Practical Guide to Seismic Restraint

Second Edition

About the Authors

James R. Tauby is chief executive engineer for Mason Industries, Inc. He is a professional engineer in over 40 states. He holds a Bachelor of Science in Mechanical Engineering from the University of Alabama. An ASHRAE Distinguished Lecturer, he regularly lectures around the world on topics ranging from vibration isolation, seismic, and wind restraint of mechanical systems to the use of elastomeric expansion joints for piping in seismic applications. He is a past chairman of ASHRAE's Technical Committee 2.7, Seismic and Wind Restraint Design. He is a member of ASHRAE's Standards Committee and currently chairs the committee revising ASHRAE Standard 171, *Method of Test of Seismic Restraint Devices for HVAC&R Equipment.* He is a member of the National Fire Protection Association's Technical Committee on Hanging and Bracing of Water-Based Fire Protection Systems (for *NFPA 13: Standard for the Installation of Sprinkler Systems*). He was also an editor on Federal Emergency Management Agency (FEMA) documents 412, 413, and 414 for the installation of seismic restraints on equipment, piping, ductwork, and electrical distribution systems.

Richard J. Lloyd is manager of Mason Industries' engineering office in California. He holds a Bachelor of Science in Engineering from California State University at Northridge. He led Mason engineering teams in inspecting earthquake-damaged equipment installations, analyzing failures, and designing retrofits. He is a coauthor of the California Office of Statewide Health Planning and Development (OSHPD) Pre-Approval OPA-0349 "Mason Industries Seismic Restraint Guidelines for Suspended Piping, Ductwork and Electrical Systems," which he used to design and supervise the installation of seismic bracing systems on many projects. He is currently a member of the Earthquake Engineering Institute (EERI), International Code Council (ICC), Applied Technology Council (ATC), and American Society of Civil Engineering (ASCE), where he participated in the development of ASCE 7-10 (Chapter 13). He has also participated in shake table testing of equipment mounted on vibration isolation systems for equipment certification. His most recent involvement in research projects includes designing vibration isolation, seismic restraints, and piping flexible connectors for the George E. Brown, Jr. Network for Earthquake Engineering Simulation (NEES) "Full-Scale Structural and Nonstructural Building System Performance During Earthquakes & Post-Earthquake Fire", the large-scale shake table test of a structure with mechanical, electrical, and plumbing nonstructural components.

Updates and errata to this publication will be posted on the ASHRAE website at www.ashrae.org/publicationupdates.

Practical Guide to Seismic Restraint

Second Edition

James R. Tauby
Richard Lloyd

ISBN 978-1-936504-18-3

© 1999, 2012 ASHRAE
1791 Tullie Circle, NE
Atlanta, GA 30329
www.ashrae.org

All rights reserved.

Printed in the United States of America on 10% post-consumer waste using soy-based inks.

Cover design by Tracy Becker

ASHRAE has compiled this publication with care, but ASHRAE has not investigated, and ASHRAE expressly disclaims any duty to investigate, any product, service, process, procedure, design, or the like that may be described herein. The appearance of any technical data or editorial material in this publication does not constitute endorsement, warranty, or guaranty by ASHRAE of any product, service, process, procedure, design, or the like. ASHRAE does not warrant that the information in the publication is free of errors, and ASHRAE does not necessarily agree with any statement or opinion in this publication. The entire risk of the use of any information in this publication is assumed by the user.

No part of this book may be reproduced without permission in writing from ASHRAE, except by a reviewer who may quote brief passages or reproduce illustrations in a review with appropriate credit; nor may any part of this book be reproduced, stored in a retrieval system, or transmitted in any way or by any means—electronic, photocopying, recording, or other—without permission in writing from ASHRAE. Requests for permission should be submitted at www.ashrae.org/permissions.

Library of Congress Cataloging-in-Publication Data

Tauby, James R.
Practical guide to seismic restraint / James R. Tauby and Richard Lloyd. -- 2nd ed.
p. cm.
Includes bibliographical references.
ISBN 978-1-936504-18-3 (pbk.)
1. Earthquake resistant design. I. Lloyd, Richard, 1952- II. Title.
TA658.44.P73 2102
693.8'52--dc23
2011049428

ASHRAE Staff

Special Publications

Mark Owen
Editor/Group Manager of Handbook and Special Publications

Cindy Sheffield Michaels
Managing Editor

James Madison Walker
Associate Editor

Elisabeth Warrick
Assistant Editor

Meaghan O'Neil
Editorial Assistant

Michshell Phillips
Editorial Coordinator

Publishing Services

David Soltis
Group Manager of Publishing Services and Electronic Communications

Jayne Jackson
Publication Traffic Administrator

Tracy Becker
Graphics Specialist

Publisher

W. Stephen Comstock

Contents

Preface

There have been many interesting developments in the design of mechanical, plumbing, and electrical systems for earthquakes since publication of the first edition of this manual in 1999. Information from these developments has been used to revise some parts of this book and to add new material of interest to design engineers, contractors, owners, code enforcement agencies, and equipment manufacturers.

Included in Chapter 2 of this second edition are updated summaries of the 2009 International Building Code® (IBC®) and the American Society of Civil Engineers' (ASCE) Standard ASCE 7-10 design requirements. Also included is a summary of additional requirements from the 2010 California Building Code (CBC). In all cases, the focus on differential system displacement is highlighted.

There are updates to the specification considerations, seismic restraint devices and connection methods, and equations for seismic wind force determination along with examples. There is also a chapter that includes photographs of correct and incorrect methods for installation of mechanical, electrical, and plumbing equipment and piping.

An important addition is the information on equipment certification in Chapter 4. In the first edition, this chapter was limited to a discussion of equipment ruggedness from a nuclear industry study and some information on shipping fragility and test methods. This information has for the most part been deleted. A summary of industry standard fragility levels has been retained as a useful general reference. But the additional information on equipment shake table testing and equipment certification may be of particular value to equipment manufacturers, who may be struggling with meeting equipment certification requirements and confused over the options of analysis, testing, or experience data.

Although the testing and analysis of equipment is important, current equipment certification standards do not address piping connections. In an earthquake, the piping connected to equipment can be subject to differential displacement, straining and rupturing vulnerable piping connections. Broken piping connections were to blame for many instances of equipment shutdown in an earthquake. Accordingly, new codes require system design to accommodate differential displacements. This second edition includes valuable new information on the advantages of different types of flexible connectors, the importance of proper connector orientation, and the need for flexible connector testing to develop accurate stiffness at operating pressure.

In Chapter 5 of the first edition, we noted that concrete anchor values were easily obtainable from ICBO reports where, with a few adjustments for spacing and edge distance, most engineers could comfortably select anchor bolts. Since that time, we have learned of the critical importance of cracked-concrete testing of anchors and complex calculations to determine various anchor failure modes. Included in the second edition are illustrations and descriptions of anchors with cracked-concrete testing and a table of allowable loads for reference.

We believe this guide remains pertinent for the design engineer, installing contractors, code officials, local inspectors, and others who are interested in seismic and wind restraint design. Although this publication is considered a design guide, it should be understood that the information it presents reflects our understanding of the current code requirements and generally accepted good engineering practices. The responsibility for satisfying local codes and regulations still remains with the design professional.

Jim Tauby
Rich Lloyd
November 2011

Acknowledgments

As with any publication, there are people and organizations that deserve to be thanked for their assistance. This publication is no different. The very first person we would like to thank is former ASHRAE President, Terry Townsend. Terry was the inspiration to ASHRAE Technical committee TC 2.7 Seismic and Wind Restraint to begin the process to have this publication written. Terry has helped coach us through this and other ASHRAE projects throughout the years. To say this publication would have not been possible without his guidance is an understatement.

We would like to thank ASHRAE Technical Committee TC 2.7. Their editing skills of the first edition and technical guidance for both editions of this publication were invaluable.

We would like to thank our original co-authors of the first edition, Todd Noce and Joep Tunnissen. They contributed to several chapters and helped edit the first edition.

We were extremely lucky to have Norm Mason edit this manual. Norm Mason gave the first ASHRAE talk on the design of Seismic Restraints in San Francisco in the early 1970s. This manual and the industry as a whole is an extension of that talk.

Most importantly, we would like to thank our wives, Maureen Tauby and Ellen Lloyd, for their support and understanding. Without their understanding and allowances into our personal time, this publication would not have been possible.

1 Fundamentals of Earthquakes

An earthquake is simply defined as shaking of the ground. This shaking can cause irreparable damage to natural and human-made structures. It has been known to change the course of rivers, cause landslides down into valleys, and create tsunamis that engulf islands. There are three natural and one human-made cause for earthquakes.

COLLAPSE EARTHQUAKES

Collapse earthquakes are the least common of the natural causes. They are caused by the sudden fall of the roof of an underground cavern or mine. Earthquakes that are caused by a landslide are also grouped into the collapse category.

VOLCANIC EARTHQUAKES

These are earthquakes that are caused by volcanic activity. Their large explosive eruptions can produce shock waves that form earthquakes. Volcanoes are usually formed on the edges of the tectonic plates and are linked, therefore, to the final natural cause, the tectonic earthquake.

TECTONIC EARTHQUAKES

This is the most common earthquake. The shifting or slipping of one tectonic plate edge adjacent to another causes most tectonic earthquakes. This slipping could be minor, as in a tremor, or major, as in a full earthquake.

EXPLOSIVE EARTHQUAKES

These are human-made quakes that are usually caused by underground nuclear explosions.

PLATES

Plates (or tectonic plates) are the outermost rock layers and cover the entire surface of the Earth. The major tectonic plates are shown in Figure 1-1. The plates themselves are relatively stable in their interiors but are unstable where they contact adjacent plates. The constant moving of the plates, as they try to slip past or under one another, can cause violent earthquakes. Most earthquakes occur at faults along the plates.

FAULTS

A fault is an offset in a geological structure. Inactive faults are faults that have not been active for thousands of years. They are well documented and their abrupt strata changes are easily identifiable. Active faults, on the other hand, move constantly, producing stress and eventually giving rise to earthquakes. Active faults are also documented, with new faults being discovered after each seismic event.

There are three basic fault types, as seen in Figure 1-2. The first is a normal fault. A normal fault is a vertical displacement, where one side slips downward, at an angle from 0° to 90° away from the other side. The second is a reverse fault, which is also a vertical displacement, where one side slides up past the other. Both normal and reverse faults are

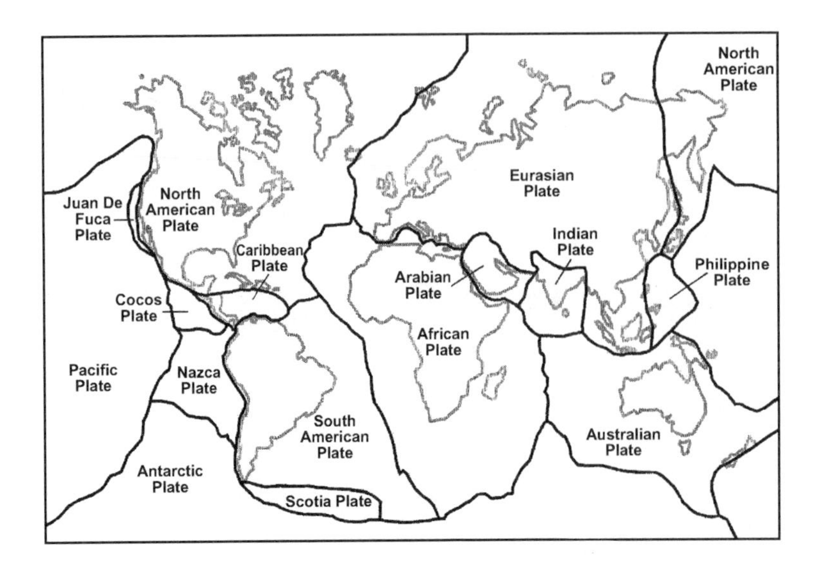

Figure 1-1 Major tectonic plates of the world.

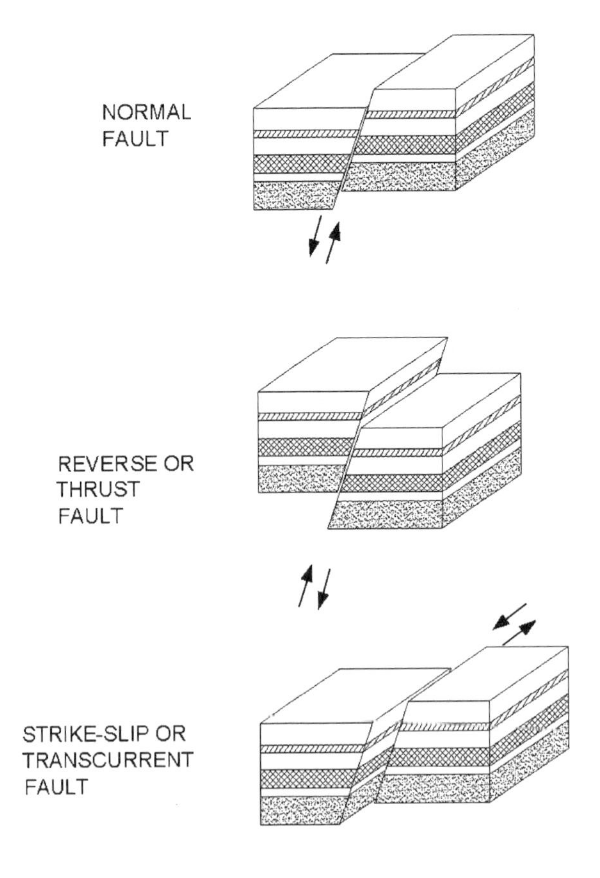

Figure 1-2 Three basic fault types.

sometimes called dip-slip faults. The last type is a strike-slip, or transcurrent, fault. These are predominantly horizontal slips and are defined by their movement as either left-lateral or right-lateral faults. It is easy to distinguish between the two. If you are standing on one side of the fault and the other side has moved from your right to your left, it is a left-lateral fault. In reality, faults usually take on characteristics of all three.

WAVES

While the four types of earthquake are different, they all have one thing in common—they emit shock or seismic waves. There are four basic waves associated with an earthquake. The fastest is the P, or primary wave, and the next is the S, or secondary, wave. Both propagate inside of rocks.

P waves have zones of compression and elongation, while S waves shear the strata perpendicular to the direction of travel. Though P waves arrive first, it is S waves, with their up-and-down and side-to-side motion, that do the most damage to structures.

The last two waves are known as surface waves because they appear only near the surface. The Love wave mimics an S wave but without the vertical component and is one of the main causes for residential buildings to be moved off their foundations. The last is the Rayleigh wave, which rolls elliptically in the direction of travel.

P and S waves emanate from the focus of the earthquake and can be refracted, or bounce off, different strata, sending the waves in many directions. The focus is at a depth below the surface, and the projection of the point to the surface is called the epicenter of the earthquake. (See Figure 1-3.)

INTENSITY AND MAGNITUDE

Intensity and *magnitude* are terms used to describe an earthquake's level of destruction. Intensity scales have changed over the years, but today's Modified Mercalli Intensity Scale estimates the strength of an earthquake based on 12 values of visual inspection of damage to buildings and ground surfaces, as well as the reaction of animals. While the scale is accurate, it is a time-consuming process that can take weeks or months. The scale is not based on mathematics but on the accumulation of information and verbal accounts of the damage. For an abridged version of the Modified Mercalli Intensity Scale, see Appendix A.

Earthquake magnitudes are also presented in terms of magnitude scales. The media generally quote the Richter magnitude, as it is available almost immediately after an earthquake. Richter magnitudes are based on seismographic readouts and gathering these data is much faster than using the Modified Mercalli Intensity method. Determining Richter magnitudes helps pinpoint the focus by determining the distance to the focus based on the time lag between P and S waves. The P waves arrive first, and the time between the P and S waves is plotted along with the maximum amplitude of the S waves. A straight line is drawn between these two values and the line intersects the Richter magnitude value on

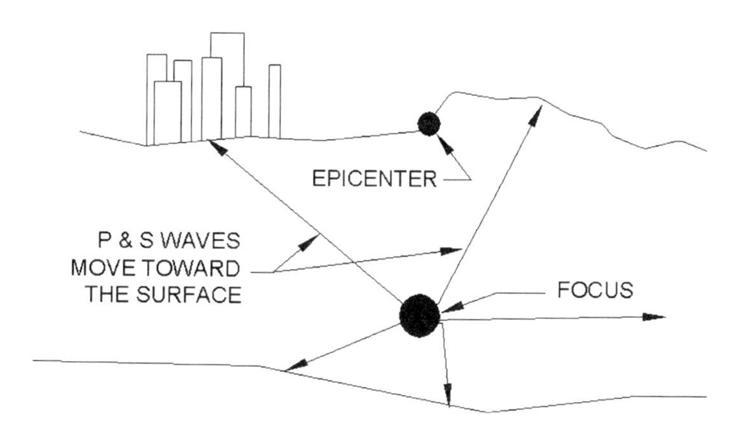

Figure 1-3 Seismic wave propagation.

coordinate forms designed for this purpose. The moment magnitude is what seismologists use for defining the size of an earthquake and is calculated by the geometry of the fault plane after the seismic event.

Magnitude values offer two distinct, related forms of reference. They are ground motion and released energy. A magnitude 6 earthquake has 10 times as much ground motion as a magnitude 5 earthquake and 30 times as much released energy along the fault than a 5. This means that a magnitude 7 earthquake has 100 times ($10 \cdot 10$) more ground motion and 900 times ($30 \cdot 30$) as much energy as a magnitude 5 earthquake. The Bikini Atomic Blast after WWII produced approximately $1 \cdot 10^{19}$ ergs, while a magnitude 5 earthquake can generate over $1 \cdot 10^{20}$ ergs of energy.

BIBLIOGRAPHY

Bolt, B.A. 1978. *Earthquakes.* New York: W.H. Freeman and Company.

2 Building Codes

Building codes have addressed seismic requirements for mechanical and electrical systems for over 35 years. It has been well documented that earthquake damage to mechanical and electrical systems accounts for over 50% of monetary damage to buildings. Mechanical and electrical systems are critical for the continued operation of essential facilities after an earthquake, such as hospitals, communication control centers, and fire and police stations.

The primary intent of seismic design for mechanical and electrical systems is life safety. This translates to complete system anchorage and functionality in essential occupancies, such as hospitals, and only complete system support in standard occupancies, such as office buildings. System anchorage ensures the system will not fall over and either hit someone or block someone's means of egress after an earthquake. System functionality ensures the building's mechanical and electrical systems will be fully operational after an earthquake.

Once adopted, building codes represent the minimum seismic requirements for design and installation. There are many participants who take part in ensuring the codes are followed. These include building owners, the design team (i.e., architects, engineers), government review boards, contractors, component manufacturers, and government inspectors. Not one or any combination of these parties has the option to compromise or ignore the adopted code.

In California, the Office of Statewide Health Planning and Development (OSHPD) and Department of the State Architect (DSA) were formed to ensure that schools and hospitals are built to meet the codes. The resulting consistent enforcement of the code along with additional published clarifications, an anchorage preapproval program, and a concentrated inspection effort with highly qualified inspectors has resulted in what many engineers believe are the safest buildings of their type in the world. But the fact is that these building are not build to higher standards; they are simply being built to published building code standards. The difference is the level of enforcement.

The International Building Code® (IBC®) is published by the International Code Council (ICC). The ICC is comprised of the three former building code entities, Building Officials Code Administration (BOCA), Southern Building Code Congress International (SBCCI), and the International Conference of Building Officials (ICBO). The language in this model code is drawn from the provisions of the National Earthquake Hazards Reduction Program (NEHRP) for new buildings. NEHRP is a division of the Building Seismic Safety Council (BSSC) and is funded by the Federal Emergency Management Agency (FEMA). The IBC is entirely drawn from the NEHRP provisions. Most U.S. jurisdictions will adopt the IBC to ensure financial backing from FEMA following an earthquake. Since the 2003 edition, the

IBC has included a direct reference to the American Society of Civil Engineers' (ASCE) Standard ASCE 7, "Minimum Design Loads for Buildings and Other Structures," for the seismic requirements of nonstructural components. Both IBC 2006 and IBC 2009 include a reference to ASCE 7-05. The next edition of IBC, scheduled for 2012, will reference ASCE 7-10. Meanwhile, in California, the 2010 California Building Code includes new requirements from ASCE 7-10. In Canada, the *National Building Code* (NBC) has been used for many years. The 2005 edition incorporates information and seismic zoning maps that differ from those utilized in earlier editions.

The design earthquake used to determine ground accelerations in these codes is based on geologic evidence gathered over the last 500 years. More specifically, the code ground accelerations are based on a 10% probability that a larger earthquake will occur in 50 years and a 90% probability that an equal or smaller earthquake will occur in 50 years.

In all codes, there is a distinction between rigidly or hard-mounted equipment, having a fundamental period less than or equal to 0.06 s, and flexibly or resiliently mounted equipment having a fundamental period greater than 0.06 s.

2006 AND 2009 INTERNATIONAL BUILDING CODE (IBC)

General Requirement

Every structure, including nonstructural components permanently attached to structures, and their support and attachments, must be designed and constructed to resist the effects of earthquake motions per ASCE 7-05.

Special attention should be paid to the design of distribution systems for seismic relative displacements. Although previous codes included some references to seismic relative displacements as they effect nonstructural components, ASCE 7-05 includes at least 10 separate requirements. These requirements are most significant as they relate to piping connections, because the mass and stiffness of connected piping can damage equipment connections in an earthquake.

This summary is presented in (4) parts:

Part 1: Seismic Design Category.
Part 2: Seismic Design Requirements.
Part 3: Seismic Design Force.
Part 4: Tests and Inspection Requirements.

Part 1: Seismic Design Category

The values of S_s, the spectral response acceleration at 5 Hz, and S_1, the spectral response acceleration at a 1 Hz, can be found at www.earthquake.usgs.gov using the project's zip code for estimated values or the project latitude and longitude for more exact values.

F_a is the site coefficient defined in Table 2-1 for specific site class A to F and S_s. Use straight-line interpolation for intermediate values of S_s. Where the soil condition is unknown, Site Class D may be used.

Table 2-1 Site Coefficient F_a

Site Class	S_S <= 0.25	S_S = 0.50	S_S < 0.75	S_S < 1.00	S_S >=1.25
A	0.8	0.8	0.8	0.8	0.8
B	1.0	1.0	1.0	1.0	1.0
C	1.2	1.2	1.1	1.0	1.0
D	1.6	1.4	1.2	1.1	1.0
E	2.5	1.7	1.2	0.9	0.9
F	site-specific information required				

S_{ms} is the maximum considered earthquake spectral response accelerations for short periods and is equal to $F_a \times S_s$. S_{ds} is the design spectral response acceleration at short period and is equal to (2/3) $S_{ms.}$

F_v is the site coefficient defined in Table 2-2 for specific site class A to F and S_1. Use straight-line interpolation for intermediate values of S_1. Where the soil condition is unknown, Site Class D may be used.

S_{m1} is the maximum considered earthquake spectral response accelerations for a 1 s period and is equal to $F_v \times S_1$. S_{d1} is the design spectral response acceleration at a 1 s period and is equal to (2/3) S_{m1}.

Both short-period spectral acceleration and 1 s period acceleration must be used to determine Seismic Design Category. Results must be compared and the worst case will determine the actual Seismic Design Category for the project.

Select the Seismic Design Category as a letter designation between A and F based on the short-period design spectral acceleration, S_{DS}, determined in Step 3 and the Occupancy Category of the building where Occupancy Category I and II include standard buildings, III includes schools and assembly halls, and IV includes essential facilities. (See Table 2-3)

Select the Seismic Design Category as a letter designation between A and F based on the 1 s period design spectral acceleration, S_{D1}, determined in Step 3 and the Occupancy Category of the building where Occupancy Category I and II include standard buildings, III includes schools and assembly halls, and IV includes essential facilities. (See Table 2-4)

Table 2-2 Site Coefficient F_v

Site Class	S_1 <= 0.1	S_1 = 0.2	S_1 < 0.3	S_1 < 0.4	S_1 >=0.5
A	0.8	0.8	0.8	0.8	0.8
B	1.0	1.0	1.0	1.0	1.0
C	1.7	1.6	1.5	1.4	1.3
D	2.4	2.0	1.8	1.6	1.5
E	3.5	3.2	2.8	2.4	2.4
F	site-specific information required				

Table 2-3 Seismic Design Category Based on S_{DS}

S_{DS}	Occupancy Category I and II	Occupancy Category III	Occupancy Category IV
$S_{DS} < 0.167g$	A	A	A
$0.167g \le S_{DS} < 0.33g$	B	B	C
$0.33g \le S_{DS} < 0.50g$	C	C	D
$0.50g \le S_{DS}$	D	D	D

Table 2-4 Seismic Design Category Based on S_{D1}

S_{D1}	Occupancy Category I and II	Occupancy Category III	Occupancy Category IV
$S_{D1} < 0.067g$	A	A	A
$0.067g \le S_{D1} < 0.133g$	B	B	C
$0.133g \le S_{D1} < 0.20g$	C	C	D
$0.20g \le S_{D1} < 0.75g$	D	D	D
$0.75g \le S_{D1}$	E	E	F

Part 2: Seismic Design Requirements

Scope

Chapter 13 of ASCE 7-05 establishes minimum design criteria for nonstructural components that are permanently attached to structures and for their supports and attachments.

General Requirements

Mechanical and electrical systems with $I_p = 1.0$ require supports and attachments to comply with the requirements of ASCE 7-05.

Mechanical and electrical systems with $I_p = 1.5$ require the component, including supports and attachments to comply with the requirements of ASCE 7-05.

The failure of an essential or nonessential component must not cause the failure of an essential component.

Component design must include consideration of both flexibility and strength.

Compliance can be shown by either of the following means:

1. Project-specific design and documentation prepared by a registered design professional.
2. Manufacturer's certification that a component has been certified by analysis, or by testing per a nationally recognized standard, or by experience data per a nationally recognized procedure.

Determine the Component Importance Factor I_p

The component importance factor I_p is 1.5 for any component required to function for life safety purposes, components containing hazardous material, or any component that is needed for the continued operation of a critical facility. For all other components, $I_p = 1.0$.

Special Certification Requirement for Designated Seismic Systems

Any active mechanical or electrical equipment that must remain operable is a "designated seismic system" that requires certification by the manufacturer based on approved shake table testing or experience data. Any component with hazardous contents must maintain containment and is a "designated seismic system" that requires certification by the supplier based on analysis, shake table testing, or experience data.

Reference Documents

Reference documents that provide a basis of design for a system or component are permitted to be used, subject to the approval of the building authority having jurisdiction. Design forces must not be less than that determined by code. Each component's seismic interactions with all other components and the structure must be considered. The component must accommodate drifts, deflections, and relative displacements.

Reference Documents for Allowable Stress Design

Reference documents, such as OSHPD OPA approvals, that provide a basis for design for a particular system or component and define acceptance criteria in terms of allowable stress, are permitted to be used. Additional loads, including dead, live, operating, and earthquake loads, must be used.

Component Seismic Design Category

For seismic design purposes, nonstructural components are assigned the same Seismic Design Category as the structure.

Seismic Design Category A

All components are exempt from seismic requirements.

Seismic Design Category B

Mechanical and electrical components are exempt from seismic requirements.

Seismic Design Category C

Mechanical and electrical components with $I_p = 1.0$ are exempt from seismic requirements. Mechanical and electrical components with $I_p = 1.5$ must meet the seismic requirements.

Seismic Design Categories D, E, and F

Mechanical and electrical components with $I_p = 1.0$ and 1.5 must meet the seismic requirements.

Mechanical and Electrical Component Support General Design Requirements

1. Supports and attachment of support to the component must be designed for forces and displacements.
2. Supports include structural members, braces, skirts, legs, saddles, pedestals, cables, guys, stays, snubbers, tethers, and elements forged or cast as part of the component.
3. Industry standard and proprietary supports must be designed for loads and stiffness.
4. Support and support attachment materials must be suitable for the application.
5. Support design and construction must insure that support engagement is maintained.
6. Reinforcement must be provided at bolted connections of a sheet metal base when required unless the base is reinforced with stiffeners.
7. Weak-axis bending of cold-formed steel supports must be specifically evaluated.
8. Components mounted on vibration isolators require horizontal and vertical restraints. Isolator housings and restraints must be made of ductile materials. An elastomeric pad of appropriate thickness must be used to limit the impact load.
9. Expansion anchors cannot be used on non-vibration-isolated equipment over 10 hp (7.5 kW).
10. The attachments to concrete for piping, boilers and pressure vessels must be designed for cyclic loads.
11. The attachment of mechanical equipment with drilled and grouted-in-place anchors for tensile load applications must use expansive cement or expansive epoxy grout.
12. Light fixtures, lighted signs, and ceiling fans not connected to duct or pipe, where supported by chains or otherwise suspended, are not required to be braced for seismic if the support is designed for $1.4 \times$ dead load and an equivalent simultaneous horizontal load. Failure cannot cause the failure of an essential component and the connection must allow a 360 degree motion.
13. Components installed inline with the duct system and having an operating weight greater than 75 lbs (34 kg) must be supported and braced independent of the duct system. Unbraced piping attached to unbraced equipment must have flexible connectors to accommodate the motion.
14. The effects of seismic relative displacements combined with displacements caused by other loads must be considered. An estimate of design drift can be taken as 1% of the story height. For example, differential motion from floor to ceiling for a 20 ft (6 m) story height is 2.4 (61mm) inches. Suspended piping, ductwork and conduit attached to floor mounted equipment must have inherent flexibility or flexible connectors to allow differential motion without overloading the component connection.
15. When $I_p = 1.5$, the local region of the support attachment to the component must be evaluated for the effect of load transfer.

Utility and Service Line Special Requirements

Utility lines must be designed for differential motions between structures. Interruption of service must be considered for critical facilities. Special attention must be given to underground systems in Site Class E or F where S_{DS} is greater than 0.33.

Supports for electrical distribution components must be designed for forces and displacement if any of the following conditions apply:

1. Supports are attached to the floor.
2. Supports include bracing to limit deflection.
3. Supports are made of rigid welded frames.
4. Attachments into concrete are nonexpanding inserts, power actuated fasteners, or cast iron embedments.
5. Attachments to steel are spot welds, plug welds, or minimum-size welds defined by the American Institute of Steel Construction (AISC).

Seismic braces are not required on ductwork fabricated and installed per an accepted standard and, for the entire run of duct when either:

1. The hangers are 12 in. (305 mm) or less in length from the top of duct to the supporting structure detailed to avoid significant bending to the hangers or their connections.
2. The cross-sectional area is less than 6 ft^2 (0.55 m^2).

Important Note: ASCE 7-05 only includes these ductwork bracing exclusions for systems with $I_p = 1.0$. However, the same bracing exclusions for ductwork with $I_p = 1.5$ have also been included in IBC 2009.

Seismic restraints can be excluded from the following when I_p=1.0:

1. Mechanical and electrical components where flexible connections are provided between the components and associated ductwork, piping and conduit, <u>and</u> the system components are mounted at 4 ft (1.2m) or less above floor level and weigh 400 lb (182 kg) or less.
2. Mechanical and electrical components weighing 20 lb (9 kg) or less <u>and</u> flexible connections are provided between the components and associated ductwork, piping, and conduit.
3. Piping, ductwork, and electrical distribution systems weighing 5 lb/ft (7.45 kg/m) or less <u>and</u> flexible connectors are provided between the component and the piping, ductwork or electrical distribution system.

Seismic braces are not required on piping when:

The piping is supported by rod hangers and the hangers in the entire run are 12 in. (305 mm) or less in length from the top of the pipe to the supporting structure; hangers are detailed to avoid bending of the hangers and their attachments; and provisions are made for piping to accommodate expected deflections.

Electrical Components with I_p = 1.5 Design Requirements

1. Design for seismic forces and displacements.
2. Design to eliminate seismic impacts between components.
3. Design for loads imposed by utilities or service lines attached to separate structures.
4. Special battery rack requirements include wrap-around restraints and spacers to protect batteries and racks evaluated for load capacity.
5. Internal coils of dry type transformers positively attached.
6. Slide-out components must have latches.
7. Cabinet design per National Electrical Manufacturers Association (NEMA). Lower shear panel field modifications must be evaluated.
8. Attachment of field added external items over 100 lb must be evaluated.

9. Components must be designed for relative displacement of connected conduit, cable trays or other distribution systems.

The supports for electrical distribution components with $I_p = 1.5$ must be designed for forces and displacement if any of the following conditions apply:

1. Conduit is larger than 2 1/2 in. (64 mm) trade size.
2. Trapeze assemblies are used to support conduit, bus duct or cable tray weighing more than 10 lb/ft (14.9 kg/m).

The following pipe bracing exemptions are dependent on project Seismic Design Category:

1. For projects in Seismic Design Category C, seismic braces are not required on piping when $I_p = 1.5$; high-deformability piping is used; provisions are made to avoid impact with larger pipe or mechanical components or to protect the pipe in the event of such impact; and the nominal pipe size is 2 in. (50 mm) diameter or less.
2. For projects located in seismic design category D, E or F, seismic braces are not required on high-deformability piping when $I_p = 1.0$; provisions are made to avoid impact with larger pipe or mechanical components or to protect the pipe in the event of such impact; and the nominal pipe size is 3 in. (75 mm) diameter or less.
3. For projects located in seismic design category D, E or F, seismic braces are not required on high-deformability piping when $I_p = 1.5$; provisions are made to avoid impact with larger pipe or mechanical components or to protect the pipe in the event of such impact; and the nominal pipe size is 1 in. (25 mm) diameter or less.

Part 3: Seismic Design Force

Where F_p is the seismic design force, S_{DS} is the short period spectral acceleration, a_p is the component amplification factor that varies from 1.0 to 2.5, R_p is the component response modification factor that varies from 1.0 to 12, I_p is the component importance factor that is 1.0 or 1.5, z is the height of point of attachment to structure with respect to the base, h is the roof height with respect to the base, and W_p is the component operation weight.

Use the following formula to determine F_p:

$$F_p = \frac{0.4 a_p S_{DS}}{(Rp/Ip)}\left(1 + 2\frac{z}{h}\right) W_p \tag{2-1}$$

F_p shall not be greater than

$$F_p = 1.6\ S_{DS}\ I_p\ W_p \tag{2-2}$$

and shall not be less than

$$F_p = 0.3\ S_{DS}\ I_p\ W_p \tag{2-3}$$

F_p shall be applied independently in at least two directions.

The component must be designed for a concurrent vertical load of

$$F_p = 0.2\ S_{DS}\ W_p \tag{2-4}$$

Where

$$S_{DS} = 2/3\ S_{MS} \tag{2-5}$$

and

$$S_{MS} = 2/3\ F_a\ S_s \tag{2-6}$$

Table 2-5 Seismic Coefficients a_p and R_p for Common System Components (see ASCE 7-05 Table 13.6.1 for a complete list)

Electrical System Components	a_p	R_p
Electrical Components		
Engines, generators, batteries, inverters, motors, transformers and other electrical components constructed of high-deformability materials	1.0	2.5
Motor control centers, panel boards, switchgear, instrumentation cabinets, and other components constructed of sheet metal framing	2.5	6.0
Communication equipment, computers, instrumentation, and controls	1.0	2.5
Lighting fixtures	1.0	1.5
Other electrical components	1.0	1.5
Vibration-Isolated Components and Systems		
Neoprene Isolated	2.5	2.5
Spring Isolated	2.5	2.0
Suspended, vibration-isolated equipment	2.5	2.5
Distribution Systems		
Electrical conduit, bus ducts, and rigidly mounted cable trays	1.0	2.5
Suspended cable trays	2.5	6.0
Air-side HVAC Mechanical System Components and Ductwork	a_p	R_p
Components		
Fans, air handlers, air-conditioning units, cabinet heaters, air distribution boxes, and other mechanical components constructed of sheet metal framing	2.5	6.0
Roof-mounted chimneys, stacks, and cooling towers laterally braced below their center of mass	2.5	3.0
Roof-mounted chimneys, stacks, and cooling towers laterally braced above their center of mass	1.0	2.5
Other components	1.0	1.5
Vibration-Isolated Components and Systems		
Neoprene Isolated	2.5	2.5
Spring Isolated	2.5	2.0
Suspended, vibration-isolated equipment	2.5	2.5
Ductwork Systems		
Ductwork, including in-line components, constructed of high-deformability materials, with joints made from welding or brazing	2.5	9.0
Ductwork, including in-line components, constructed of high- or limited-deformability materials, with joints made from means other than welding or brazing	2.5	4.5
Ductwork, including in-line components, constructed of low-deformability materials such as cast iron, glass and non-ductile plastic	2.5	3.0
Wet-side HVAC, Mechanical, and Plumbing and Piping	a_p	R_p
Components		
Boilers, furnaces, tanks, chillers, water heaters, heat exchangers, evaporators, air separators, other components constructed of high-deformability materials	1.0	2.5
Engines, turbines, pumps, and compressors	1.0	2.5
Other components	1.0	1.5
Vibration-Isolated Components and Systems		
Neoprene Isolated	2.5	2.5

Table 2-5 Seismic Coefficients a_p and R_p for Common System Components (see ASCE 7-05 Table 13.6.1 for a complete list) (Continued)

Wet-side HVAC, Mechanical, and Plumbing and Piping (Continued)	a_p	R_p
Spring Isolated	2.5	2.0
Suspended, vibration-isolated equipment	2.5	2.5
Piping Systems		
Piping, including in-line components, constructed of high-deformability materials, with joints made from welding or brazing	2.5	9.0
Piping, including in-line components, constructed of high- or limited-deformability materials, with joints made from threading, bonding, compression couplings or grooved couplings	2.5	4.5
Piping constructed of low-deformability materials such as cast iron, glass and nonductile plastic	2.5	3.0
Plumbing	1.0	2.5

Consider additional F_p factors:

1. Where a reference document provides allowable stress or allowable loads, F_p can be reduced by multiplying by a factor 0.7.
2. Anchors embedded in concrete shall be designed for 1.3 times the force developed due to F_p, or the maximum force that can be transferred to the anchor by the component and its supports, whichever is least.
3. The value of R_p used to determine anchor loads cannot exceed 1.5 unless the anchorage load is limited by the bending of a ductile steel element or postinstalled anchors are prequalified by American Concrete Institute (ACI) 355.2 or cast-in-place anchors are designed per code.
4. Components mounted on vibration isolation systems shall have bumper restraint or snubber in each horizontal direction. If the maximum clearance (air gap) between the equipment support frame and the restraint is greater than 1/4 in. (6 mm), the design force shall be taken as $2 \times F_p$. If the maximum clearance specified on the construction documents is 1/4 in. (6 mm), the design force may be taken as F_p.

Important note: IBC 2006 references ACI 318-05 for anchor design in concrete. ACI 318-08 includes several important changes and is referenced in IBC 2009.

Following are examples illustrating the calculation of the seismic forces for a rigidly mounted boiler in an office building and a flexibly mounted pump in a hospital. Each system is calculated at grade and on the roof level.

Example 1

A boiler is rigidly mounted to a concrete slab located on grade in a 200 ft (61 m) tall office building (Occupancy Category II). The building is located where the value of S_S is 1.00 and the site class is D. (See Figure 1.)

W_p = 10,000 lb (4536 kg)

I_p = 1.0

z = 0 ft (0 m)

h = 200 ft (61 m)

a_p = 1.0

R_p = 2.5

F_a = 1.1

S_{DS} = $(2/3)S_{MS} = (2/3)F_a \times S_S = (2/3)(1.1)(1.00) = 0.73$

$$F_p = \frac{0.4a_p S_{DS} I_p}{R_p}\left[1 + 2\frac{z}{h}\right]W_p$$

$$F_p = \frac{0.4(1.0)(0.73)(1.0)}{2.5}\left[1 + 2\frac{0}{200}\right]10,000$$

$$F_p = 1168 \text{ lb } (5.2 \text{ kN})$$

except F_p shall not be greater than

$$F_p = 1.6S_{DS} I_p W_p = 1.6(0.73)(1.0)(10{,}000 \text{ lb}) = 11{,}680 \text{ lb } (52 \text{ kN})$$

and shall not be taken as less than

$$F_p = 0.3S_{DS} I_p W_p = 0.3(0.73)(1.0)(10{,}000 \text{ lb}) = 2190 \text{ lb } (9.8 \text{ kN})$$

Therefore,

$$F_p = 2190 \text{ lb } (9.8 \text{ kN})$$

Example 2

The same boiler from Example 1 is rigidly mounted to a concrete slab located on the roof of a 200 ft (61 m) tall office building (Occupancy Category II). The building is located where the value of S_S is 1.00 and the site class is D. (See Figure 1.)

W_p = 10,000 lb (4536 kg)
I_p = 1.0
z = 200 ft (61 m)
h = 200 ft (61 m)
a_p = 1.0
R_p = 2.5
F_a = 1.1
S_{DS} = $(2/3)S_{MS} = (2/3)F_a \times S_S = (2/3)(1.1)(1.00) = 0.73$

$$F_p = \frac{0.4a_p S_{DS} I_p}{R_p}\left[1 + 2\frac{z}{h}\right]W_p$$

$$F_p = \frac{0.4(1.0)(0.73)(1.0)}{2.5}\left[1 + 2\frac{200}{200}\right]10,000$$

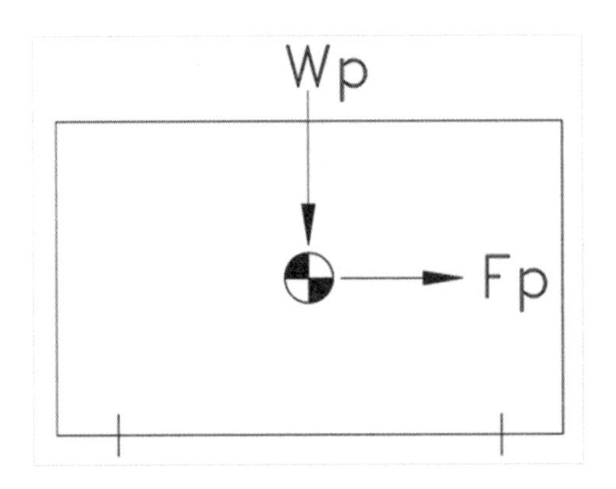

Figure 2-1 Typical rigidly mounted equipment.

$$F_p = 3504 \text{ lb } (15.5 \text{ kN})$$

except F_p shall not be greater than

$$F_p = 1.6S_{DS} I_p W_p = 1.6(0.73)(1.0)(10{,}000 \text{ lb}) = 11{,}680 \text{ lb } (52 \text{ kN})$$

and shall not be taken as less than

$$F_p = 0.3S_{DS} I_p W_p = 0.3(0.73)(1.0)(10{,}000 \text{ lb}) = 2190 \text{ lb } (9.8 \text{ kN})$$

Therefore,

$$F_p = 3504 \text{ lb } (15.5 \text{ kN})$$

Example 3

A pump is spring vibration isolated with separate seismic snubbers and mounted to a concrete slab located on grade in a 100 ft (31 m) tall hospital (Occupancy Category IV). The hospital is located where the value of S_S is 0.50 and the site class is E. (See Figure 2-1.)

W_p = 2000 lb (907 kg), including concrete inertia base
I_p = 1.5
z = 0 ft (0 m)
h = 100 ft (31 m)
a_p = 2.5
R_p = 2.0
F_a = 1.7
S_{DS} = $(2/3)S_{MS} = (2/3)F_a \times S_S = (2/3)(1.7)(0.50) = 0.57$

Seismic snubbers will be used ($F_p \times 2$).

$$F_p = \frac{0.4a_p S_{DS} I_p}{R_p}\left[1 + 2\frac{z}{h}\right]W_p$$

$$F_p = \frac{0.4(2.5)(0.57)(1.5)}{2.0}\left[1 + 2\frac{0}{200}\right]2000 \times 2$$

$$F_p = 1710 \text{ lb } (7.6 \text{ kN})$$

except F_p shall not be greater than

$$F_p = 1.6S_{DS} I_p W_p \times 2 = 1.6(0.57)(1.5)(2000 \text{ lb}) \times 2 = 5472 \text{ lb } (24.4 \text{ kN})$$

and shall not be taken as less than

$$F_p = 0.3S_{DS} I_p W_p \times 2 = 0.3(0.57)(1.5)(2000 \text{ lb}) \times 2 = 1026 \text{ lb } (4.6 \text{ kN})$$

Therefore,

$$F_p = 1710 \text{ lb } (7.6 \text{ kN})$$

Example 4

The same pump is vibration isolated with separate seismic snubbers and mounted to a concrete slab located on the roof of a 100 ft (31 m) tall hospital (Occupancy Category IV). The hospital is located where the value of S_S is 0.50 and the site class is E. (See Figure 2-2.)

W_p = 2000 lb (907 kg), including concrete inertia base

I_p = 1.5
z = 100 ft (31 m)
h = 100 ft (31 m)
a_p = 2.5
R_p = 2.0
F_a = 1.7
S_{DS} = $(2/3)S_{MS} = (2/3)F_a \times S_S = (2/3)(1.7)(0.50) = 0.57$

Seismic snubbers will be used ($F_p \times 2$).

$$F_p = \frac{0.4a_pS_{DS}I_p}{R_p}\left[1 + 2\frac{z}{h}\right]W_p$$

$$F_p = \frac{0.4(2.5)(0.57)(1.5)}{2.0}\left[1 + 2\frac{100}{100}\right]2000 \times 2$$

$$F_p = 5130 \text{ lb } (22.8 \text{ kN})$$

except F_p shall not be greater than

$$F_p = 1.6S_{DS}I_p\,W_p \times 2 = 1.6(0.57)(1.5)(2000 \text{ lb}) \times 2 = 5472 \text{ lb } (24.4 \text{ kN})$$

and shall not be taken as less than

$$F_p = 0.3S_{DS}I_p\,W_p \times 2 = 0.3(0.57)(1.5)(2000 \text{ lb}) \times 2 = 1026 \text{ lb } (4.6 \text{ kN})$$

Therefore,

$$F_p = 5130 \text{ lb } (22.8 \text{ kN})$$

ASCE 7-10

According to the IBC, every structure, including nonstructural components, must be designed and constructed to resist the effects of earthquake motions per ASCE 7. ASCE 7 establishes minimum design criteria for nonstructural components that are permanently attached to structures and for their supports and attachments. ASCE 7-10 will be the basis of seismic requirements in IBC 2012. It will serve as a good reference for IBC 2009-based current code installations.

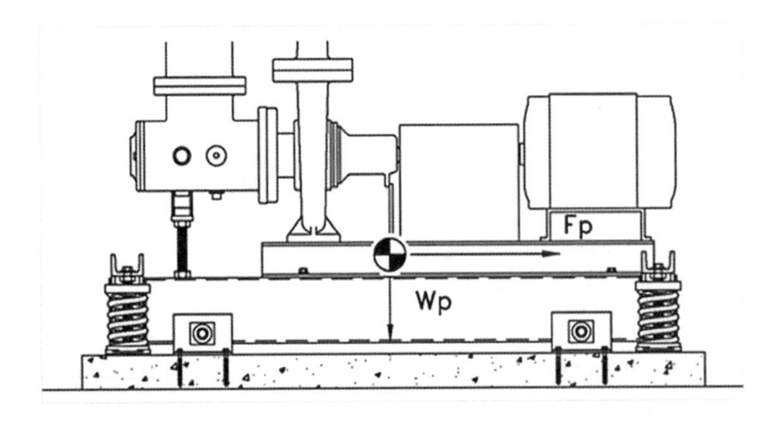

Figure 2-2 Typical vibration isolation and seismically restrained equipment.

Special attention should be paid to the design of distribution system connections for seismic relative displacements. While previous codes included some references to seismic relative displacements on nonstructural components ASCE 7-10 includes at least 10 separate requirements. These requirements are most significant as they apply to piping connections, because the mass and stiffness of connected piping can damage equipment connections in an earthquake.

The following summary of information includes the method to determine the project Seismic Design Category from ASCE 7-10 Chapter 11 and the seismic design demands and mechanical and electrical component requirements from Chapter 13.

Seismic Design Category:

Follow these steps to determine the project Seismic Design Category:

Find the values of S_s, the spectral response acceleration at 5 Hz, and S_1, the spectral response acceleration at a 1Hz, at www.earthquake.usgs.gov using the project's zip code for estimated values or the project latitude and longitude for more exact values.

Determine site coefficient F_a knowing the Site Class A to F, using straight-line interpolation for intermediate values of S_s. Where the soil condition is unknown Site Class D may be used. (See Table 2-6)

S_{ms} is the maximum considered earthquake spectral response accelerations for short periods and is equal to $F_a \times S_s$.

S_{ds} is the design spectral response acceleration at short period and is equal to $(2/3)S_{ms}$.

Determine the site coefficient F_v knowing Site Class A to F. Use straight-line interpolation for intermediate values of S_1. Where the soil condition is unknown Site Class D may be used. (See Table 2-7)

S_{m1} is the maximum considered earthquake spectral response accelerations for a 1 s period and is equal to $F_v \times S_1$.

S_{d1} is the design spectral response acceleration at a 1 s period and is equal to $(2/3)S_{m1}$.

Find the Seismic Design Category as a letter designation between A and D based on the short-period spectral design acceleration S_{ds} and the Risk Category of the building.

Risk Categories I and II include standard buildings, III includes schools and assembly halls, and IV includes essential facilities.

Table 2-6 Site Coefficient F_a

Site Class	S_S<= 0.25	S_S = 0.50	S_S = 0.75	S_S = 1.00	S_S>=1.25
A	0.8	0.8	0.8	0.8	0.8
B	1.0	1.0	1.0	1.0	1.0
C	1.2	1.2	1.1	1.0	1.0
D	1.6	1.4	1.2	1.1	1.0
E	2.5	1.7	1.2	0.9	0.9
F	site-specific information required				

Table 2-7 Site Coefficient F_v

Site Class	S_1 <= 0.1	S_1 = 0.2	S_1 = 0.3	S_1 = 0.4	S_1 >=0.5
A	0.8	0.8	0.8	0.8	0.8
B	1.0	1.0	1.0	1.0	1.0
C	1.7	1.6	1.5	1.4	1.3
D	2.4	2.0	1.8	1.6	1.5
E	3.5	3.2	2.8	2.4	2.4
F	site-specific information required				

Find the Seismic Design Category as a letter designation between A and D based on the 1 s period design spectral acceleration, S_{d1}, and the Risk Category of the building. (See Table 2-8)

Risk Categories I and II include standard buildings, III includes schools and assembly halls, and IV includes essential facilities.

Compare the Seismic Design Categories based on S_{ds} and S_{d1}. The worst-case Seismic Design Category must be used for the project. (See Table 2-9)

Knowing the project Seismic Design Category, the specific seismic requirements of nonstructural components can be determined.

For seismic design purposes, nonstructural components are assigned the same Seismic Design Category as the structure.

Component Importance Factor

1. The Component Importance Factor I_p is 1.5 for any component required to function for life safety purposes, components containing hazardous material, or any component that is needed for the continued operation of a critical facility.
2. For all other components, $I_p = 1.0$.

General Seismic Requirements Based on Component Importance Factor

1. Unless specifically noted, the supports and attachments for all components with $I_p = 1.0$ must meet these seismic requirements.
2. Unless specifically noted, mechanical and electrical components with $I_p = 1.5$ must meet these seismic requirements.

Based on the project Seismic Design Category, the following mechanical and electrical components are exempt from the seismic requirements of ASCE 7-10:

1. All components are exempt from seismic requirements in Seismic Design Category A.
2. Mechanical and electrical components are exempt from seismic requirements in Seismic Design Category B.
3. Mechanical and electrical components with $I_p = 1.0$ are exempt from seismic requirements in Seismic Design Category C.

Table 2-8 Seismic Design Category Based on S_{DS}

S_{DS}	Risk Category I or II or III	Risk Category IV
$S_{DS} < 0.167g$	A	A
$0.167g \leq S_{DS} < 0.33g$	B	C
$0.33g \leq S_{DS} < 0.50g$	C	D
$0.50g \leq S_{DS}$	D	D

Table 2-9 Seismic Design Category Based on S_{D1}

S_{D1}	Risk Category I or II or III	Risk Category IV
$S_{D1} < 0.067g$	A	A
$0.067g \leq S_{D1} < 0.133g$	B	C
$0.133g \leq S_{D1} < 0.20g$	C	D
$0.20g \leq S_{D1}$	D	D

Based on the project Seismic Design Category, the following mechanical and electrical components must meet seismic requirements of ASCE 7-10:

1. Mechanical and electrical components with I_p = 1.5 must meet the seismic requirements in Seismic Design Category C.
2. Mechanical and electrical components with I_p = 1.0 and 1.5 must meet the seismic requirements in Seismic Design Categories D, E, and F.

Mechanical and electrical components are exempt from these requirements in Seismic Design Categories D, E, or F where all of the following are true:

1. The component importance factor is 1.0;
2. The components is positively attached to the structure;
3. Flexible connectors are provided between the component and associated ductwork, piping and conduit and either
   a. The component weighs 400 lb (1780 N) or less and has a center of mass located 4 ft (1.22 m) or less above adjacent floor level; or
   b. The component weighs 20 lb (89 N) or less or, in the case of a distributed system 5 lb (73 N/m) or less.

Reference Documents

Reference documents that provide a basis of earthquake-resistant design of a particular type of nonstructural component are permitted to be used, subject to the approval of the building authority having jurisdiction. Design forces must not be less than that determined by code. Each component's seismic interactions with all other components and the structure must be considered. The component must accommodate drifts, deflections and relative displacements. Nonstructural component anchorage must meet the requirements of this chapter.

Reference Documents for Allowable Stress Design

Reference documents such as OSHPD OPA approvals that provide a basis for design for a particular system or component and define acceptance criteria in terms of allowable stress, are permitted to be used. Additional loads, including dead, live, operating, and earthquake loads, must be used. Earthquake loads determined in accordance with this chapter can be multiplied by 0.7. The component must also accommodate the relative displacements specified in this chapter.

General Design Requirements for Mechanical and Electrical Components

Compliance can be shown by either of the following means:

1. Project-specific design and documentation submitted to the authority having jurisdiction after review and acceptance by a registered design professional.
2. Submittal of a manufacturer's certification that a component is seismically qualified by at least one of the following:
   a. Analysis
   b. Testing based on a nationally recognized standard such as AC-156, acceptable to the authority having jurisdiction
   c. Experience data per nationally recognized procedures acceptable to the authority having jurisdiction

Special Certification Requirement for Designated Seismic Systems

Certification shall be provided for designated seismic systems assigned to Seismic Design Category C through F as follows:

1. Active mechanical and electrical equipment that must remain operable following the design earthquake ground motion shall be certified by the manufacturer as operable whereby active parts or energized components shall be certified exclusively on the basis of approved shake table testing or experience data unless it can be shown that the component is inherently rugged by comparison with similar seismically qualified components.
2. Components with hazardous substances and assigned an I_p of 1.5 shall be certified by the manufacturer as maintaining containment following the design earthquake ground motion by analysis, shake table testing, or experience data.

Evidence demonstrating compliance shall be submitted for approval by the authority having jurisdiction after review and acceptance by the registered design professional.

Current test standards and certifications do not address design of distribution system connections for seismic relative displacements. See Chapter 4 (of ASCE7-10) for recommendations regarding acceptable nozzle loads and flexible connector design. These requirements are most significant as they apply to piping connections, because the mass and stiffness of connected piping can damage equipment connections in an earthquake.

Determine Seismic Demands

1. Using the project S_{ds} and component I_p, there are still several factors required before the seismic design force, F_p, can be determined.
2. Table 2-10 is adapted from Table 13.6.1 of ASCE 7-10 and includes the a_p and R_p values for most common HVAC and electrical components.
3. In general, $a_p = 1.0$ for rigid components and rigidly attached components and $a_p = 2.5$ for flexible components and flexibly attached components.
4. Please refer to the table in ASCE 7-10 for the complete list.

Seismic Design Force

The horizontal seismic design force F_p must be applied at the component center of mass and distributed according to the component mass distribution using the following equation:

$$F_p = \frac{0.4 a_p S_{DS} I_p}{R_p}\left(1 + 2\frac{z}{h}\right) W_p$$

where

F_p = seismic design force

S_{ds} = short period spectral acceleration,

a_p = component amplification factor that varies from 1.0 to 2.5

R_p = Component Response Modification Factor that varies from 1.0 to 12

I_p = Component Importance Factor that is 1.0 or 1.5

Z = height of point of attachment to structure with respect to the base

h = roof height with respect to the base

W_p = the component operation weight

Also consider these additional seismic force requirements:

1. F_p shall not be greater than 1.6 S_{ds} I_p W_p and shall not be less than 0.3 S_{ds} I_p W_p.
2. F_p shall be applied independently in at least two orthogonal horizontal directions and added to service loads.
3. For vertically cantilevered systems, the force F_p must be taken in any horizontal direction.
4. The component must be designed for a concurrent vertical load of 0.2 S_{ds} W_p.

Table 2-10 Mechanical and Electrical Systems Coefficients

Mechanical and Electrical System Components	a_p	R_p
Air-side HVAC, fans, air handlers, air-conditioning units, cabinet heaters, air distribution boxes, and other mechanical components constructed of sheet metal framing	2.5	6.0
Wet-side HVAC, boilers, furnaces, atmospheric tanks and bins, chillers, water heaters, heat exchangers, evaporators, air separators, and other mechanical components constructed of high-deformability materials	1.0	2.5
Engines, turbines, pumps, compressors, and pressure vessels not supported on skirts	1.0	2.5
Skirt-supported pressure vessels	2.5	2.5
Generators, batteries, inverters, motors, transformers, and other electrical components constructed of high-deformability materials	1.0	2.5
Motor control centers, panel boards, switchgear, instrumentation cabinets and other components constructed of sheet metal framing	2.5	6.0
Communication equipment, computers, instrumentation, and controls	1.0	2.5
Roof-mounted chimneys, stacks, and cooling towers laterally braced below their center of mass	2.5	3.0
Roof-mounted chimneys, stacks, and cooling towers laterally braced above their center of mass	1.0	2.5
Lighting fixtures	1.0	1.5
Other mechanical or electrical components	1.0	1.5
Vibration-Isolated Components and Systems		
Components and systems using neoprene elements and neoprene-isolated floors with built-in or separate elastomeric snubbing devices or resilient perimeter stops	2.5	2.5
Spring-isolated components and systems and vibration-isolated floors closely restrained using built-in or separate elastomeric snubbing devices or resilient perimeter stops	2.5	2.0
Internally isolated components and systems	2.5	2.0
Suspended, vibration-isolated equipment, including inline duct devices and suspended, internally isolated components	2.5	2.5
Distribution Systems		
Piping and tubing not in accordance with ASME B31, including in-line components, constructed of high-deformability materials, with joints made by welding or brazing	2.5	9.0
Piping and tubing not in accordance with ASME B31, including in-line components, constructed of high- or limited-deformability materials, with joints made from threading, bonding, compression couplings or grooved couplings	2.5	6.0
Piping and tubing constructed of low-deformability materials such as cast iron, glass and nonductile plastics	2.5	3.0
Ductwork, including in-line components, constructed of high-deformability materials, with joints made from welding or brazing	2.5	9.0
Ductwork, including in-line components, constructed of high- or limited-deformability materials, with joints made from means other than welding or brazing	2.5	6.0
Ductwork, including in-line components, constructed of low-deformability materials such as cast iron, glass and nonductile plastics	2.5	3.0
Electrical conduit and cable trays	2.5	6.0
Bus ducts	1.0	2.5
Plumbing	1.0	2.5

5. Where a reference document provides allowable stress or allowable loads, F_p can be reduced by multiplying by a factor 0.7.
6. An R_p greater than 6 cannot be used to determine the component attachment loads.
7. Components mounted on vibration isolation systems shall have bumper restraint or snubber in each horizontal direction. If the maximum clearance (air gap) between the equipment support frame and the restraint is greater than 1/4 in. (6 mm), the design force shall be taken as $2F_p$. If the maximum clearance specified on the construction documents is 1/4 in. (6 mm), the design force may be taken as F_p.

Seismic Relative Displacements

1. The effects of seismic relative displacements must be considered.
2. For two connection points on the same structure, the relative displacement can be calculated as the difference between the allowable story drifts at the points of connection.
3. For two connection points on different structures, the relative displacement can be calculated as the sum of the allowable story drifts at the points of connection.
4. The calculated relative displacement must be added to displacements due to other loads, and the result multiplied by the seismic importance factor of the building, I_e, to determine the total seismic relative displacement. $I_e = 1.0$ for Risk Category I and II, 1.25 for Risk Category III, and 1.5 for Risk Category IV structures. I_e is only used in this chapter for seismic relative displacements.

Nonstructural Component Anchorage

1. Nonstructural components and their supports must be attached or anchored to the structure per these requirements.
2. Component must be bolted, welded, or otherwise positively attached without consideration of friction due to gravity.
3. There must be a continuous load path of sufficient strength and stiffness between the component and the structure.
4. Local elements of the structure must be designed for the component forces.
5. Component forces do not need to be increased due to anchorage considerations.
6. Design documents must include all information related to the attachments.
7. Anchors in concrete must be designed per ACI 318 Appendix D.
8. Anchors in masonry must be designed per TMS 402/ACI 503/ASCE 5. Anchors capacity must be governed by the tensile or shear strength of a ductile steel element or additional design factors must be used.
9. Postinstalled anchors in concrete must be prequalified per ACI 355.2 or other approved procedures. Postinstalled anchors in masonry must be prequalified by approved procedures.
10. Forces in attachments must consider eccentricity and prying.
11. Anchorage of attachments with multiple anchors must be done per ACI 318 or other methods to account for the stiffness and ductility of the component.
12. Power actuated fasteners in steel or concrete cannot be used in SDC D, E, or F without additional considerations.
13. Friction clips cannot be used in SDC D, E, or F without additional considerations.

Design Requirements for Mechanical Components with $I_p = 1.5$

1. Provisions must be made to eliminate impact for components vulnerable to impact, components constructed of nonductile materials and where ductility is reduced due to service conditions.
2. Evaluate loads on components due to attached service or utility lines due to differential movement of supports on separate structures.

3. Components must be designed to accommodate the relative seismic displacements where piping or HVAC ductwork components are attached to structures that could displace relative to each other.

Other General Requirements

1. The failure of an essential or nonessential component must not cause the failure of an essential component.
2. Component design must include consideration of both flexibility and strength.

Mechanical and Electrical Component Support General Design Requirements

1. Supports and attachment of support to the component must be designed for forces and displacements.
2. Supports include structural members, braces, skirts, legs, saddles, pedestals, cables, guys, stays, snubbers, tethers, and elements forged or cast as part of the component.
3. Industry standard and proprietary supports must be designed for loads and stiffness.
4. Support and support attachment materials must be suitable for the application.
5. Support design and construction must insure that support engagement is maintained.
6. Reinforcement or Bellville washers must be provided at bolted connections of a sheet metal base when required unless the base is reinforced with stiffeners.
7. Weak-axis bending of cold-formed steel supports must be specifically evaluated.
8. Components mounted on vibration isolators require horizontal and vertical restraints. Isolator housings and restraints must be made of ductile materials. An elastomeric pad of appropriate thickness must be used to limit the impact load.
9. Expansion anchors on non-vibration-isolated equipment over 10 hp (7.45 kW) must be qualified by ACI 355.2.
10. The attachments to concrete for piping, boilers, and pressure vessels must be suitable for cyclic loads.
11. The attachment of mechanical equipment with drilled and grouted-in-place anchors for tensile load applications must use expansive cement or expansive epoxy grout.
12. Light fixtures, lighted signs, and ceiling fans not connected to duct or pipe, where supported by chains or otherwise suspended, are not required to be braced for seismic if the support is designed for 1.4 × dead load and an equivalent simultaneous horizontal load. Failure can not cause the failure of an essential component and the connection must allow 360° motion.

Utility and Service Line Special Requirements

1. Utility lines must be designed for differential motions between structures.
2. Interruption of service must be considered for critical facilities.
3. Special attention must be given to underground systems in Site Class E or F where S_{ds} is greater than 0.33.

Unless ductwork is designed to carry toxic, highly toxic, or flammable gases or used for smoke control, design for seismic forces and displacements is not required on ductwork when:

1. Trapeze assemblies are used to support ductwork and the total weight of the ductwork supported by the trapeze assemblies is less than 10 lb/ft (146 N/m) or,
2. The ductwork is supported by hangers and each hanger in the duct run is 12 in. (305 mm) or less in length from the duct support point to the supporting structure. Where rod hangers are used, they shall be equipped with swivels to prevent inelastic bending of the rod.

3. Provisions are made to avoid impact with other ducts or mechanical components or to protect the ducts in the event of such impact and HVAC ducts have a cross-sectional area of 6 ft^2 (0.557 m^2) or less or weigh 17 lb/ft (248 N/m) or less.

Special Rules for Inline Duct Components

1. Components installed that are installed in-line with the duct system and have an operating weight greater than 75 lb (334 N) must be supported and braced independent of the duct system.
2. In-line components that weigh 75 lb (334 N) or less can remain unbraced in unbraced ductwork. Where the ductwork is braced, the additional weight of in-line components must be braced by the duct bracing.
3. Unbraced piping attached to in-line equipment of any weight, braced or unbraced, must have flexible connectors to seismic relative displacements.

Design of piping systems and attachments for the seismic forces and relative displacements shall not be required where one of the following conditions apply:

1. Trapeze assemblies are used to support piping whereby no single pipe exceeds the limits set forth in 3a, 3b, or 3c below and the total weight of the piping supported by the trapeze assemblies is less than 10 lb/ft (146 N/m).
2. The piping is supported by hangers and each hanger in the piping is 12 in. (305 mm) or less in length from the top of the pipe to the supporting structure. Where pipes are supported on a trapeze, the trapeze shall be supported by hangers having a length of 12 in. (305 mm) or less. Where rod hangers are used, they shall be equipped with swivels, eye nuts, or other devices to prevent bending in the rod.
3. Piping having an R_p in Table 13.6-1 of 4.5 or greater is used and provisions are made to avoid impact with other structural or nonstructural components or to protect the piping in the event of such impact and where the following size requirements are satisfied:
   a. For Seismic Design Category C where I_p is 1.5, the nominal pipe size shall be 2 in. (50 mm) or less.
   b. For Seismic Design Categories D, E, or F and I_p is 1.5, the nominal pipe size shall be 1 in. (25 mm) or less.
   c. For Seismic Design Categories D, E, or F where I_p is 1.0, the nominal pipe size shall be 3 in. (80 mm) or less.

Piping system design when $I_p = 1.5$

1. Unless otherwise exempt from seismic considerations, piping systems must be designed for seismic forces and seismic relative displacements.
2. Where other applicable standards or recognized design manuals are not used, piping must be designed so that seismic and other design loads limit the pipe stress to 90% yield strength for piping made with ductile materials, 70% yield strength for threaded connected ductile piping, 10% yield strength for nonductile piping, and 8% yield strength for threaded connected nonductile piping.
3. Piping connections to other components must be designed to accommodate relative seismic displacements.

Flexible Connectors for Piping

Piping not designed to accommodate seismic relative displacements at the connections to other components must be provided with connections having sufficient flexibility to avoid failure of the connection to the other component. This is most often achieved with flexible connectors to limit the effects of the relative displacement.

Electrical components with $I_p = 1.5$ design requirements

1. Design for seismic forces and displacements.
2. Design to eliminate seismic impacts between components.
3. Design for loads imposed by utilities or service lines attached to separate structures.
4. Special battery rack requirements include wraparound restraints and spacers to protect batteries and racks evaluated for load capacity.
5. Internal coils of dry type transformers positively attached.
6. Slide-out components must have latches.
7. Cabinet design per NEMA. Lower shear panel field modifications must be evaluated.
8. Attachment of field added external items over 100 lb (445 N) must be evaluated.
9. Components must be designed for relative displacement of connected conduit, cable trays, or other distribution systems.

Electrical Components with $I_p = 1.0$ or 1.5 design requirements:

1. Conduit, cable tray, and other electrical distribution systems must be designed for forces and relative displacements.
2. Conduit greater than 2 1/2 in. (63.5 mm) trade size and attached to panels, cabinets, or other equipment and subject to seismic relative displacement must be provided with flexible connectors or designed for forces and displacements.
3. Design for seismic forces and displacements are not required when
   a. Trapeze assemblies are used to support electrical distribution systems and the total weight of the system supported by trapeze assemblies is less than 10 lb/ft (146 N/m).
   b. The distribution system is supported by hangers and each hanger in the run is less than 12 in." (305 mm) or less in length from the distribution system support point to the supporting structure. When rods are used they must have swivels.
   c. Conduit is less than 2 1/2 in. (63.5 mm) trade size.

CALIFORNIA BUILDING CODE (CBC 2010)

The 2010 California Building Code is based on IBC 2009 and ASCE 7-05, but has been modified to include many requirements for mechanical and electrical components from ASCE 7-10. The modifications are included in CBC Chapters 16 and 16A.

This summary includes the modified requirements of CBC 2010 as they apply to the design of mechanical and electrical systems. For general requirements, refer to the IBC 2006-2009 summary. For exact requirements, please refer to CBC 2010.

Chapter 16A includes the design requirements for buildings regulated by DSA, including public elementary and secondary schools, as well as community colleges and state-owned or state-leased essential services buildings and projects regulated by OSHPD, including hospitals, skilled-nursing facilities, intermediate-care facilities, and correctional-treatment centers.

Chapter 16 includes the design requirements for all other buildings in California including general use commercial and residential buildings. Chapter 16 also includes the design requirements for community colleges regulated by DSA-SS/CC, some single-story OSHPD skilled-nursing and intermediate-care facilities, and OSHPD-licensed clinics and outpatient facilities. However, for the purposes of this summary, the requirements for mechanical and electrical components listed for Chapter 16A are appropriate for these buildings.

2010 CBC CHAPTER 16A SUMMARY

Mechanical and electrical components are exempt from seismic requirements when

1. The component is positively attached to the structure; and
2. Flexible connectors are provided between the component and associated ductwork, piping and conduit and either

a. The component weighs 400 lb (1780 N) or less and has a center of mass located 4 ft (1.22 m) or less above adjacent floor or roof level that directly supports the component, except where Special Seismic Certification Requirements for Designated Seismic Systems apply, or

b. The component weighs 20 lb (89 N) or less or, in the case of a distributed system, 5 lb/ft (73 N/m) or less, except for the attachment of equipment with hazardous content, which must be shown on the plan regardless of weight.

Seismic Relative Displacement

Use the building importance factor I in the calculation to determine the displacement requirements of displacement-sensitive components.

Nonstructural Component Anchorage

1. The design force for the component anchorage must be based on a maximum $R_p = 6$.
2. Anchors in concrete must be designed per ACI 318 Appendix D.
3. Anchors in masonry must be designed per ACI 503. Anchor capacity must be governed by the tensile or shear strength of a ductile steel element or additional design factors must be used.
4. Postinstalled anchors in concrete must be prequalified per ACI 355.2, ICC-ES AC193, or ICC-ES AC308. Postinstalled anchors in masonry must be prequalified per ICC-ES AC01, AC58, or AC106.
5. Adhesive anchors are not permitted in overhead installations or in applications with continuous tension loads.
6. Anchors prequalified for seismic applications do not need to be governed by the steel strength of a ductile steel element.
7. Power-actuated fasteners in steel or concrete cannot be used in SDC D, E or F without additional considerations.

Ductwork Design Requirements

Ductwork that carries toxic, highly toxic, or flammable gases or is used for smoke control must be designed for seismic forces and seismic relative displacements.

Design of ductwork for seismic forces and relative displacements is not required when either

1. Trapeze assemblies are used to support ductwork and the total weight of the ductwork supported by the trapeze assemblies is less than 10 lb/ft (146 N/m) or,
2. The ductwork is supported by hangers and each hanger in the duct run is 12 in. (305 mm) or less in length from the duct support point to the supporting structure. Where rod hangers are used with a diameter greater than 3/8 in. (9.5 mm), they shall be equipped with swivels to prevent inelastic bending of the rod.
3. Provisions are made to avoid impact with larger ducts or mechanical components or to protect the ducts in the event of such impact and HVAC ducts have a cross-sectional area of 6 ft^2 (0.557 m^2) or less or weigh 10 lb/ft (146 N/m) or less.

Duct Construction Requirements

HVAC duct systems fabricated and installed per approved standards meet the requirements of this section.

Special Requirements for In-line Duct Components

1. Components that are installed in-line with the duct system and have an operating weight greater than 75 lb (334 N) must be supported and braced independent of the duct system.
2. In-line components that weigh 75 lb (334 N) or less can remain unbraced in unbraced ductwork. Where the ductwork is braced, the additional weight of in-line components must be braced by the duct bracing.
3. Unbraced piping attached to in-line equipment of any weight, braced or unbraced, must have flexible connectors to seismic relative displacements.

Design of piping systems and attachments for the seismic forces and relative displacements shall not be required where one of the following conditions apply:

1. Trapeze assemblies are used to support piping whereby no single pipe exceeds the limits set forth in 3a or 3b below and the total weight of the piping supported by the trapeze assemblies is less than 10 lb/ft (146 N/m).
2. The piping is supported by hangers and each hanger in the piping is 12 in. (305 mm) or less in length from the top of the pipe to the supporting structure. Where pipes are supported on a trapeze, the trapeze shall be supported by hangers having a length of 12 in. (305 mm) or less. Where rod hangers are used with a diameter greater than 3/8in. (9.5 mm), they shall be equipped with swivels, eye nuts, or other devices to prevent bending in the rod.
3. Piping having an R_p in Table 13.6-1 of 4.5 or greater is used and provisions are made to avoid impact with other structural or nonstructural components or to protect the piping in the event of such impact and where the following size requirements are satisfied:
   a. For Seismic Design Categories D, E, or F where $I_p = 1.5$, the nominal pipe size shall be 1 in. (25 mm) or less.
   b. For Seismic Design Categories D, E, or F where $I_p = 1.0$, the nominal pipe size shall be 3 in. (80 mm) or less.

Piping System Design Requirements when $I_p = 1.5$

1. Unless otherwise exempt from seismic considerations, piping systems must be designed for seismic forces and seismic relative displacements.
2. Where other applicable standards or recognized design manuals are not used, piping must be designed so that seismic and other design loads limit the pipe stress to 90% yield strength for piping made with ductile materials, 70% yield strength for threaded connected ductile piping, 10% yield strength for nonductile piping, and 8% yield strength for threaded connected nonductile piping.
3. Piping connections to other components must be designed to accommodate relative seismic displacements.

Flexible Connectors for Piping

Piping not designed to accommodate seismic relative displacements at the connections to other components must be provided with connections having sufficient flexibility to avoid failure of the connection to the other component.

This is most often achieved with flexible connectors to limit the effects of the relative displacement.

Electrical Components with $I_p = 1.0$ or 1.5 Design Requirements

1. Conduit, cable tray, and other electrical distribution systems must be designed for forces and relative displacements.

2. Conduit greater than 2 1/2in. (63.5 mm) trade size and attached to panels, cabinets, or other equipment and subject to seismic relative displacement must be provided with flexible connectors or designed for forces and displacements.
3. Design for seismic forces and displacements are not required when:
   a. Trapeze assemblies are used to support electrical distribution systems and the total weight of the system supported by trapeze assemblies is less than 10 lb/ft (146 N/m).
   b. The distribution system is supported by hangers and each hanger in the run is less than 12 in. (305 mm) or less in length from the distribution system support point to the supporting structure. When rods hangers with a diameter greater than 3/8 in. (9.5 mm) are used they must have swivels to prevent inelastic bending of the rod.
   c. Conduit is less than 2 1/2 in. (63.5 mm) trade size regardless of I_p.

2010 CBC CHAPTER 16 SUMMARY

HVAC ductwork with $I_p = 1.5$

Seismic supports are not required for HVAC ductwork with $I_p = 1.5$ if either of the following conditions is met for the full length of each duct run:

1. HVAC ducts are suspended for hangers 12 in. (305 mm) or less in length with hangers detailed to avoid significant bending of the hangers and their attachments, or
2. HVAC ducts have a cross-sectional area of less than 6 ft^2 (0.557 m^2).

Based on the project Seismic Design Category, the following mechanical and electrical components are exempt from the seismic requirements:

1. All components are exempt from seismic requirements in Seismic Design Category A.
2. Mechanical and electrical components are exempt from seismic requirements in Seismic Design Category B.
3. Mechanical and electrical components with $I_p = 1.0$ are exempt from seismic requirements in Seismic Design Category C.

Mechanical and electrical components are exempt from these requirements in Seismic Design Categories D, E, or F where all of the following are true:

1. The component importance factor I_p is 1.0;
2. The components is positively attached to the structure;
3. Flexible connectors are provided between the component and associated ductwork, piping and conduit and either
   a. The component weighs 400 lb (1780 N) or less and has a center of mass located 4 ft (1.22 m) or less above adjacent floor level; or
   b. The component weighs 20 lb (89 N) or less or, in the case of a distributed system, 5 lb/ft (73 N/m) or less.

2005 NATIONAL BUILDING CODE OF CANADA

According to the 2005 *National Building Code of Canada*, seismic designs must be considered for machinery, fixtures, equipment, ducts, tanks, and pipes in all seismic regions of Canada. The magnitude of forces and damage are represented in seismic zoning maps that are based on a statistical analysis of the earthquakes that have been experienced in Canada and adjacent regions.

Lateral Force Factor V_p

Mechanical and electrical equipment and its anchorage shall be designed for a lateral force V_p applied through the center of mass,

$$V_p = 0.3\, F_a\, S_a(0.2)\, I_E\, S_p W_p \tag{2-7}$$

where

F_a = the acceleration-based site coefficient from table 2-12

I_E = the importance factor for the building from table 2-13

S_a (0.2) = the spectral response acceleration value at 0.2 s

Detailed spectral response accelerations, S_a at 0.2 s can be found in Division B, Appendix C, Table C-2 of the code for many locations in Canada.

S_p is equal to $C_p\, A_r\, A_x \,/\, R_p$, where

C_p = the component factor from Table 2-14

A_r = the component force amplification factor from Table 2-14

A_x = the height factor $(1 + 2\ h_x/h_n)$, where

h_x = the height level of the component within the building

h_n = the total number of levels of the building

R_p = the component response modification factor from Table 2-14

W_p = the weight of the equipment

The following also apply to Equation 2-7:

- Earthquake forces shall be applied in the horizontal direction that results in the most critical loading for design.
- For buildings other than postdisaster buildings, where $I_E F_a S_a(0.2)$ is less than 0.35, the requirements of equation 2-5 need not apply to categories 6 thru 21 of table 2-14.
- When the mass of a tank plus its contents is greater than 10% of the mass of the supporting floor, the lateral forces shall be determined by rational analysis.
- Connections to the structure of elements and components listed in table 2-14 shall be designed to support the component or element for gravity loads, shall be in accordance with equation 2-5, and shall satisfy the following:
  - Friction due to gravity shall not be considered to provide resistance to seismic forces.
  - R_p for nonductile connections, such as adhesives or power-actuated fasteners, shall be taken as 1.00.
  - R_p for anchorage using shallow expansion anchors, chemical, epoxy or cast in place anchors shall be 1.5, where shallow anchors are those with a ratio of embedment length to diameter of less than 8.
  - Power-actuated fasteners and drop-in anchors shall not be used for tension loads.
- Lateral deflections of elements or components shall be based on the loads from equation 2-5 and lateral deflections obtained from an elastic analysis shall be multiplied by R_p/I_E to give realistic values of the anticipated deflections
- Seismic restraint for suspended pipes, equipment, ducts, electrical cable trays, etc. shall be designed to meet the force and displacement requirements of this section and be constructed in a manner that will not subject the hanger rods to bending
- Isolated suspended equipment and components, such as pendant lights, may be designed as a pendulum system provided that adequate chains or cables capable of supporting 2.0 times the weight of the suspended component are provided and the deflection requirements are satisfied

Example 5

A boiler is rigidly mounted to a concrete slab connected to steel pipe, located on grade. The boiler is in a 12-story office building. The building is located in Alberni, BC.

W_p = 10,000 lb (4536 kg)

Table 2-11 Site Classification for Seismic Site Response

Site Class	Ground Profile Name	Average Properties in Top 30 m, as per code: Average Shear Wave Velocity V_s, m/s (ft/s)	Average Standard Penetration Resistance, N_{60}	Soil Undrained Shear Strength S_u, kPa (psi)
A	Hard Rock	$V_s > 1500$ (4500)	n/a	n/a
B	Rock	760 (2500) $< V_s \leq$ 1500 (4500)	n/a	n/a
C	Very Dense Soil and Soft Rock	360 (1200) $< V_s <$ 760 (2500)	$N_{60} > 50$	$S_u > 100$ (14.5)
D	Stiff Soil	180 (600) $< V_s <$ 360 (1200)	$15 \leq N_{60} \leq 50$	50 (7.25) $< S_u \leq$ 100 (14.5)
E	Soft Soil	$V_s < 180$ (600)	$N_{60} < 15$	$S_u < 50$ (7.25)
		Any profile with more than 3 m (10 ft) of soil with the following characteristics: • Plasticity index PI > 20, • Moisture content W ≥ 40%, and • Undrained shear strength: $S_u < 25$ kPa (3.6 psi).		
F	Other Soils[1]	Site Specific Evaluation Required		

Notes:
1. See the code for explanation and examples of "other soils"

Table 2-12 Values of F_a as a Function of Site Class and S_a (0.2)

Site Class	Values of F_a: S_a (0.2) ≤ 0.25	S_a (0.2) = 0.50	S_a (0.2) = 0.75	S_a (0.2) = 1.00	S_a (0.2) ≥ 1.25
A	0.7	0.7	0.8	0.8	0.8
B	0.8	0.8	0.9	1.0	1.0
C	1.0	1.0	1.0	1.0	1.0
D	1.3	1.2	1.1	1.1	1.0
E	2.1	1.4	1.1	0.9	0.9
F	(1)	(1)	(1)	(1)	(1)

Table 2-13 Importance Factor I_E

Importance Category	Importance Factor I_E: ULS	SLS
Low	0.8	(Note 1)
Normal	1.0	
High	1.3	
Postdisaster	1.5	

Notes:
1. See Appendix A.

Table 2-14 Elements of Nonstructural Components and Equipment

Category	Equipment or other Nonstructural Component	C_p	A_r	R_p
5	Towers, chimneys, smokestacks, and penthouses when connected to or forming part of a building	1.00	2.50	2.50
7	Suspended ceilings, light fixtures, and other attachments to ceilings with independent vertical support	1.00	1.00	2.50
11	Machinery, fixtures, equipment, ducts, and tanks (including contents)			
	That are rigid and rigidly connected[1]	1.00	1.00	1.25
	That are flexible or flexibly connected[1]	1.00	2.50	2.50
12	Machinery, fixtures, equipment, ducts, and tanks (including contents) containing toxic or explosive materials, materials having a flash point below 38°C (100°F) or firefighting fluids			
	That are rigid and rigidly connected[1]	1.50	1.00	1.25
	That are flexible or flexibly connected[1]	1.50	2.50	2.50
13	Flat-bottom tanks (including contents) attached directly to a floor at or below grade within a building	0.70	1.00	2.50
14	Flat-bottom tanks (including contents) attached directly to a floor at or below grade within a building containing toxic or explosive materials, materials having a flash point of 38°C (100°F) or firefighting fluids	1.00	1.00	2.50
15	Pipes, ducts, cable trays (including contents)	1.00	1.00	3.00
16	Pipes, ducts (including contents) containing toxic or explosive materials	1.50	1.00	3.00
17	Electrical cable trays, bus ducts, and conduits	1.00	2.50	5.00
18	Rigid components with ductile materials and connections	1.00	1.00	2.50
19	Rigid components with nonductile materials and connections	1.00	1.00	1.00
20	Flexible components with ductile materials and connections	1.00	2.50	2.50
21	Flexible components with nonductile materials and connections	1.00	2.50	1.00

Notes:
1. Elements or components shall be assumed to be flexible or flexibly connected, unless it can be shown that the fundamental period of the element or component is less than or equal to 0.06 s, in which case the element or component is classified as being rigid or rigidly connected.

$S_a(0.2) = 0.75$
$I_E = 1.0$
$F_a = 1.1$
$R_p = 1.25$
$C_p = 1.0$
$h_x = 0$
$h_n = 12$
$A_r = 1.0$
$A_x = 1.0 + 2\, h_x/h_r = 1.0 + 2(0/12) = 1.0$
$S_p = C_p A_r A_x / R_p = (1.0)(1.0)(1.0)/1.25 = 0.8$

$$V_p = 0.3\, F_a\, S_a(0.2)\, I_E\, S_p\, W_p = (0.30)(1.1)(0.75)(1.0)(0.8)(10{,}000 \text{ lb}) = 1980 \text{ lb } (8.8 \text{ kN})$$

Example 6

A boiler is rigidly mounted to a concrete slab connected to steel pipe, located on the roof. The boiler is in a 12-story office building. The building is located in Alberni. BC.

W_p = 10,000 lb (4536 kg)
S_a (0.2)= 0.75
I_E = 1.0
F_a = 1.1
R_p = 1.25
C_p = 1.0
h_x = 12
h_n = 12
A_r = 1.0
A_x = $1.0 + 2\, h_x/h_r = 1.0 + 2(12/12) = 3.0$
S_p = $C_p\, A_r\, A_x\, /R_p = (1.0)(1.0)(3.0)/1.25 = 2.4$

$$V_p = 0.3\, F_a\, S_a(0.2)\, I_E\, S_p\, W_p = (0.30)(1.1)(0.75)(1.0)(2.4)(10{,}000 \text{ lb}) = 5940 \text{ lb } (26.4 \text{ kN})$$

Example 7

A pump is vibration isolated, mounted to concrete slab connected to steel pipe, and located on grade. The pump is in a six-story hospital. The hospital is located in Quebec City.

W_p = 2000 lb (907 kg), including concrete inertia base
S_a (0.2)= 0.59
I_E = 1.5
F_a = 1.3
R_p = 2.50
C_p = 1.0
h_x = 0
h_n = 12
A_r = 2.5
A_x = $1.0 + 2\, h_x/h_r = 1.0 + 2(0/12) = 1.0$
S_p = $C_p\, A_r\, A_x\, /R_p = (1.0)(2.5)(1.0)/2.50 = 1.0$

$$V_p = 0.3\, F_a\, S_a(0.2)\, I_E\, S_p\, W_p = (0.30)(1.3)(0.59)(1.5)(1.0)(2{,}000 \text{ lb}) = 690 \text{ lb } (3.1 \text{ kN})$$

Example 8

A pump is vibration isolated, mounted to concrete slab connected to steel pipe, and located on the roof. The pump is in a six-story hospital. The hospital is located in seismic zone 3.

W_p = 2000 lb (907 kg), including concrete inertia base
S_a (0.2 = 0.59
I_E = 1.5
F_a = 1.3
R_p = 2.50
C_p = 1.0
h_x = 12
h_n = 12

A_r = 2.5

A_x = $1.0 + 2\,h_x/h_r = 1.0 + 2(12/12) = 3.0$

S_p = $C_p\,A_r\,A_x\,/R_p = (1.0)(2.5)(3.0)/2.50 = 3.0$

$$V_p = 0.3\,F_a\,S_a(0.2)\,I_E\,S_p\,W_p = (0.30)(1.3)(0.59)(1.5)(3.0)(2{,}000\text{ lb})$$
$$= 2071\text{ lb }(9.2\text{ kN})$$

BIBLIOGRAPHY

ACI. 2008. *318-08, Building Code Requirements for Structural Concrete*. Farmington Hills, MI: American Concrete Institute.

ASCE. 2005. *Minimum Design Loads for Buildings and Other Structures, ASCE 7-05*. Reston,VA: American Society of Civil Engineers.

ASCE. 2010. *Minimum Design Loads for Buildings and Other Structures, ASCE 7-10*. Reston,VA: American Society of Civil Engineers.

CBSC. 2007. *California Building Code*. Sacramento, CA: California Building Standards Commission.

CBSC. 2010. *California Building Code*. Sacramento, CA: California Building Standards Commission.

ICC. 2006. *2006 International Building Code® (IBC®)*. Washington, DC: International Code Council.

ICC. 2009. *2009 International Building Code® (IBC®)*. Washington, DC: International Code Council.

NRCC. 2005. *National Building Code of Canada*. Ottawa, ON: National Research Council of Canada.

3 Specification Guidelines

One of the most important ways to help ensure that a project in a seismic area is constructed as planned is to specify it correctly. A properly prepared set of construction documents (drawings and specifications), detailed review and approval of shop drawings, and field enforcement are essential. For example, problems could arise if the vibration isolation and the seismic restraint system are not specified together. Separating the two specifications could leave a gap in responsibility and system design, and, therefore, they must be appropriately coordinated. The restraint system should be looked at as an engineered system and not a selection of hardware. The system should take into account the complete load path from equipment or system to structure, as the design is only as strong as its weakest link.

Note to design professionals: The following are basic guidelines for consideration when writing a seismic restraint specification. This is not a complete specification. This guideline covers force levels, systems that should be restrained, product specifications, and general notes. Seismic restraint specifications may be added to the vibration isolation specification or have its own section. It is not recommended that the seismic restraint specifications be added to the specification section for the individual pieces of equipment. It is recommended that the specification sections for the individual pieces of equipment include a requirement that the equipment manufacturer supply a certification of its seismic capability. This is required by the 2009 International Building Code® (IBC®), published by the International Code Council (ICC), and by other model codes for certain buildings.

IDENTIFICATION OF APPLICABLE CODES

See Chapter 2 for a guide to the building codes and their seismic restraint levels. Specifications should require seismic restraints for code-required force levels or may specify higher force levels if appropriate for the application and/or the owner's requirements. The specification should identify the applicable codes and building classification.

This information can be obtained from the project structural engineer. The mechanical and electrical engineers must identify which systems require special seismic attachments. These are typically life safety systems but may vary due to local code or owner requirements. These systems must be clearly identified and defined. Life safety systems generally require both mechanical and electrical systems, except in the case of fire alarms. Therefore, there must be a coordinated effort between the two disciplines.

Table 3-1 Equipment that Should Be Restrained

Electrical Equipment		
Battery racks	Electrical raceways	Switch gear
Bus ducts	Generators	Transformers
Cable trays	Light fixtures	Unit substations
Conduit	Motor control centers	Variable-frequency drives
Electrical panels		
Mechanical Equipment		
A/C units	Computer room A/C units	Heat exchangers
Air distribution boxes	Condensers	Piping
Air-handling units	Condensing units	Pumps (all types)
Air separators	Cooling towers	Rooftop units
Boilers	Ductwork	Sound attenuators
Cabinet heaters	Fans (all types)	Tanks (all types)
Chillers	Fan-coil units (suspended)	Unit heaters
Compressors	Fan terminal units	Water heaters

EQUIPMENT AND SYSTEMS THAT SHOULD BE RESTRAINED

Table 3-1 consists of partial tables of equipment and systems that should be restrained.

Life Safety Systems

1. All systems involved with fire protection, including sprinkler piping, fire pumps, jockey pumps, fire pump control panels, service water supply piping, water tanks, fire dampers, smoke dampers, and smoke exhaust systems.
2. All systems involved with and/or connected to emergency power supply, including all generators, transfer switches, transformers, all power and control flow paths to fire protection, smoke control, emergency lighting systems, and other life safety systems.
3. All systems carrying flammable, toxic, or other hazardous substances.
4. All medical and life support systems.
5. All air-handling systems that are part of the building smoke control and removal systems, including emergency controls, conduit, duct, dampers, etc.
6. The specification for a specific facility should identify the life safety systems that are to remain operational during and after a seismic event.

The following are examples.

Ventilation systems:

Supply Fan ________, serving ______________, together with all ductwork and appurtenances from ______________ to ______________. This includes all power and controls required for system operation.

Exhaust Fan ________, serving ______________, together with all ductwork and appurtenances from ________________ to ______________. This includes all power and controls required for system operation.

Heating systems:

Boiler ____________, serving ______________, together with pumps ________________, distribution piping, terminal units, and appurtenances. This includes all power and controls required for system operation.

Cooling systems:
Chiller ____________, serving ____________, together with pumps ____________________, distribution piping, terminal units, and appurtenances. This includes all power and controls required for system operation.
Electrical power systems:
Services ____________, serving ____________, together with feeders ________________, transformers ____________, switchboards ______________, mains ________________, distribution panels ____________________, and appurtenances.
Alarm systems:
System ______________, serving ____________, including the following equipment __________ ______________________________.
Control and monitoring systems (including BMS or SCADA):
System ______________________, serving ______________________, including the following equipment __________________.
Other systems or equipment:
____________________________________.

An example of how the preceding would be filled in is as follows:

Supply Fan SF-23, serving staircase B pressurization, together with all ductwork and appurtenances from outside air intake to discharge registers. This includes all power and controls required for system operation.

SUBMITTAL REQUIREMENTS AND SEISMIC CERTIFICATION

Descriptive Data

1. Catalog cuts or data sheets on vibration isolators and specific restraints detailing compliance with the specification.
2. Detailed schedules of flexibly and rigidly mounted equipment, showing vibration isolators and seismic restraints by referencing numbered descriptive drawings.

Shop Drawings

1. Submit fabrication details for equipment bases, including dimensions, structural member sizes, and support point locations.
2. Provide all details of suspension and support for equipment hung from the ceiling.
3. Where walls, floors, slabs, or supplementary steel work are used for seismic restraint locations, details of acceptable attachment methods for ducts, conduit, and pipe must be included and approved before the condition is accepted for installation. Restraint manufacturers' submittals must include spacing, static loads, and seismic loads at all attachment and support points.
4. Provide specific details of seismic restraints and anchors; include number, size, and locations for each piece of equipment.

Seismic Certification and Analysis

See Chapter 4 on fragility levels and their use.

Note to design professionals: Many states have laws that require project drawings and calculations be signed and sealed by a professional engineer licensed in the state where the project is located. The designer may choose to perform these detailed calculations as part of the scope of work. However, since actual equipment dimensions, weights, and mounting details may not be known until the equipment is acquired, the designer may require that the contractor provide the seismic calculations. An engineer on the seismic restraint vendor's staff, or a specialty consultant, would perform these calculations and submit them with the restraint submittals and installation details. These calculations would either be forwarded to the appropriate authority having jurisdiction or become part of the project records. Check local laws before writing a specification or include this requirement.

Note to design professionals: For an individual piece of equipment or system, either section 1 or 2 below may be used. Sections 3 and 4 are required with either section 1 or 2.

1. The contractor shall retain a specialty consultant or equipment manufacturer to develop a seismic restraint system and perform seismic calculations in accordance with the state and local codes and additional requirements specified in this section. Calculations, restraint selections, and installation details shall be done by a professional engineer experienced in seismic restraint design and installation and licensed in the state where the project is located.

   The seismic restraint design, consisting of calculations, restraint selection, installation details, and other documentation, shall be submitted. This submittal shall be signed and sealed by a professional engineer, as stated above. This submittal will become part of the project design calculations, included in the project records, and when required, will be submitted to the authority having jurisdiction.

   The seismic restraint design shall clearly indicate the attachment points to the building structure and all design forces (in X, Y, and Z direction) at the attachment points. The seismic restraint engineer shall coordinate all attachments with the building's structural engineer of record, who shall verify the attachment methods and the ability of the building structure to accept the loads imposed.

   The seismic restraint design shall be based on actual equipment data (dimensions, weight, center of gravity, etc.) obtained from submittals or the manufacturers. The equipment manufacturer shall verify that the attachment points on the equipment can accept the combination of seismic, weight, and other loads imposed. For life safety systems and other systems that must remain operational during and after an earthquake, the manufacturer shall provide certification that the equipment can accept the loads imposed and remain operational. (See Chapter 4, "Equipment Fragility Levels.")
2. Seismic restraint calculations are not required if one of the following conditions is met.
   a. The engineer of record designs the restraint system based on specified equipment and provides complete details on drawings, and the contractor supplies and installs the specified equipment as detailed.
   b. The engineer of record designs the restraint system based on specified equipment and provides complete details on drawings, and the contractor selects alternative equipment where the dimensions and weights are within the tolerance allowed in the engineer's design.
3. Analysis should include calculated dead loads, static seismic loads, and capacity of materials utilized for the connection of the equipment or system to the structure. Analysis should detail anchoring methods, bolt diameter, embedment, and/or welded length. All seismic restraint devices should be designed to accept, without failure, the forces detailed above in "Identification of Application Codes" in this chapter acting through the equipment or system's center of gravity.
4. All seismic restraints and combination isolator/restraints should have verification of their seismic capabilities. Manufacturers may verify their capabilities by testing that is witnessed by an independent professional engineer. A manufacturers' association is currently developing a uniform set of test standards. Independent approval can also be obtained by agencies such as the Office of Statewide Health, Planning and Development (OSHPD) from the State of California, NES, ICC ES, Factory Mutual, Underwriters Lab, Seismic Source International, etc.

Note to design professionals: Seismic requirements should be included in the contract documents as follows (example uses CSI specification format):

Seismic specifications may be in Section 13080, Special Construction—Vibration Control and Seismic Restraints, or in Section 15241, Mechanical—Vibration Control and Seismic Restraints. If both 13080 and 15241 are used, a cross reference to the other should be added to both.

Section 15010, Basic Mechanical Requirements, should reference section 13080 or 15241.

Section 16010, Basic Electrical Requirements, should reference section 13080 or 15241.

Drawing(s) of "Typical Seismic Restraint Details" should be included in the design. The drawing(s) may contain a note stating that the details are typical and may change as the result of the seismic restraint design to be performed under specification Section 15241 (or 13080).

PRODUCT SPECIFICATION

Note to design professionals: The following are descriptions for typical seismic restraint types. Please refer to Chapter 48 of the 2011 *ASHRAE Handbook—HVAC Applications* for complete descriptions and selection charts for vibration isolation. These seismic restraint types or combination vibration isolator/seismic restraints should be used in conjunction with or replace the vibration isolators in Chapter 48.

Elastomeric pads designed for the application. Pads may be either a single layer or two layers separated horizontally by a 16 gage (1.5 mm) galvanized shim. Load distribution plates shall be used as required. If bolting through the pad is required, type 4 bushings should be used. The neoprene should be compounded to bridge bearing specifications. See Figure 3-1.

Neoprene mountings having all-directional seismic capability. The mount should consist of a ductile iron casting or welded steel housing containing a molded neoprene element. The element should prevent the central threaded sleeve and attachment bolt from contacting the housing during normal operation. The neoprene should be compounded to bridge bearing specifications. See Figure 3-2.

Neoprene bushing assembly for restraining electric panelboards. The assemblies should have both neoprene and seismic restraint capabilities in all directions. The assemblies should have steel sections to spread loads over the neoprene area and to prevent the panel from cutting through the assembly. The assembly should have a positive steel stop to allow

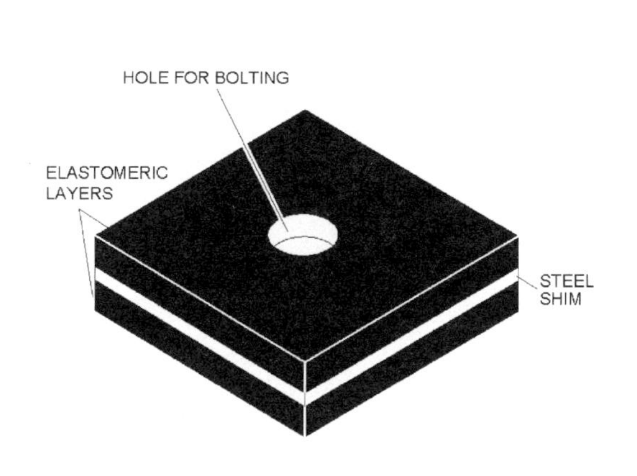

Figure 3-1 Layered neoprene pads with steel shim.

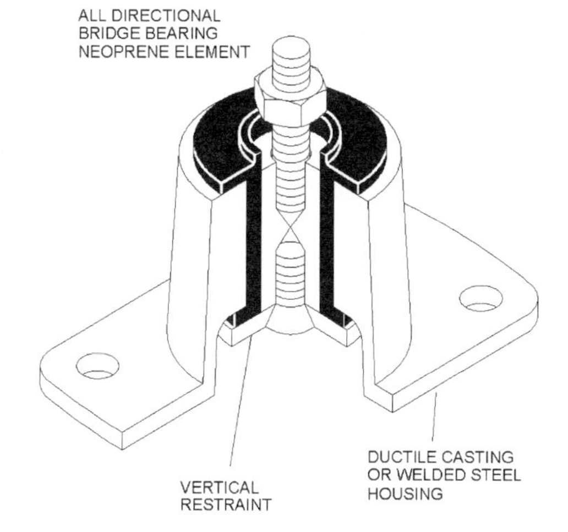

Figure 3-2 All-directional bridge bearing neoprene element.

tightening of the through-bolt without short-circuiting it. The neoprene should be compounded to bridge bearing specifications. See Figure 3-3.

A one-piece molded bridge bearing neoprene washer/bushing. The bushing should surround the anchor bolt and have a flat washer face to avoid metal-to-metal contact. See Figure 3-4.

Spring mountings, as in the 2011 *ASHRAE Handbook*, Chapter 48, type 3, should be built into a ductile casting or welded steel housing to provide all-directional seismic snubbing. The snubber should be adjustable vertically and allow a maximum of 1/4 in. (6 mm) travel in all directions before contacting the resilient snubbing collars. See Figures 3-5 A, B, C, and D.

Seismic cable restraints should consist of steel cables sized to resist seismic loads with a minimum safety factor of 2 and arranged to provide all-directional restraint. Cables should be prestretched to achieve a certified minimum modulus of elasticity. Cable end connections should be steel assemblies that rotate to the final installation angle and utilize two clamping bolts to provide proper cable engagement. Alternatively, 45° bent steel plates, with holes for attachment to the structure and for steel cable loops with thimbles and wire rope clamps, are acceptable. A minimum of two wire rope clamps are required at each end of the cable assembly. See Figures 3-6 A and B.

Seismic solid braces should consist of steel angles, channels, or strut channels to resist seismic loads with a minimum safety factor of 2 and arranged to provide all-directional restraint. Seismic solid brace end connectors should be steel assemblies that swivel to final installation angle and utilize two through-bolts to provide proper attachment. Alternatively, 45° bent plates with holes for attaching to the structure and to the steel brace section are acceptable. See Figure 3-7.

Steel angles or strut channels sized to prevent buckling should be clamped to vertical support rods utilizing a minimum of two clamps at each restraint location when required. Clamp assemblies may be ductile casting or strut channels assemblies. See Figures 3-8 A and B.

Pipe clevis cross bolt braces should be required at all restraint locations. They may be special purpose preformed channels deep enough to be held in place by bolts passing over the cross bolt or pipe sections installed over the cross bolt with a minimum of 1/8 in. (3 mm) wall thickness. See Figure 3-9.

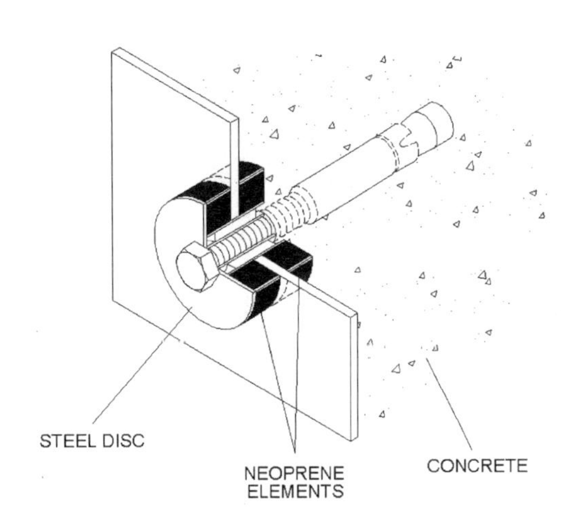

Figure 3-3 Bushing.

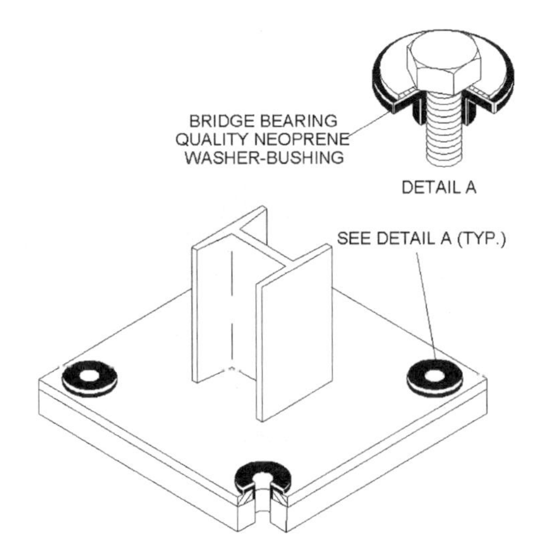

Figure 3-4 Neoprene bushing.

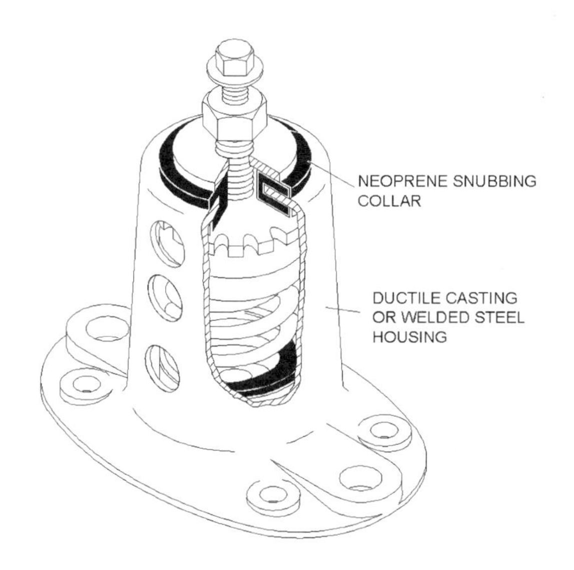

Figure 3-5A All-directional seismic spring mount.

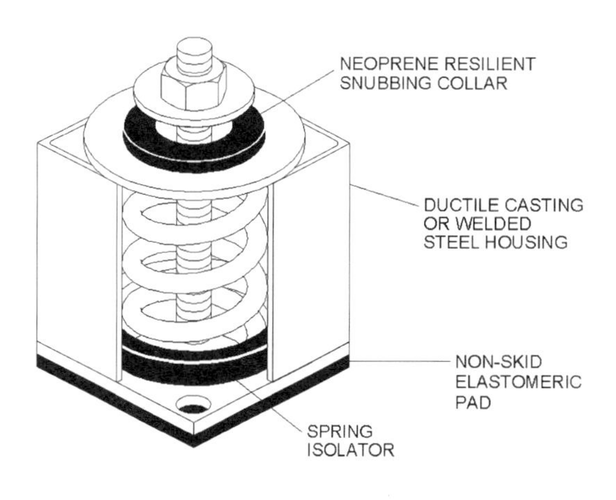

Figure 3-5B All-directional seismic spring mount.

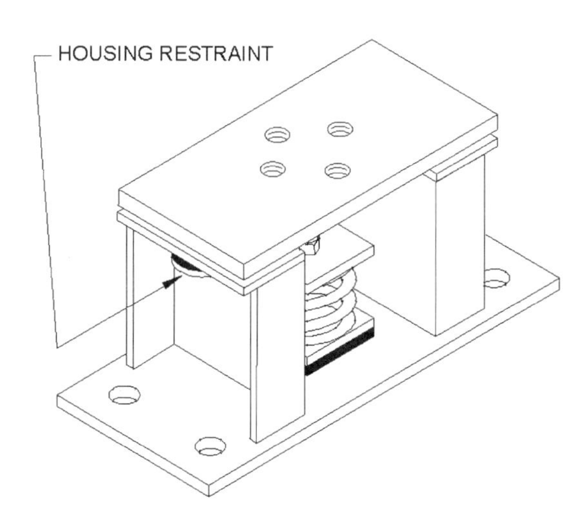

Figure 3-5C Restrained spring isolator.

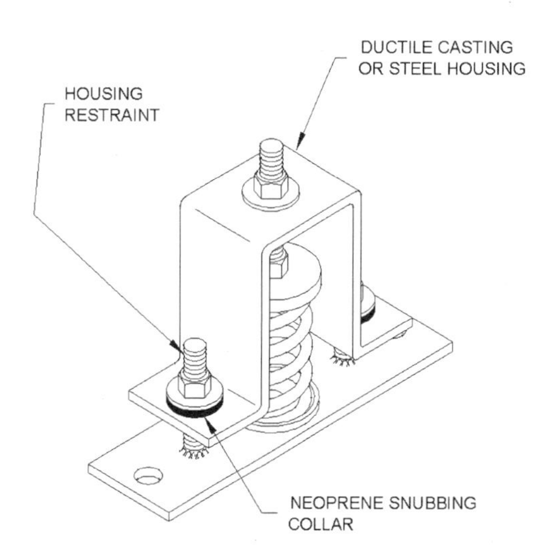

Figure 3-5D Restrained spring isolator.

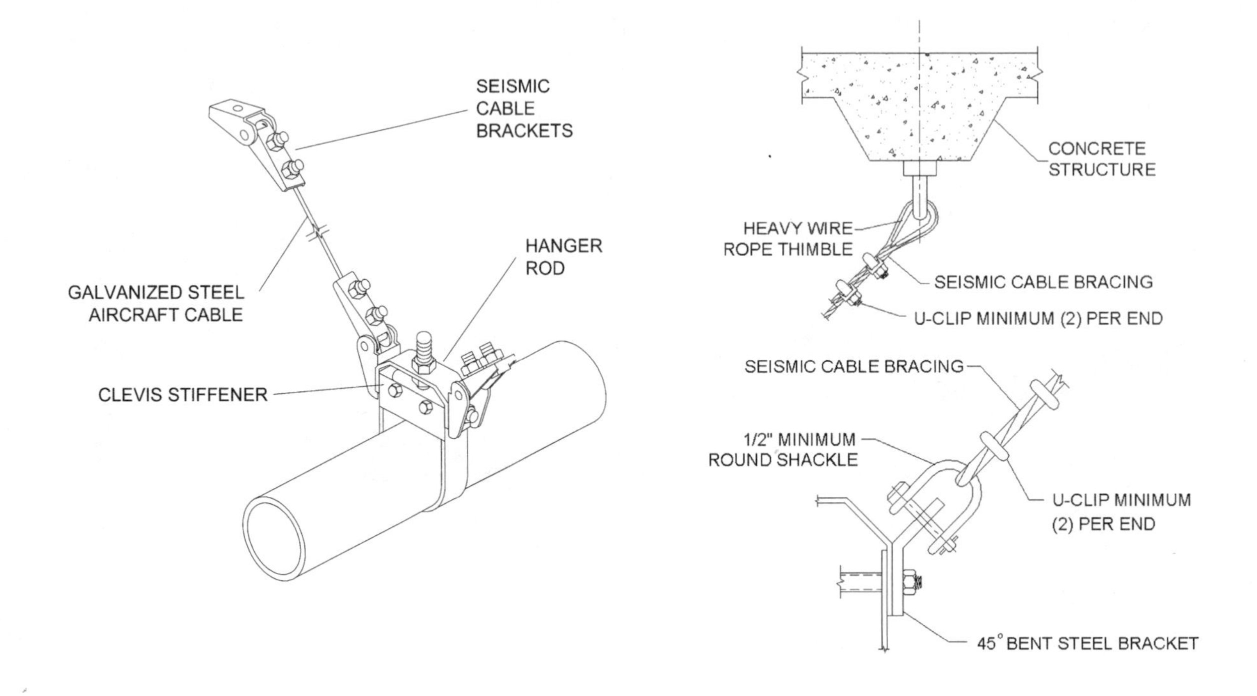

Figure 3-6A Cable restraints.

Figure 3-6B Cable restraints.

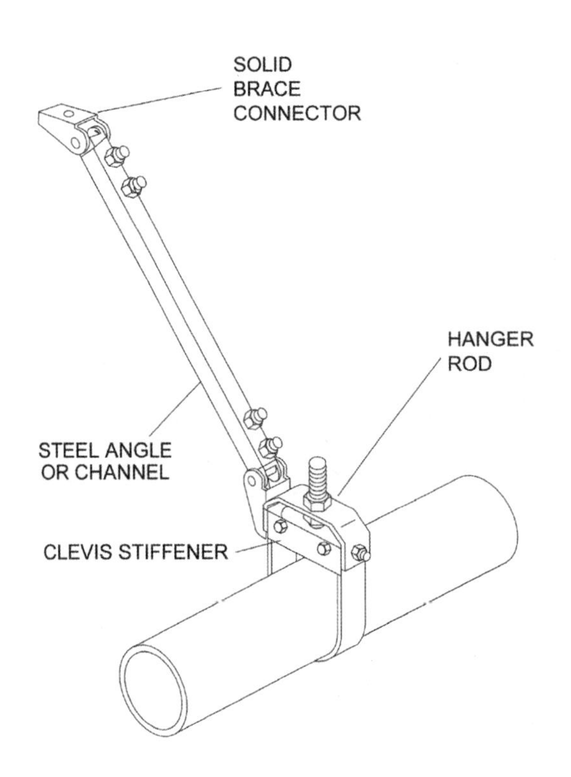

Figure 3-7 Rigid brace.

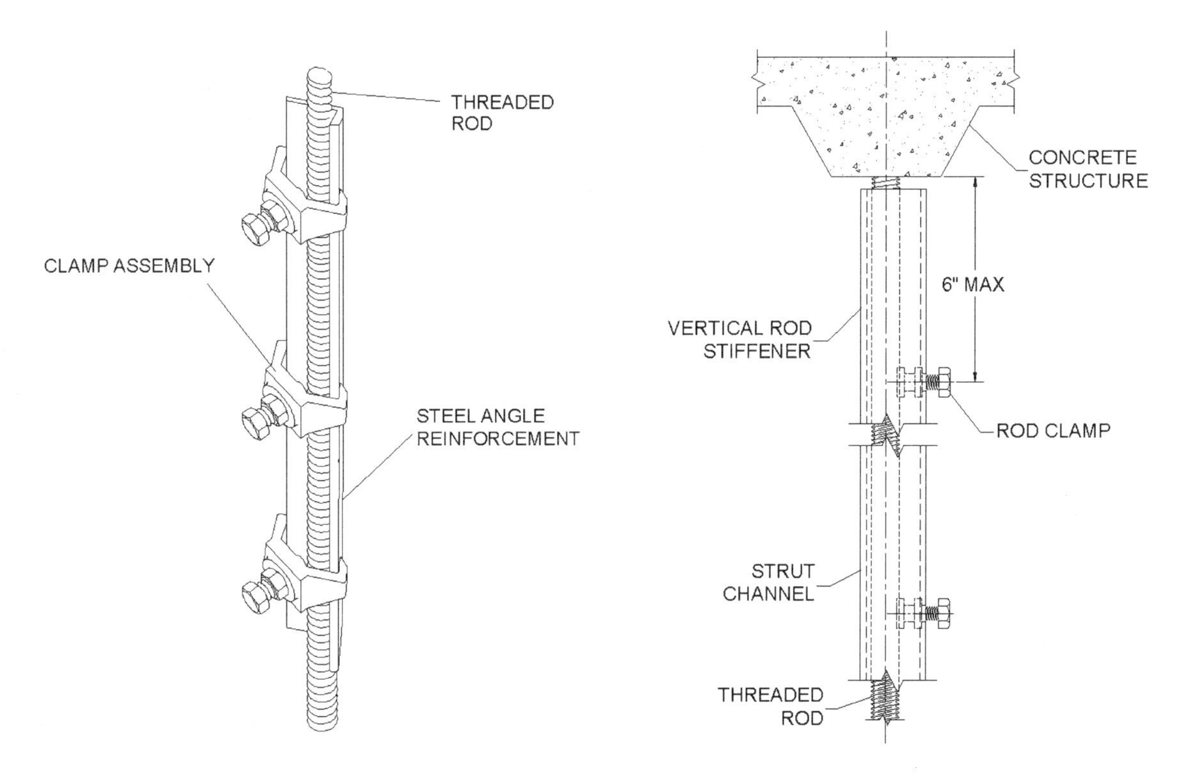

Figure 3-8A Vertical rod stiffener.

Figure 3-8B Rod stiffener.

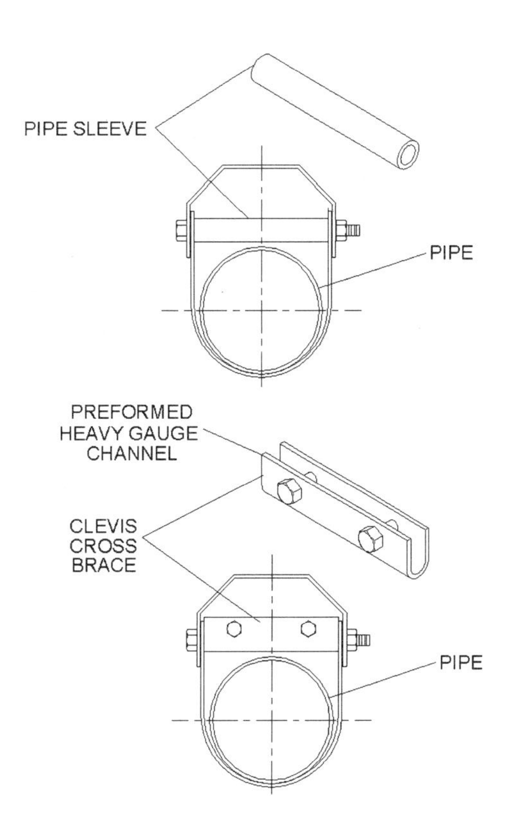

Figure 3-9 Clevis cross brace.

Seismic snubbers should consist of interlocking steel members restrained by molded neoprene bushings or pads of bridge bearing neoprene. Bushings or pads should be replaceable and a minimum of 1/4 in. (6 mm) thick. A minimum air gap of 1/8 in. (3 mm) should be incorporated in the snubber design before contact is made between the rigid and resilient surfaces. Snubbers must have a minimum of two bolt holes for attachment to the structure. The neoprene should be compounded to bridge bearing specifications. See Figures 3-10 A, B, C, and D.

All-directional seismic snubbers should consist of interlocking steel members restrained by shock absorbent neoprene material compounded of bridge bearing neoprene. Neoprene should be a minimum of 3/4 in. (19 mm) thick. Snubbers should be manufactured with an air gap between hard and resilient material of not less than 1/8 in. (3 mm) or more than 1/4 in. (6 mm). Snubbers should be installed with factory set clearances. Submittals should include the load deflection curves in the X, Y, and Z planes. The neoprene should be compounded to bridge bearing specifications. See Figure 3-11.

All-directional acoustical pipe anchors, consisting of two sizes of steel tubing, pipes, or plates should be separated by a minimum 1/2 in. (13 mm) thick neoprene. Vertical restraint should be provided by similar material arranged to prevent vertical travel in either direction. The design should be balanced for equal resistance in any direction. The neoprene should be compounded to bridge bearing specifications. See Figure 3-12.

Pipe guides should consist of an acoustically telescopic arrangement of two sizes of steel tubing or pipes separated by a minimum 1/2 in. (13 mm) thickness of neoprene. The guides should be preset with a device for the setting of the height to allow vertical motion due to pipe expansion or contraction. Guides should be capable of ±1 5/8 in. (41 mm) motion or to meet location requirements. The neoprene should be compounded to bridge bearing specifications. See Figure 3-13.

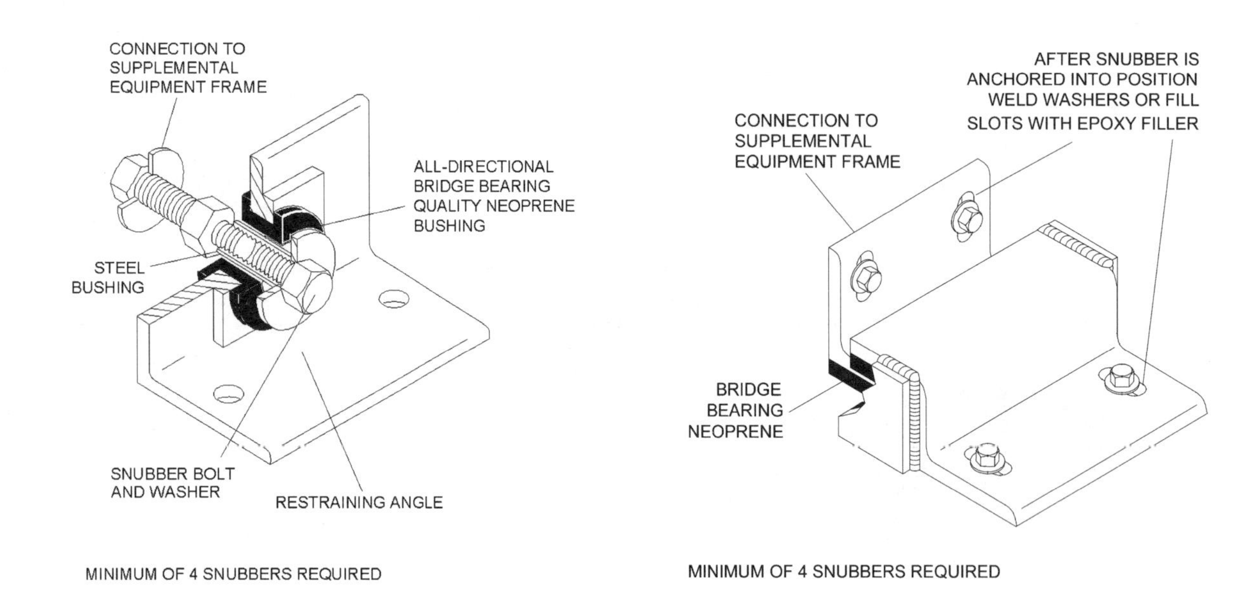

Figure 3-10A All-directional seismic snubber.

Figure 3-10B Two-directional seismic snubber.

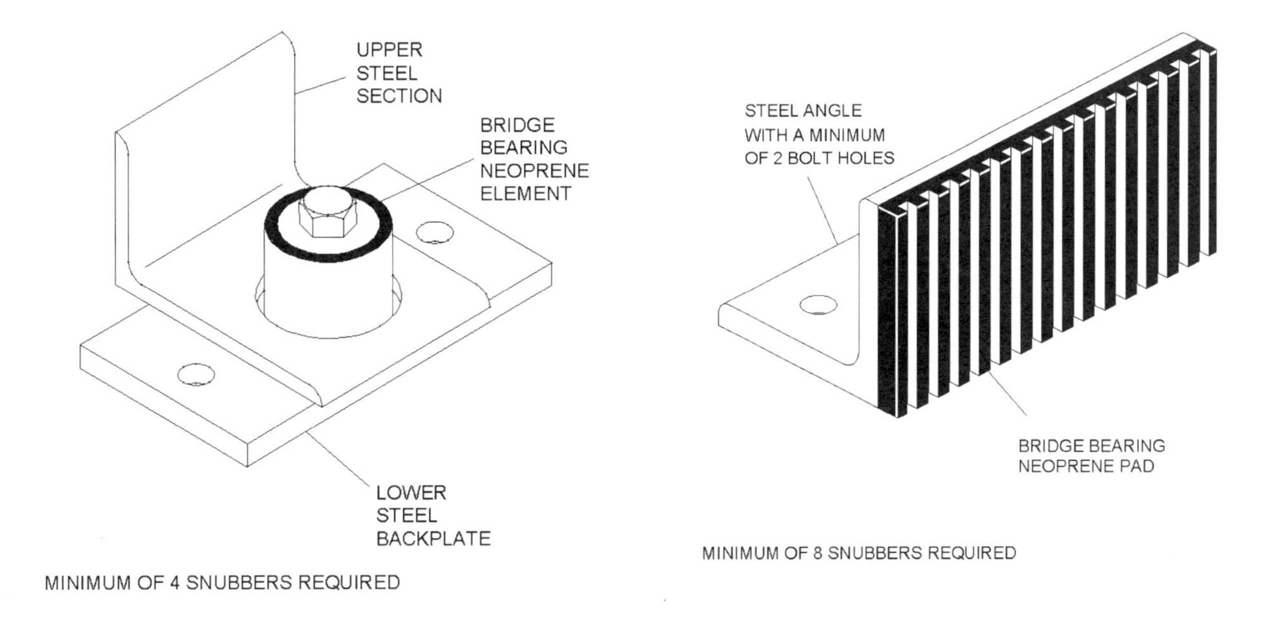

Figure 3-10C Horizontal seismic snubber.

Figure 3-10D Single direction seismic snubber.

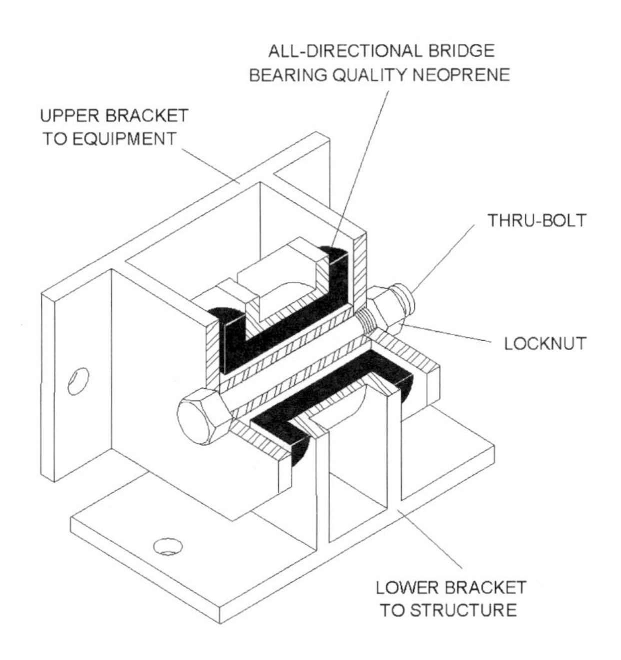

Figure 3-11 All-directional seismic snubber.

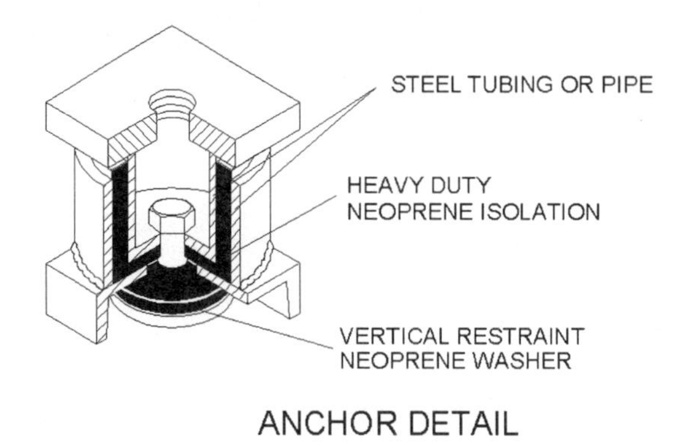

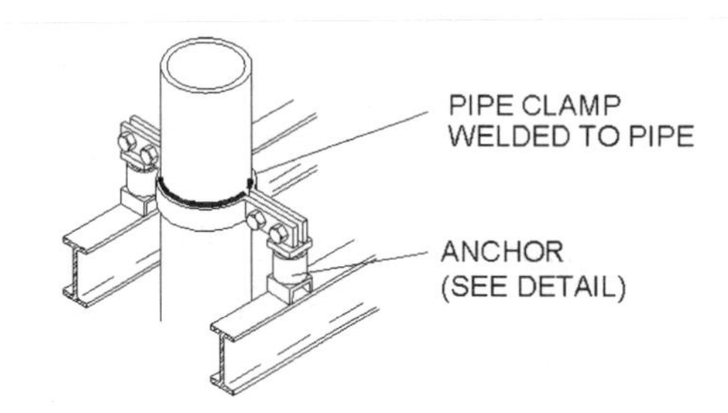

Figure 3-12 All-directional anchor.

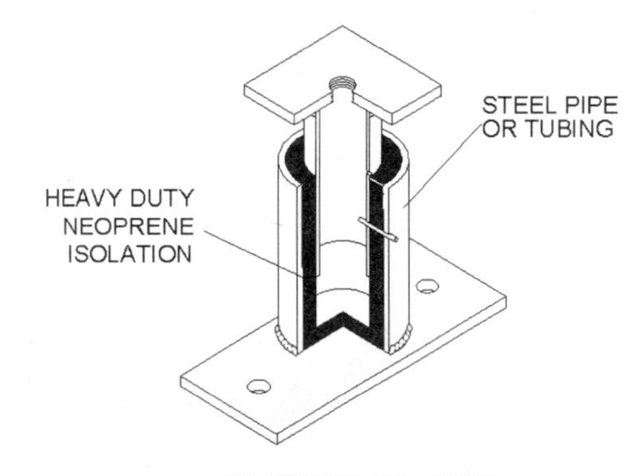

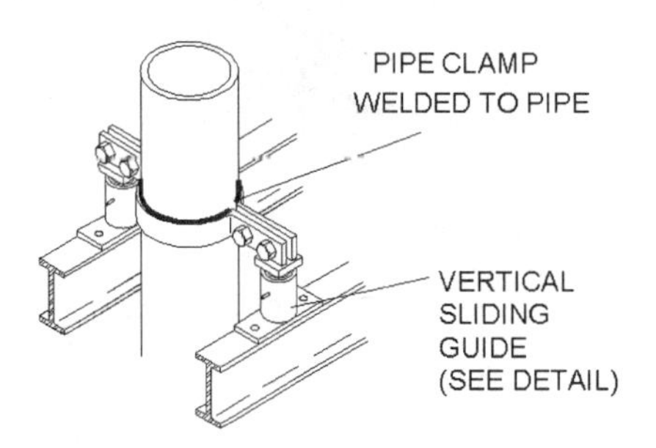

Figure 3-13 Sliding glides.

GENERAL SPECIFICATION AND DESIGN NOTES

1. All seismic restraint systems should be installed in strict accordance with the manufacturer's written instructions and all certified submittal data.
2. Installation of seismic restraints should not cause any change of position of equipment, piping, or ductwork, resulting in stresses or misalignment.
3. No rigid connections between equipment and the building structure should be made that degrade the noise and vibration-isolation system specified.
4. The contractor shall not install any equipment, piping, duct, or conduit that makes rigid connections with the building unless isolation is not specified. "Building" includes, but is not limited to, slabs, beams, columns, studs, and walls.
5. Coordinate work with other trades to avoid rigid contact with the building.
6. Any conflicts with other trades that will result in rigid contact with equipment or piping due to inadequate space or other unforeseen conditions should be brought to the architect's/engineer's attention prior to installation.
7. Prior to installation, bring to the architect's/engineer's attention any discrepancies between the specifications and the field conditions or changes required due to specific equipment selection.
8. Overstressing of the building structure should not occur because of overhead support of equipment. Contractor should submit loads to the structural engineer of record for approval. Generally, bracing may occur from
   a. flanges of structural beams,
   b. upper truss cords in bar joist construction, and
   c. cast-in-place inserts or wedge-type drill-in concrete anchors.
9. Type 6 (figures 3-6A & 3-6B) cable restraints should be installed slightly slack to avoid short-circuiting the isolated suspended equipment, ductwork, piping, or conduit.
10. Type 6 (figures 3-6A & 3-6B) cable assemblies should be installed taut on nonisolated systems. Type 7 (figure 3-7) seismic solid braces may be used in place of cables on rigidly attached systems only.
11. Cables should not be installed over sharp corners.
12. At locations where type 6 or 7 restraints are located, the support rods should be braced when necessary to accept compressive loads with type 8 (figure 3-8) braces. Welding of compression braces to the vertical support rods is not acceptable.
13. At all locations where type 6 or 7 restraints are attached to a pipe clevis, the clevis cross bolt should be reinforced with type 9 (figure 3-9) braces.
14. The vibration-isolation manufacturer shall furnish integral structural steel bases as required. Independent steel rails should not be permitted.
15. Post-installed concrete anchors should be as specified in Chapter 5, "Connections."
16. When vertical pipe risers are flexibly supported to accommodate thermal motion and/or pipe vibration concerns, the pipe shall be guided with type 13 (figure 3-13) pipe guides, located to maintain pipe stability and provide horizontal seismic restraint. Where necessary, the riser shall also be anchored with type 12 (figure 3-12) pipe anchors, located to provide thermal control and vertical seismic restraint. See Chapter 9 of this manual for more information.
17. Seismic restraints should be mechanically attached to the system. It is not sufficient to loop restraints around the system.
18. Piping crossing building seismic joints, passing from building to building, or supported from different portions of the building shall be installed to allow differential support displacements without damaging the pipe, equipment connections, or support connections. Flexible Hose, pipe offsets, loops, anchors, and guides shall be installed as shown on the plans or as required to provide required motion capability and limit motion of adjacent piping. See chapter 8.
19. Avoid crossing seismic separations. If necessary, cross at the lowest possible floor and provide for flexibility in piping that exceeds the anticipated movement as defined by the structural engineer of record. See chapter 8.
20. Avoid locating equipment (especially water tanks) on the roofs of buildings.

21. Water tanks should be secured to their saddles by welding or proper concrete attachment, and those saddles should be properly attached to the structure.
22. Brace all air distribution terminal units with water coils. Design flexibility into the piping/coil connection.
23. Design piping systems into zones that can be isolated during emergencies. Provide sectional shutoff valves and readily accessible drainage ports.
24. Do not brace a system to two different structures, such as a wall and a ceiling.
25. Provide appropriate openings in walls, floors, and ceilings, or design flexibility into the system for the anticipated movements.

BIBLIOGRAPHY

ASHRAE. 2011. *ASHRAE Handbook—HVAC Applications*, chapter 48, Noise and vibration control. Atlanta: American Society of Heating, Refrigerating and Air-Conditioning Engineers, Inc.

ASHRAE. 2011. *ASHRAE Handbook—HVAC Applications*, chapter 55 Seismic- and Wind-Resistant Design. Atlanta: American Society of Heating, Refrigerating and Air-Conditioning Engineers, Inc.

ICC. 2009. *International Building Code® (IBC®)*, chapter 16, Structural design requirements. Washington, D.C.: International Code Council.

Mason, N.J. 2004. *SVCS—Complete Seismic HVAC Engineering Specifications, Parts 1 and 2*. New York: Mason Ind., Inc.

4 Equipment Seismic Certification

The code requirements for equipment certification are not new, but have changed since their widespread publication and acceptance in IBC 2000. According to IBC, a manufacturer's certificate of compliance is not required for all equipment but is required for all designated seismic systems. Designated seismic systems are defined as all equipment with an importance factor greater than 1.0. But IBC 2003 and 2006 directly reference ASCE 7, which includes more specific certification requirements, with special attention to active mechanical and electrical equipment that must function after an earthquake, and for all components containing hazardous contents.

IBC 2000 Section 1708.5 required the manufacturer of designated seismic system components to test or analyze the component and submit a certificate of compliance based on shake table testing, shock testing, analysis, or experience data. That was the only requirement.

A similar general requirement was included in IBC 2003 Section 1708.5, but IBC 2003 also references requirements in ASCE 7. ACSE 7-02 Section 9.6.3.6 specifically required mechanical and electrical equipment that must remain operable, to demonstrate operability by shake table testing or analysis, and required a manufacturer's certificate of compliance indicating compliance.

The equipment certification general requirement was included in IBC 2006 Section 1708.5, which required equipment manufacturers to test or analyze equipment and submit a certificate of compliance, but the referenced ASCE 7-05 Section 13.2.2 required active mechanical and electrical equipment to be certified based on shake table tests or experience data.

The general requirement paragraph has been renumbered to 1708.4 but otherwise remains the same in IBC 2009.

QUALIFICATION BY TESTING

Before 2000, there was no defined standard for testing equipment to meet building code requirements. ICC ES AC156 was developed by the TS8 committee of the Building Seismic Safety Council (BSSC) working with ICBO (now ICC ES) with industry partners and became the first test standard to translate the building code design requirements into a testing protocol. As such, it was specifically developed to include the acceleration demands of the National Earthquake Hazards Reduction program (NEHRP) provisions and Federal Emergency Management Agency (FEMA) 302,303,450-1 and 2. AC156 is referenced in ASCE7-05 Section 13.2.5 and is in wide use.

AC156 is of great benefit to the equipment manufacturer, as it includes important criteria for test sample determination and justification of testing for product lines that can greatly reduce the cost of multiple unit tests. AC156 also includes appropriate information for development of a test plan, data collection, report requirements, and the critically all-important pass/fail criteria for equipment performance after shake table testing.

If an equipment supplier chooses to use other testing protocols, the design professional reviewing the test results is in the unenviable position to make the appropriate comparisons with the building code requirements, including justification of the shake table motion, include appropriate factors to account for structural interaction, account for variations in equipment size and construction, and define post-test equipment functionality requirements.

When an equipment vendor uses AC-156 in their testing for seismic compliance, the design professional is assured that a recognized standard was used and that code compliance is virtually guaranteed.

QUALIFICATION BY EXPERIENCE DATA

The code allows equipment qualification by use of experience data. The use of experience data was prominent in the nuclear power industry in the 1980s. The program focused primarily on the electrical equipment found in nuclear power plants and may have very little information of use to manufacturers attempting to qualify modern HVAC equipment. The experience data itself may not include the higher seismic demand requirements of the newer codes or projects in more active seismic areas.

To qualify equipment by experience data today, the equipment must be included in a database of equipment that has been subjected to known earthquakes, has detailed studies of performance and anchorage evaluation, and has the root cause of any failures determined by experts. It must be shown that the current seismic requirement for the project is lower than that included in the database, and the new equipment must be of equal or better construction.

QUALIFICATION BY ANALYSIS

Although the code technically limits the use of analysis for equipment qualification, some jurisdictions allow analysis where no other method is practical. The sheer size of some components is often a concern, but some shake table laboratories have multiple shake tables that can be operated in unison to increase capacity for the largest common HVAC components. A complete investigation of shake table lab capabilities may be warranted before considering the use of analysis.

There are many types of analysis, and many opinions of the appropriateness of different analysis types. In general, the analysis would evaluate the linear elastic performance of the component, so if the analysis confirms the linear elastic performance of the component frame from the support attachment to the active or energized elements, the performance of the component may be assured. However, tests have confirmed that most component frames constructed of sheet metal sections with bolted or screwed frames will see some sort of nonlinear load characteristics, invalidating a linear analysis.

It has been suggested that the cost of a state-of-the–art-analysis that will satisfy the most knowledgeable reviewing engineers or building officials, along with added cost of shake table testing the smaller active or energized components, may be more expensive than shake table testing the entire piece of equipment.

The overriding difficulty with qualification by analysis is operational verification, unless the active components have known capabilities that can be compared with analysis results. Unlike qualification by testing, qualification by analysis offers no realistic pass/fail operational test that can be performed on the component.

RESPONSE SPECTRUM DESIGN

Knowing the maximum acceleration or displacement of any particular earthquake is not enough to define the severity of an earthquake for equipment certification. Instead, a response spectrum is used to better represent the damaging effects of earthquakes. A response spectrum is nothing more than a series of values, acceleration being the most common, representing the maximum response of a series of simple oscillators, each with a different natural frequency, subjected to the expected ground motion. In simplest terms, equipment will tend to have its greatest motions and largest accelerations when its natural frequencies are excited. A response spectrum design allows the test lab technician and design engineers the opportunity to guarantee that the equipment is subjected to the appropriate table motion to result in the appropriate equipment response.

Damping is an important consideration. All systems have some degree of damping. Using a response spectrum based on a realistic evaluation of the system damping is a critical part of the design. For most general earthquake design, 5% damping is assumed appropriate, but it is also possible to determine actual system damping and use it in the system seismic design. Damping and response spectrums can be confusing. In a classic flexible system, as the damping is increased, the system will see less displacement and higher force. But response spectrums include the accelerations at resonant frequencies, so the higher the damping, the lower the forces. Displacement and acceleration will also decrease.

SHAKE TABLE TESTING

The shake table lab creates a series of artificial time histories and subjects the equipment to table motions that have been developed to result in a test response spectrum (TRS) that matches the required response spectrum (RRS).

Equipment is not a single-degree-of-freedom simple oscillators. It is a multiple-degree-of-freedom system with complexities that require the attention of an experienced shake table lab. Before the actual seismic tests, a series of initial tests is run. The primary purpose of the initial tests is to determine the actual system resonant frequencies and damping. This is often done as a sine sweep, white noise, or pulse test.

This information is used to adjust the table motion to ensure that the RRS is met and not exceeded significantly. During the shake table test, the feedback loop automatically adjusts the table motion within certain operating limitations.

The RRS is the target. The actual TRS is available after the test is complete and is based on the generated shake table motions. For a successful test, the TRS must meet or exceed the RRS at all system natural frequencies, within the requirements of the test acceptance criteria.

EQUIPMENT ANCHORAGE AND SUPPORT DESIGN

One of the most critical aspects of equipment certification is certifying the equipment anchorage and support capability. Historically, most equipment failures are directly attributable to the poor design of equipment anchorage and supports, including integral steel rails and brackets that provide equipment attachment points to the structure. This was confirmed by the nuclear energy findings, which concluded that the primary cause of damage to electrical equipment during earthquakes was loss of anchorage. It is the responsibility of the equipment manufacturer to qualify the equipment frame all the way down to and including the attachment points to the structure.

It is also critically important for the manufacturer to define the support requirements of the equipment and duplicate them in the testing. This is especially true of equipment supports for equipment that will be supported on vibration isolation or project-supplied steel framing. Equipment certification testing or analysis can ensure a reduction of anchorage failures caused by problems with the equipment itself and leave the project engineer to deal with external causes of anchorage failure, such as improper anchorage, improperly installed anchors, or problems with housekeeping pad design or attachment to structure.

EQUIPMENT CERTIFICATION FOR CBC 2007 CALIFORNIA HOSPITALS

With the adoption of CBC 2007, IBC-based equipment certification was required in California for the first time. The Office of Statewide Health Planning and Development (OSHPD) developed Code Application Notice (CAN) 2-1708A.5 to clarify the certification requirements of nonstructural components in acute-care hospitals.

The CAN was the developed by OSHPD and reviewed by the Hospital Building Safety Board (HBSB), an advisory body to OSHPD that includes engineers, architects, hospital owner representatives, and related personnel. Specific items were addressed in meetings of the Ad Hoc Committee on Non-Structural Components and included discussion among OSHPD personnel, several HBSB members, and several other interested parties, including equipment and seismic bracing manufacturers. These meetings included very thoughtful discussions and decisions.

CAN 2-1708A.5 is mandatory reading for anyone wishing to be involved in equipment certification on California hospitals and is an excellent resource for information on equipment certification in all states and jurisdictions. The CAN offers some insight into the future, with language added that was eventually to be included in ASCE 7-10. Several engineers at OSHPD participate in ASCE and ICC code committees and are responsible for the development of the code. As such, their enforcement of the code should be taken very seriously. Not only do they know what the code says, they know its intent.

OSHPD has removed the confusion from the code and defined two types of certification. Because of the confusion with the terms *seismic certification (SC)* and *seismic qualification*, OSHPD added a third term, *special seismic certification (SSC)*, to identify active components that must remain operable. The CAN includes a list of equipment and components that require SSC and also includes a list of equipment and components that are considered rugged and do not require SSC.

SCis required of all equipment and components and is limited to anchorage and the equipment's ability to remain anchored. It can be performed on a project-by-project basis by calculation or testing, or submitted for preapproval. Most vibration isolation and seismic restraint vendors have participated in the preapproval program since its inception and have current preapprovals on many restrained mounts, snubbers and seismic braces. Several medical equipment manufacturers have preapprovals that include anchorage of their equipment.

SSCis required of all active equipment necessary for the continued operation of a hospital.

The CAN includes a list of components requiring SSC, including several major HVAC components including the following:

- Smoke control fans
- Built-up or field-assembled mechanical equipment
- Air-conditioning units
- Air-handling units
- Chillers used for HVAC
- Cooling towers designed as components

There are several HVAC equipment components that are considered rugged and do not require SS C:

- Valves (not in cast-iron housings, except for ductile cast iron)
- Motors and motor operators
- Horizontal and vertical pumps (including vacuum pumps)
- Air compressors
- Equipment and components weighing not more than 20 lb supported directly on the structures (and not mounted on any other equipment or components) with supports and attachments in accordance with Chapter 14, ASCE/SEI 7-05, as modified by Section 1614A, CBC 2007

See the CAN for the complete lists. OSHPD also adds a note to the list of rugged equipment saying that the exemptions in this section are for factory-assembled, discrete equipment and components only and do not apply to site-assembled or field-assembled equipment or equipment anchorage.

OSHPD allows SSC by testing and requires testing to be done by an accredited lab or under the responsible charge of a California-licensed engineer, with test reports reviewed and stamped by a California-licensed structural engineer. This allows manufacturers to do testing in their facilities, but with the proper independent engineering witness and certification.

OSHPD allows SSC by analysis but only where active parts or energized components are certified exclusively on the basis of shake table testing or experience data, unless the component is inherently rugged. OSHPD does allow analysis on multicomponent systems where active parts or energized components are certified by test or rugged and only the connecting supports justified by analysis. Whereas ASCE 7 does not allow analysis, OSHPD allows analysis but only for the connecting elements, attachments and supports. Active parts or energized components must be certified by shake table testing or experience data.

OSHPD allows SSC by experience data per ASCE 7-05 Sections 13.2 to 13.6, based on nationally recognized procedures. OSHPD lists two such procedures for the Uniform Facilities Criteria (UFC) 3-310-04 (2007), including equipment that has been certified as *mission critical (MC)* with accompanying peer review reports submitted for review, and equipment experience data per Appendix F of the UFC. Some electrical equipment may have such data available, but the availability of such data for more common HVAC equipment is doubtful. OSHPD does allow the use of experience data to certify active mechanical and electrical equipment and components with hazardous content. However, they do require proof that the experience data includes equipment with similar structural integrity. They also require the owner of the SSC to maintain a quality assurance program to continually evaluate new information. Although OSHPD includes experience data, they also reference the use of Appendix F of UFC 3-310-04 (2007) for examples of SSC by experience data. This is a very intensive procedure and will not work for the vast majority of standard HVAC equipment. OSHPD has also expressly rejected reference to Table 4-1 of this chapter. Table 4-1 in still included in the chapter as a good general reference, and may be acceptable to some engineers and jurisdictions.

Manufacturers' equipment certificates and accompanying documentation can be reviewed and accepted by OSHPD on a project-by-project basis, but this is a time consuming process and can cause delays in project design and construction. Submittal of the certificate and accompanying documents for preapproval is preferred. Once accepted, OSHPD will issue an OSHPD SSSC preapproval number (OSP) and list the certification on the OSHPD Web site.

OSHPD allows SSC with submittal of report approved by the International Code Council—Evaluation Service (ICC-ES) based on ICC AC-156. OSHPD will accept a third party certification from a listing agency accredited by ICC-IAS with several additional inspection and quality assurance requirements and review and acceptance of the SSC documents by a California-registered structural engineer. Additionally, attempts at providing SSC by analysis must meet the added OSHPD requirements for analysis per the CAN.

Although CAN 1708A.5 is not valid for CBC 2011 projects, it remains in effect for CBC 2007 projects and is a very useful reference for equipment certification on all projects in California and in other jurisdictions.

EQUIPMENT CERTIFICATION FOR CBC 2010 OSHPD HOSPITALS

Beginning on January 1, 2011, all new OSHPD projects are subject to the requirements of CBC 2010. The equipment certification clarifications from CAN 1708A.5 are no longer valid, because a number of clarifications and additional requirements have been included in CBC 2010 Section 1708A.4. These clarifications and additional requirements match those found in ASCE 7-10, which was not published in time to be included in the code requirements.

Included is a statement that active or energized components require certification based on shake table testing or experience data unless the component is inherently rugged by comparison with similar certified components. This expressly rules out the possibility of certification of active or energized components by analysis. This section also includes a statement that a minimum of two tests are required, with a few exceptions, and allows the tests to be performed on different-sized components to qualify a range of products. Analysis can be used for component certification but only on connecting elements, attachments or supports where active or energized components have been certified by test or experience data.

Unlike the CAN, the new code does not include a list of rugged components, but does include an expanded list of systems, equipment, and components that require certification including:

- Emergency and standby power systems, including generators, turbines, fuel tanks and automatic transfer switches
- Elevator equipment (excluding elevator cabs)
- Components with hazardous contents (excluding pipes, ducts, and underground tanks)
- Smoke control fans
- Exhaust fans
- Switchgear
- Motor control centers
- X-ray machines in fluoroscopy rooms
- Computerized tomography (CT) scanners
- Air-conditioning units
- Air-handling units
- Chillers
- Cooling towers (excluding cooling towers designed as nonbuilding structures).
- Transformers
- Electrical substations
- UPS (Inverters) and associated batteries
- Distribution panels, including electrical panel boards
- Control panels, including fire alarm, fire suppression, preaction, and auxiliary or remote power supplies

EQUIPMENT MANUFACTURERS' CAPABILITIES AND TESTING

The IBC and ASCE 7 require manufacturers to certify their equipment. There is no requirement for an independent certifier, third-party certifier, or an ICC-accredited certifier. A manufacturer may have the engineering capabilities to design their own certification program and make arrangements at a shake table lab to run the required tests. Other manufacturers may not have the in-house engineering capabilities to develop their own certification programs. There are a number of professional engineering consultants who have developed a specialty in equipment certification and are able to provide this service to manufacturers.

As professional consulting engineers, they will work with the manufacturer and shake table lab to develop the certification program. Professional engineering consultants offer the best possible service to their client, the manufacturer, and can recommend the most effective vibration isolation and restraints for the testing process.

For equipment certification to meet OSHPD requirements, a manufacturer should obtain the services of a professional consulting engineer that has already completed testing programs and obtained OSP approvals from OSHPD. The manufacturer and the professional consulting engineer can then meet with OSHPD before a single test run and obtain a preapproved test procedure and a clear definition of the equipment variations and models requiring tests.

Completing testing or analyses with either in-house engineering or with others, especially those without extensive experience of working with OSHPD, can be a very costly mistake.

SHAKE TABLE EXPERIENCE

The equipment manufacturer should make every effort to have a factory engineer on site for tests along with the consulting engineer who is supervising the test. It is also a good idea to have an engineer for the vibration and seismic restraint manufacturer there to work with the test lab mechanics and insure that the isolators and restraints are installed properly and will provide representative results during the testing.

One example of test experience resulting in design modifications occurred when a HVAC chiller was mounted on restrained isolators at the supports, which were inboard from the ends of the chiller, resulting in significant equipment cantilevers at each end. While insignificant from a static calculation viewpoint, the cantilevered mass resulted in dramatic load increases during the shake table test, and higher accelerations on the equipment. Once the supports were moved to the end of the chillers, the loads and accelerations were reduced significantly.

Equipment setup is the most time-consuming aspect of shake table testing. When equipment is installed on vibration isolation systems, setup times can become excessive unless the professional consulting engineer and vibration isolation supplier can work together to create a series of interchangeable supports, fully coordinated and preadjusted for quick change-out. Isolators and snubbers should be mounted on load cells to confirm support calculations and help in future design.

Care should be taken when using water in equipment being tested. The test lab may have rules specifically banning water or limiting or the amount of water in equipment being tested. In some cases, sandbags or other nonliquid weights can be used to represent water loads. In at least one lab, bags of grout are used instead, to eliminate potential sand and grit damage to hydraulic actuator cylinders.

COMPARISON OF SHAKE TABLE TEST RESULTS

There is a great deal of theory regarding the amplification of accelerations on vibration-isolated systems, but until recently, very few data have been collected. One of the problems with comparisons is the control of the test system to insure a valid comparison. Differences in air gaps and installation issues, including lack of restraint centering, have made previous test data somewhat questionable. And the inclusion of too many variables has made results difficult if not impossible to compare.

Figures 4-1 and 4-2 are 5 s segments of 30 s long time histories from recent shake table tests. The tests were based on an RRS with the maximum flexible system response acceleration (A_{flex}) of 2.5g per AC 156 requirements to meet code requirements. The shake table displacement time histories are shown, and are, in fact, identical in each test. But the resulting acceleration time history of the equipment support frame is very, very different.

The acceleration time history of the equipment support frame used in this comparison provides very valuable data, specifically for comparison to the shake table acceleration time history as a measure of the restraint system amplification. Because equipment has widely different construction and corresponding variations in stiffness and damping, any comparison with the acceleration time history at the equipment center of gravity includes added variables that greatly complicate comparison, and in fact almost defy comparison.

Figure 4-1 illustrates the results of the test on code-compliant restrained spring mounts with 1/4 in. (6.4 mm) air gaps and 1/4 in. (6.4 mm) thick neoprene bushings. The resulting amplification of the table acceleration can be seen quite prominently. The worst case occurs at about the 15.3 s mark, with a table acceleration of 1g resulting in an equipment support frame acceleration of nearly 8g. The amplification factor is 8.

Figure 4-2 illustrates the results of the test on a system consisting of open springs and separate snubbers with 1/4 in. (6.4 mm) air gaps and 3/4 in. (19 mm) thick neoprene bushings. At the 15.3 s mark, the table acceleration is again 1g, but the acceleration of the equipment support frame is about 2.8 g. The amplification factor is less than 3, showing the importance of a properly designed spring and resilient snubber system.

EQUIPMENT PIPING CONNECTIONS

Failures of equipment piping connections are a major cause of problems in earthquakes. Although not currently included in shake table testing requirements, the evaluation of equipment piping connections would be an extremely valuable addition to the certification. Evaluation of piping flexibility at equipment connections is required by code, but currently left to the design engineer to determine on a project by project basis. The evaluation begins with a call to the equipment supplier to obtain acceptable nozzle loads, which are the maximum piping loads the equipment manufacturer will allow on the equipment connections. The initial response from the equipment manufacturer is that the allowable nozzle loads are "zero". This is an unrealistic and relatively useless response. At some point, realistic loads

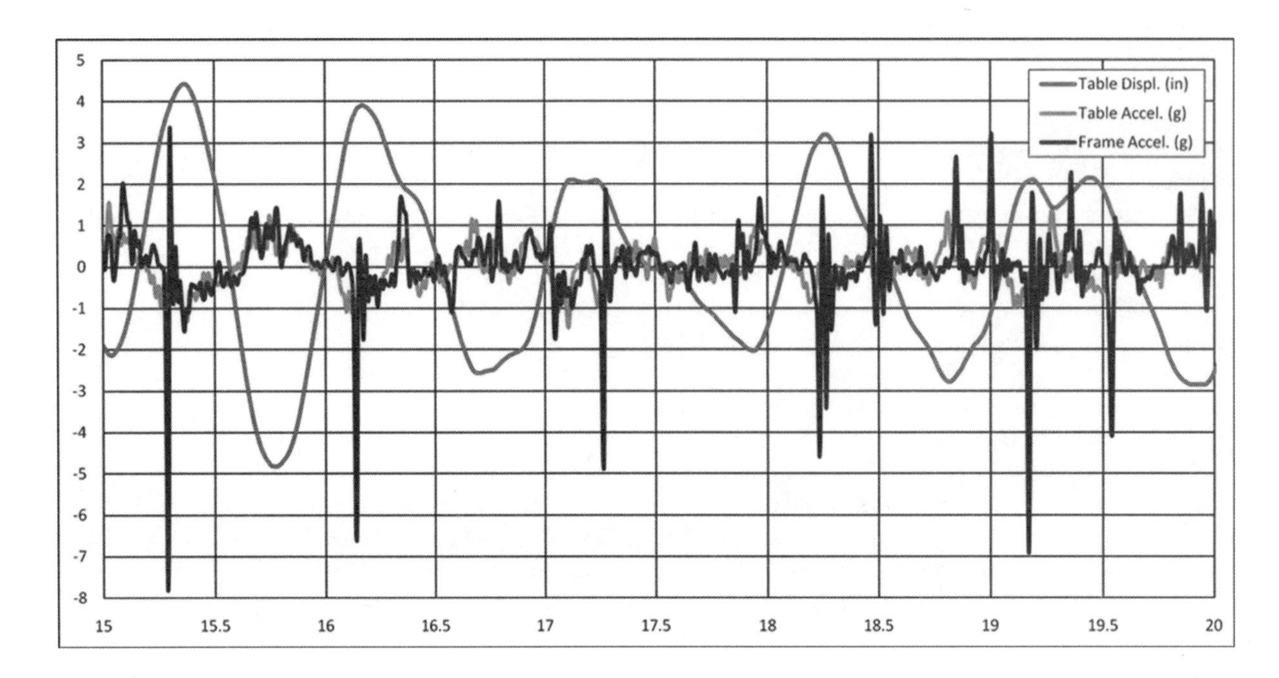

Figure 4-1 Shake table test results for restrained spring mounts with 1/4 in. (6.4 mm) air gaps and 1/4 in. (6.4 mm) thick neoprene bushings.

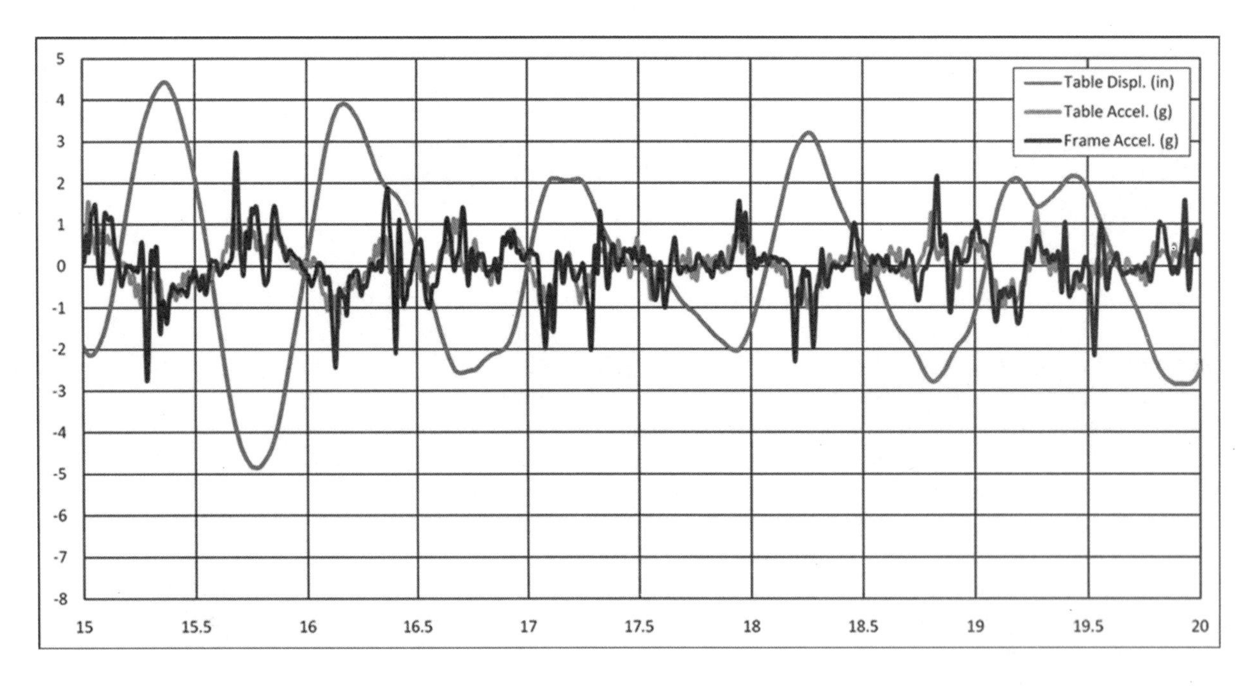

Figure 4-2 Shake table test results for open springs and separate snubbers with 1/4 in. (6.4 mm) air gaps and 3/4 in. (19 mm) thick neoprene bushings.

are discussed and the design engineer can proceed with the piping calculations. Because pipe is very stiff, the most common design includes flexible connectors, selected to accommodate expected differential motion and keep loads on the equipment connections below the acceptable nozzle load values.

The unknown effects of rigid piping attached to equipment and the potential for damage require additional study. Flexible-connection stiffness and motion capabilities must be provided by the manufacturer to be used in testing or analysis to guarantee the operability of equipment.

BIBLIOGRAPHY

ASCE. 2005. SEI/ASCE 7-05, *Minimum Design Loads for Buildings and Other Structures*. Reston, VA: American Society of Civil Engineers.

ASCE. 2010. ASCE 7-10, *Minimum Design Loads for Buildings and Other Structures*. Reston, VA: American Society of Civil Engineers.

CBSC. 2007. *2007 California Building Code*. Sacramento, CA: California Building Standards Commission.

ICC. 2006. 2006 International Building Code. Washington, DC: International Code Council.

ICC. 2009. 2009 International Building Code. Washington, DC: International Code Council.

5 Connections

The installation of connections for seismically restrained HVAC systems is not as simple as a contractor locating a hole in an equipment leg and putting in an anchor. Complexities associated with equipment attachment require coordination between the equipment manufacturer, contractor, mechanical engineer, and structural engineer. An improper equipment attachment will only become evident during a seismic event, when it is too late to implement corrective modifications.

ANCHORS IN CONCRETE

Because of great variation in equipment design and structural materials, post-installed anchors are the most practical connectors for equipment and systems to concrete floors and walls. While it is possible to install anchors prior to setting equipment, small dimensional errors can create significant problems. Once installed, anchors cannot be relocated or adjusted, and it is impossible to add other anchors within certain spacing, greatly reducing the chances for a simple connection. For these reasons, post-installed anchors are most efficiently used when installed after the equipment is located and locations can be exactly marked.

There are three different types of relatively common post-installed anchors, which include wedge-type expansion anchors and two other nonexpansion anchors, including screw anchors and adhesive anchors. Embedded bolts are a preinstalled anchor option. Figure 5-1 illustrates these anchors and Table 5-1 includes a comparison of anchor attributes.

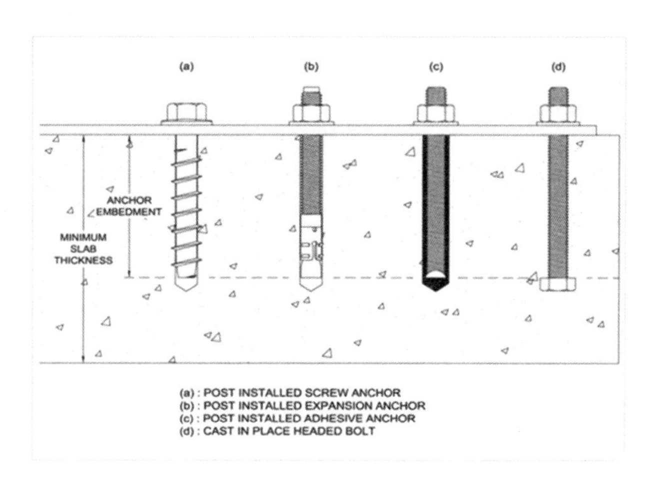

Figure 5-1 Anchor bolts.

Table 5-1 Anchor Bolt Comparison

Anchor Type	Advantages	Disadvantages
Screw	Good load capacity; can be installed with equipment in place; can be removed and reinstalled; no torque set required during installation.	Some limitations in overhead use; some limits in removal, reinstallation, and when headroom is insufficient for bolt length.
Wedge Expansion	Inexpensive; good load capacity; good reserve capacity; good vibration, shock, and cracked-concrete use.	Equipment must be lifted over studs in some situations, such as when there is insufficient headroom to drill through baseplates of equipment mounting holes. Fairly high slip potential. Must be torque set.
Adhesive	High load capacity; closer allowable edge distances and spacing, no slip; good vibration, shock, and cracked-concrete use. No torque set required during installation.	Temperature sensitive; storage limitations; special installation requirements.
Embedded Bolt	High load capacity; good for vibration, shock, and cracked-concrete use; low slip.	Must be accurately located during concrete pour; hole-to-hole clearance limitations.

SCREW ANCHORS

The screw anchor is a relatively new anchor and has yet to gain wide acceptance. However, some manufacturers have full approvals and the screw anchor is a good option for many installations. It must be installed after equipment is in place, so access is an issue, but it is also the only anchor that can be removed and reinstalled.

WEDGE-STYLE ANCHORS

The wedge-style expansion anchor is the best expansion anchor, based on performance. The stud wedge is ideal for hard-mounted (nonisolated) equipment where access is available for installation with the equipment in place. Because the stud extends above the concrete, installation is difficult where the anchors must be installed and the equipment must be lifted over the anchors and set in place.

ADHESIVE ANCHORS

Adhesive anchors consist of a threaded rod, pushed or drilled into a hole filled with adhesive. Many types of adhesives are available, including epoxy, polyurethane, methyl methacrylate, and vinyl ester. Premanufactured two-part epoxy is available inside a glass tube. These tubes also contain a small amount of sand. Once the hole is drilled, the tube is inserted. The threaded rod is fixed in a drill chuck and drilled into the hole. The rotating rod grinds up and mixes the glass, the epoxy, and the sand. This method is 99% foolproof. Problems can occur if the hole is too shallow. Rather than drilling added depth, an installer may break off the extra length of glass tube and install the anchor without proper embedment. The anchor will appear proper after installation but may not have the required design strength.

Another adhesive anchor style uses field-mixed, two-part epoxy. This material is poured into a predrilled hole and the rod inserted using a rotating motion to ensure an even bond. If the epoxy does not set because of field conditions or improper mix, the anchor should be reset with new epoxy after the hole is cleaned and re-drilled, slightly oversized to obtain clean concrete. According to most manufacturers' recommendations, adhesive anchors cannot be installed in the underside of decks or slabs, severely limiting their use. In all cases, the anchor manufacturers' recommendations and ICC-ES report requirements must be followed.

EMBEDDED BOLTS

Embedded bolts may be used, but the exact location of equipment connections must be identified and anchors installed before or during the concrete pour. This process is extremely difficult and often results in slight errors in location, requiring the relocation of isolators, restraints, or equipment anchor bolts. As with preinstalled stud anchors, embedded bolts extend beyond the top of the concrete and require isolators, restraints, or equipment to be lifted over the top of the bolts for attachment. Removal of isolators, restraints, or equipment is an extremely difficult, labor-intensive process.

Embedded bolts can be used for housekeeping pad anchorage to the floor. Unlike exact locations required for equipment anchorage, housekeeping pad sizes and cast-in-place anchor locations can be estimated before the floor is poured. After the floor is poured and pad sizes are confirmed, the reinforcing steel can be accurately installed, and the pads can be poured using the existing cast-in-place anchors to provide connections between the pads and the floor.

Where pad locations are added after the floor is complete, the equipment seismic restraint designer should provide a complete pad design including reinforcing steel and adhesive anchor requirements. See Chapter 6 for more information.

ANCHOR PLATES

The use of anchor plates and embedded anchors is often the best solution for large equipment connections, where installation of many large diameter expansion anchors would be very time consuming or impossible due to space limitations. Anchor plates are usually set in place before or during the concrete pour. Oversized anchor plates offer the installer increased dimensional flexibility in locating isolators, restraints, or equipment, which can be field-welded to the anchor plate at the exact location required. Anchor plates can be installed with post-installed anchors after the concrete is poured to increase dimensional flexibility in locating isolators, restraints, or equipment or to provide multiple anchor attachments when needed because of thin slabs or lightweight concrete. This is very important for overhead connections to the underside of slabs or decks where the lower anchor bolt allowable loads rule out single-anchor connection for all but the smallest loads.

OTHER ANCHORS

Other, less-attractive anchor options not included in this chapter include undercut anchors, through-bolts and specialty inserts. The undercut anchor has the highest load capacities of any post-installed anchor. However, undercut anchors require special drill bits and complex, time-consuming installation procedures. Both the drill bits and the anchors are prohibitively expensive and impractical on most HVAC systems. Undercut anchors should only be considered where all other types of anchors cannot be used and everyone is aware of the added cost. Through-bolts can be used to provide attachment to thin concrete slabs or decks where it is impossible to get proper embedment of expansion anchors. Most through-bolts are used with anchor plates. Backup plates must be used to maximize the bolt tension capacity. Since through-bolts can be rated for very high loads, the structural engineer should verify the capacity of the floor to accept the loads. Specialty anchors can be used in some situations; however, there are currently no ICC test criteria and no ICC-ES Reports. Manufacturer's test data can be used for specialty inserts with the understanding and acceptance of the project structural engineer.

ANCHOR BOLT LOAD RATINGS

The most widely accepted anchor ratings are those published by the International Code Council Evaluation Service (ICC-ES). Independent tests, performed by ICC-accredited test labs, are subject to uniform standards, and published ratings are based on consistent safety factors regardless of manufacturer. Installation procedures are clearly described. While the ICC-ES report is necessary, anchor selection cannot be made directly from the report. Instead, anchor selection requires additional calculations per ACI 318 Appendix D.

ACI 318 APPENDIX D

Some of the most significant changes in seismic design have been in the design requirements of anchor bolts. Codes based on IBC 2006 and IBC 2009 require the use of ACI 318 Appendix D for anchor design. Anchor bolt capacities can no longer be selected or interpolated from a chart. Instead, anchor capacities must be calculated for seven potential failures modes rather than by comparison with tested values. Failure modes for anchors in tension include steel failure, bolt pullout, concrete breakout, and side face blowout. Failure modes for anchors in shear include steel failure, concrete pry-out, and concrete breakout.

In addition to the design requirements, post-installed anchors must also be prequalified for seismic applications in accordance with ACI 355.2 by testing for seismic tension and shear loads in cracked concrete in accordance with AC193 for mechanical anchors and

AC308 for adhesive anchors. Prequalified anchors include expansion anchors, undercut anchors, screw anchors, and adhesive anchors.

Testing for use in cracked concrete is very important in seismic situations where loads on floors and walls may cause cracks. It is extremely important where suspended equipment, piping, ductwork, etc., are attached to the underside of a concrete slab or a concrete-filled metal deck. A system of tensile cracks develops because the underside of the slab or deck is in tension. There can be catastrophic results if the attachment is prone to failure in cracked concrete.

Every properly tested anchor will have a valid ICC-ES Report detailing the tested values and modification factors to use in Appendix D calculations, though on occasion a special installation condition exists that requires additional input from the anchor manufacturer to ensure proper design. The load capacity of cast-in-place anchors can also be calculated using ACI 318 Appendix D; however, cast-in-place anchors do not need to be tested, and ICC-ES reports are not available or required.

Although both IBC 2006 and 2009 reference ACI 318-05, significant changes have been made to ACI 318-08, which should be used for design. The most significant is the addition of the $y_{h,v}$ factor. This is a modification factor for the shear capacity of anchors in thin slabs, and it is in the newer version of the standard for very good reason. When using ACI 318-05, the shear breakout strength of a concrete slab can actually decrease with increasing edge distance and the edge distance required for full capacity can be overly conservative. As an extreme example, a group of four anchors calculated using ACI 318-05 was governed by concrete breakout up to 81 ft (24.7 m) from the corner of the slab. When ACI discovered the problem, they added the modification factor $y_{h,v,}$ which effectively reduces the 81 ft (24.7 m) to 4 1/2 ft (1.37 m).

Slab thickness is now an important part of the anchor calculation. Depending on the exact type of anchor and manufacturer's testing, the installation may require from 4 to 14 in. (100 to 350 mm) of concrete slab thickness. The minimum thickness required for currently available anchors is 3 1/4 (82.6 mm). Using additional manufacturer's test data, it may be possible to install anchors in thinner slabs, but the anchor capacities are reduced significantly.

Allowable tension is influenced by anchor embedment, edge distance, and proximity to other anchors. To achieve the greatest possible tension for any given anchor embedment, edge distance should be 1.5x the effective embedment, and the distance to any other anchors should be 3x the effective embedment.

Allowable shear is influenced by anchor embedment, edge distance, and proximity to other anchors. Groups of anchors perform poorly in shear, with the result that small edge distances often cannot be used. It is important that housekeeping pads not be poured before anchor design confirms there is adequate edge distance. As a general rule, housekeeping pad design should include reinforcing bars near the perimeter of the pad to increase anchor shear capacity.

ANCHORS IN PIERS

Anchor bolts in narrow concrete piers and concrete curbs using ACI 318 Appendix D have very low allowable loads. Post-installed anchors yield especially low allowable loads or require very wide piers. Instead, the use of embedded bolts tied into the reinforcement, as well as using hairpins to eliminate potential failures, may be required. For these reasons, the design of piers for equipment requires a cooperative effort with the project structural engineer.

ANCHOR EMBEDMENT

Anchor embedment is not exactly what it seems. The embedment used in calculating anchor capacities is the "effective" anchor embedment. This effective embedment is not the required depth of the hole in the concrete or the length of the bolt embedded in the concrete or the length of the embedded bolt above the wedge locking mechanism, but is an assigned length for calculations that can be found in the ES report and manufacturer's data. The actual length of the bolt embedded in the concrete is the "nominal" embedment, which may be a valid reference to include on the plans for installation personnel and inspection.

ANCHORS EXPOSED TO WEATHER

Anchors installed on rooftops and other exposed areas may need special consideration. The standard requirement listed in the ES report states: "Use of zinc-coated carbon steel anchors is limited to dry interior locations", which would seem to require stainless steel anchors for exposed areas. This requirement is almost universally ignored, especially in drier climates. Anchor bolt manufacturers claim that this note is overly conservative and may provide additional information or a letter supporting the use of zinc-plated anchors in exposed areas.

ANCHORAGE DUCTILITY REQUIREMENTS IN IBC 2006 AND 2009

IBC 2006 Section 1908.1.16 references ACI 318, D.3.3.4, which requires the design of anchors in concrete to be governed by the steel strength of a ductile steel element. Section D.3.3.5 provides an alternative to D.3.3.4, allowing for design of ductile yielding in the attachment that the anchor is connecting to the structure. When ductile yielding cannot be achieved, the design strength of the anchors must be reduced to 40% of the nonductile strength as defined in section D.3.3.6. Because of the nature of post-installed concrete anchors, it is rarely possible to design a nonstructural component attachment for ductile yielding, so the design strength of the anchor must be reduced. With this reduction, anchorage design became extremely difficult, requiring significant changes to equipment anchorage and housekeeping pads.

The difficulties of anchoring equipment to meet these requirements were recognized almost immediately and the 2007 Supplement to the IBC introduced the following exception. "Anchors in concrete designed to support nonstructural components in accordance with ASCE 7 Section 13.4.2 need not satisfy Section D.3.3.4 [or D.3.3.5 of ACI 318]." This is a very significant exception and should be used on all nonstructural component anchor designs.

The 2009 IBC also includes the exception for nonstructural components in Section 1908.1.9.

ANCHORAGE DUCTILITY REQUIREMENTS IN CBC 2007 AND 2010

Chapter 19 of the 2007 California Building Code (CBC) lists no exception to the ACI 318 requirement for nonductile yielding, but Chapter 19A for Office of Statewide Health Planning and Development (OSHPD) and Division of State Architects (DSA) projects incorporates the language of the 2007 IBC Supplement as does Chapter 19 of the 2010 CBC.

In the 2010 CBC, Chapter 19A does not include the exemption found in the 2007 version, but instead includes the exemption in section 1615A.1.14: "Anchors pre-qualified for seismic applications need not be governed by the steel strength of a ductile steel element." The end result is that the 2010 CBC anchor requirements are nearly identical to CBC 2007 Section 1908A.1.47, which included the ductility exception to ACI 318-05 Sections D.3.3.4 & D.3.3.5.

With the exception of the OSHPD chapters of 2010 CBC, ASCE 7-05 is a standard reference in the current IBC-based codes. Chapter 13.4.2 of that document requires a 1.3x increase of the demand load on anchors in concrete. The 2010 CBC, Chapter 19A does not directly reference ASCE 7, but instead includes language from the ASCE 7-10. The only difference is that the 1.3 factor has been replaced by a maximum R_p value of 6, instead of a fixed 1.3 factor, so that for current OSHPD jobs the multiplier varies from 1 to 2 depending on R_p value in ASCE 7-05 Chapter 13.

CONCRETE ANCHOR CAPACITIES

The allowable loads for several anchor types and installation conditions are summarized in the following tables. Allowable loads for screw anchors in a concrete slab are shown in Table 5-2. Allowable loads for wedge-type expansion installed in a concrete slab are shown in table 5-3A and in the underside of a concrete filled metal deck are shown in Table 5-3B. Allowable loads for adhesive anchors in concrete slabs are shown in Tables 5-4A and 5-4B.

Table 5-2 Screw Anchor Tension and Shear Values (3000 psi normal weight concrete slab)

Anchor Diameter (in. *mm*)	Anchor Embedment (in. *mm*)	Minimum Slab Thickness (in. *mm*)	Minimum Edge Distance (in. *mm*)	Allowable Tension (lb *kN*)	Allowable Shear (lb *kN*)
3/8 *10*	3 1/4 *83*	5 *127*	4 *102*	1010 *4.49*	820 *3.65*
1/2 *13*	4 *102*	7 *178*	6 *152*	1640 *7.30*	1660 *7.38*
5/8 *16*	4 1/2 *114*	7 *178*	8 *203*	1980 *8.81*	2430 *10.81*
3/4 *19*	5 1/2 *140*	9 *229*	10 *254*	2270 *10.10*	3800 *16.90*

Notes:
1. Values in the table are for single anchors only and are not valid for anchors installed in groups.
2. Each anchor is considered to be in the corner of a slab with the minimum edge distance present on two sides.
3. Values are for ASD Seismic loads in cracked concrete for nonstructural component attachments.
4. Calculations assume supplementary reinforcement Condition B and minimum #4 bar between the anchor and edge of the slab.
5. Values used for anchor design must be determined using ACI 318 Appendix D in conjunction with the appropriate ICC ES Report.

Table 5-3A Expansion Anchor Tension and Shear Values (3000 psi normal weight concrete slab)

Anchor Diameter (in. *mm*)	Anchor Embedment (in. *mm*)	Minimum Slab Thickness (in. *mm*)	Minimum Edge Distance (in. *mm*)	Allowable Tension (lb *kN*)	Allowable Shear (lb *kN*)
3/8 *10*	2 *51*	4 *102*	6 *152*	485 *2.16*	710 *3.16*
	2 7/8 *73*	5 *127*	6 *152*	1035 *4.60*	820 *3.65*
1/2 *13*	2 3/4 *70*	5 *127*	6 *152*	1070 *4.76*	1055 *4.69*
	3 7/8 *98*	6 *152*	8 *203*	1395 *6.21*	2100 *9.34*
5/8 *16*	3 3/8 *86*	6 *152*	8 *203*	1450 *6.45*	2155 *9.59*
	5 1/8 *130*	8 *203*	8 *203*	2575 *11.45*	2750 *12.23*
3/4 *19*	4 1/8 *105*	8 *203*	10 *254*	1665 *7.41*	3425 *15.24*
	5 3/4 *146*	9 *229*	10 *254*	3005 *13.37*	3930 *17.48*
1 *25*	5 1/4 *133*	9 *229*	10 *254*	2435 *10.83*	4195 *18.66*

Notes:
1. Values in the table are for single anchors only and are not valid for anchors installed in groups.
2. Each anchor is considered to be in the corner of a slab with the minimum edge distance present on two sides.
3. Values are for ASD Seismic loads in cracked concrete for nonstructural component attachments.
4. Calculations assume supplementary reinforcement Condition B and minimum #4 bar between the anchor and edge of the slab.
5. Values used for anchor design must be determined using ACI 318 Appendix D in conjunction with the appropriate ICC ES Report.

Table 5-3B Expansion Anchor Tension and Shear Values (Underside of 3000 psi lightweight-concrete-filled metal deck)

Anchor Diameter (in. *mm*)	Anchor Embedment (in. *mm*)	Minimum Spacing (in. *mm*)	Allowable Tension (lb *kN*)	Allowable Shear (lb *kN*)
3/8 *10*	3 3/8 *86*	9 *229*	761 *3.39*	1588 *7.06*
1/2 *13*	4 1/2 *114*	12 *305*	932 *4.15*	2084 *9.27*
5/8 *16*	4 5/8 *117*	12 *305*	1342 *5.97*	2302 *10.24*

Notes:
1. Minimum 20 gage profile metal deck with min. 4 1/2 in. (114 mm) wide x maximum 3 in. (76 mm) deep flutes.
2. Anchors installed in the lower flute with up to 1 in. (25 mm) offset from the center.
3. Values used for anchor design must be determined using ACI 318 Appendix D in conjunction with the appropriate ICC ES Report.

Table 5-4A Adhesive Anchor Tension and Shear Values (3000 psi normal weight concrete slab)

Anchor Diameter (in. mm)	Anchor Embedment (in. mm)	Minimum Slab Thickness (in. mm)	Minimum Edge Distance (in. mm)	Allowable Tension (lb kN)	Allowable Shear (lb kN)
1/2 *13*	4 *102*	7 *178*	8 *203*	1885 *8.38*	1595 *7.09*
5/8 *16*	5 *127*	9 *229*	8 *203*	2035 *9.05*	2540 *11.30*
3/4 *19*	6 *152*	10 *254*	10 *254*	3275 *14.57*	3755 *16.70*
1 *25*	8 *203*	13 *330*	12 *305*	6460 *28.74*	6790 *30.20*

Notes:
1. Values in the table are for single anchors only and are not valid for anchors installed in groups.
Each anchor is considered to be in the corner of a slab with the minimum edge distance present on two sides.
2. Values are for ASD Seismic loads in cracked concrete for nonstructural component attachments.
3. Calculations assume supplementary reinforcement Condition B and minimum #4 bar between the anchor and edge of the slab.
4. Values used for anchor design must be determined using ACI 318 Appendix D in conjunction with the appropriate ICC ES Report.

Table 5-4B Adhesive Anchor Tension and Shear Values (3000 psi normal weight concrete slab)

Anchor Diameter (in. *mm*)	Anchor Embedment (in. *mm*)	Minimum Slab Thickness (in. *mm*)	Minimum Edge Distance (in. *mm*)	Allowable Tension (lb *kN*)	Allowable Shear (lb *kN*)
3/8 *10*	2 *51*	3 *76*	4 *102*	1365 *6.07*	615 *2.74*
	3 *76*	5 *127*	4 *102*	1445 *6.43*	860 *3.83*
1/2 *13*	2 *51*	4 *102*	6 *152*	1365 *6.07*	1160 *5.16*
	4 *102*	6 *152*	6 *152*	2565 *11.41*	1630 *7.25*
5/8 *16*	3 *76*	5 *127*	6 *152*	2510 *11.17*	1500 *6.67*
	5 *127*	8 *203*	6 *152*	4015 *17.86*	2105 *9.36*
3/4 *19*	4 *102*	6 *152*	8 *203*	3865 *17.19*	2455 *10.92*
	6 *152*	9 *229*	8 *203*	5770 *25.67*	3260 *14.50*
7/8 *22*	5 *127*	8 *203*	8 *203*	5400 *24.02*	3105 *13.81*
	7 *178*	11 *279*	8 *203*	6445 *28.67*	3895 *17.33*
1 *25*	6 *152*	9 *229*	10 *254*	7100 *31.58*	4445 *19.77*
	8 *203*	12 *305*	10 *254*	8725 *38.81*	5435 *24.18*

Notes:
1. Values in the table are for single anchors only and are not valid for anchors installed in groups.
2. Each anchor is considered to be in the corner of a slab with the minimum edge distance present on two sides.
3. Values are for ASD Seismic loads in cracked concrete for nonstructural component attachments.
4. Calculations assume supplementary reinforcement Condition B and minimum #4 bar between the anchor and edge of the slab.
5. Values used for anchor design must be determined using ACI 318 Appendix D in conjunction with the appropriate ICC ES Report.

These loads should only be used for general reference. And since the allowable loads are very dependent on installation conditions, careful review of the notes accompanying each table is recommended.

The actual ICC-ER Report should be used to obtain the exact information for the specific anchor, including information for specific embedment lengths, edge distance, spacing, installation procedures. Then the calculations required by ACI 318 Appendix D can be completed.

ANCHOR INSPECTION

Special inspection of anchors is required by IBC. The specific inspection requirements can be found in the ICC-ES Report for the specific anchor. The frequency of periodic inspection should be noted in the statement of special inspection. The extent of the inspection should cover proper installation of the anchors. That is, the inspector should verify that the specified anchor is being used, that the base material is appropriate, that the hole was drilled and cleaned properly, and that the anchor has the correct embedment, spacing, edge distance, and torque, when required.

ANCHOR BOLT RECOMMENDATIONS

The different types of anchors have some specific advantages and disadvantages. Expansion anchors need to be torqued during installation. This can short-circuit installations on neoprene pad isolation. Instead, adhesive anchors should be considered. Because adhesive anchors do not need to be torqued during installation, the bolt isolation bushing and steel washer can remain flexible under a finger-tight nut, allowing the pad to function properly. Screw anchors have the same potential, because they do not need to be torqued tight during installation, and screw anchors can be removed and reinstalled. As an added advantage, bolts that are not torqued often have smaller edge distance requirements than expansion anchors for equivalent loads.

ANCHOR BOLT LIMITATIONS FOR EQUIPMENT ANCHORAGE

Expansion anchors are not allowed for direct attachment of non-vibration-isolated mechanical equipment rated over 10 hp (7.5 kW) per Section 13.6.5.5 of ASCE 7-05. Screw anchors and adhesive anchors are acceptable.

ANCHOR BOLT DESIGN FOR PRYING

Prying action will increase tension loads on any type of anchor bracket. Great care should be taken in checking the geometry of the bracket and designing a bracket to minimize this action. A good example is the 45° bent brackets used on seismic bracing for suspended piping, ducts, or equipment. These brackets can easily double the tension load on the anchor bolt. This is particularly bad on concrete attachments into the underside of concrete-filled metal decks or slabs because tension is the weakest mode. Figure 5-2 illustrates several types of commercially available brackets and lists their prying factors. It should be noted that these factors are for comparison only; bending of weak brackets can increase the factors. In all cases, the lower the prying factor, the lower the calculated anchor tension.

Although most equipment connections are made to concrete, many other connections are made directly to building steel members or supplementary steel members. In many structures, especially those with lightweight roofs, the suspension and seismic bracing of suspended equipment, piping, ductwork, and electrical systems can only be made to steel or wood members. In other buildings, equipment mounted on steel mezzanine areas or steel roofing systems must be attached to structural steel. Suspended systems are often supported on threaded rods and attached to steel beams with suspension brackets.

These brackets are available in many forms in cast or formed steel construction and have an optional retaining hook to prevent slippage perpendicular to the beam. For large loads, brackets that attach to both sides of a steel beam are preferred to avoid eccentric loading of the steel members. In general, seismic connections at the lower flange of building steel members should be avoided unless specifically allowed by the project structural engineers. This is especially true of connections that load the beam perpendicular to their length. All connections should be made near the top flange of building steel members that are keyed into the floor deck. Where attachment to open beams is necessary and significant loads are present, connections should be made adjacent to cross braces to avoid twisting the steel beam. While beam clips are allowable for dead load supports, seismic connections must be either bolted or welded to the beam, except where the seismic load is perpendicular to the beam and will not cause the clamp to slide. In many situations, welding a transition plate or welded "U" is the most practical means of attachment. Bolted steel connections are often limited, because drilling holes through structural steel members is often prohibited. In most cases, a welded transition plate, welded "U," or supplementary steel provide the only possible bolted connection locations.

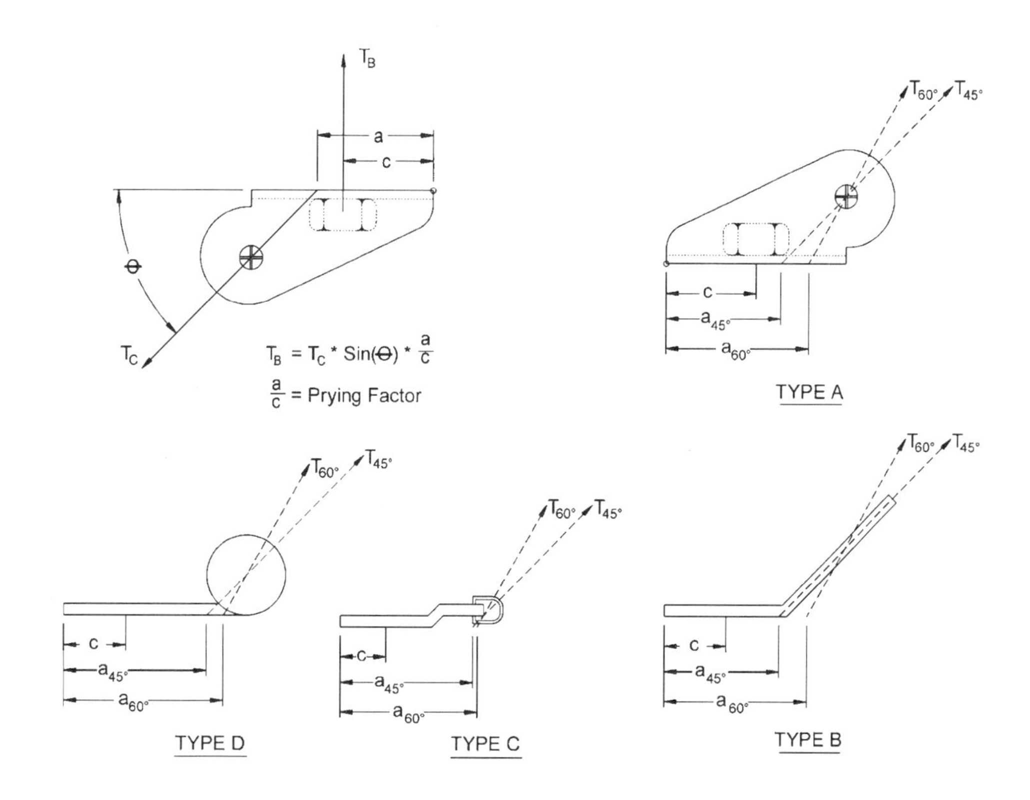

Bracket Type and Size	45°					60°				
	a		c		Prying Factor	a		c		Prying Factor
	in.	mm	in.	mm		in.	mm	in.	mm	
A-1	1.33	34	1.00	25	1.4	1.60	41	1.00	25	1.7
A-2	1.75	45	1.38	35	1.3	2.20	56	1.38	35	1.6
A-3	2.19	56	1.75	45	1.3	2.80	71	1.75	45	1.6
A-4	3.38	86	2.50	64	1.5	4.20	107	2.50	64	1.7
B-1	2.50	64	1.25	32	2.0	3.38	86	1.25	32	2.7
B-2	3.00	76	1.50	37	2.0	4.00	102	1.50	37	2.7
B-3	4.00	102	2.00	51	2.0	5.40	137	2.00	51	2.7
B-4	5.00	127	2.50	64	2.0	6.75	172	2.50	64	2.7
C-1	3.25	83	1.19	30	2.7	3.75	95	1.19	30	3.2
D-1	2.13	54	0.50	13	4.3	2.25	57	0.50	13	4.5

Figure 5-2 Seismic bracket types and prying factors.

STEEL BOLTS

Actual tensile and shear stresses on the bolts should be calculated from the anchor bolt reactions using the bolt areas listed in the American Institute of Steel Construction's *Manual of Steel Construction*, 9th edition, pp. 4-147. Allowable stresses on bolts should be obtained from AISC 9th edition, Table J3.2. For bolts subject to combined tension and shear, the interaction formula should be used from Table J3.3. A one-third increase is allowed for wind or seismic loads in accordance with Section J3.5. The tensile stress areas and shear stress areas for 3/8 in. to 1 1/4 in. bolts are listed in Table 5-5A and Table 5-5B.

Using these areas, the allowable stresses can be calculated as follows:

$$\text{Tensile stress, } f_t = T_{bolt} / A_t \text{ and}$$

$$\text{Shear stress, } f_v = V_{bolt} / A_k$$

where A_t and A_k are the tensile and minimum root areas of the bolt.

$$\text{Allowable sheer stress, } F_v = 10{,}000 \text{ psi} \times 133^* = 13{,}333 \text{ psi}$$

(*1/3 allowable stress increase for seismic loads.)

Allowable tensile stress, F_t:

If $26{,}000 - 1.8f_v \leq 20{,}000$ psi,

then, $F_t = (26{,}000 - 1.8f_v) \times 1.33$ psi;

else, $F_t = 20{,}000 \times 1.33 = 26{,}667$ psi.

If $f_t \leq F_t$ and $f_v \leq F_v$, then anchor bolts are satisfactory.

Table 5-5A Tensile Stress Area and Root Area for Steel Bolts

Bolt Diameter (in.)	Tensile Stress Area (in.^{2})	Root Area (in.^{2})
3/8	0.078	0.068
1/2	0.142	0.126
5/8	0.226	0.202
3/4	0.334	0.302
7/8	0.462	0.419
1	0.606	0.554
1 1/8	0.763	0.693
1 1/4	0.969	0.890

Table 5-5B Tensile Stress Area and Root Area for Steel Bolts

Bolt Diameter (mm)	Tensile Stress Area (mm^2)	Root Area (mm^2)
10	50.3	43.8
13	91.6	81.2
16	145.8	130.3
19	215.5	194.8
22	298.0	270.3
25	390.9	357.4
29	492.2	447.0
32	625.1	574.1

LAG SCREWS

Equipment connections to wood structures are most often made using lag screws. Lag screw connection design can be found in the *National Design Specification for Wood Construction*. The withdrawal capacity of lag screws is dependent on the lag screw diameter D, length of thread penetration p, and the specific gravity of the wood G. Withdrawal capacity is reduced if the lag screw is inserted into end grain. The diameter used for withdrawal calculations is the diameter of the unthreaded shank of the lag screw. The length of thread penetration is the length of threads penetrating the main member, not including the length of the tapered tip. The specific gravity is a measure of the wood density; the higher the specific gravity, the denser and stronger the wood.

Lateral design values for lag screws are dependent on the lag screw root diameter D_r, screw strength F_{yb}, side plate strength F_{es} and thickness t, and the specific gravity of the wood (G). Further influencing factors are edge distance, end distance, screw spacing, and penetration depth. Root diameter of a lag screw is the diameter of the shank within the threaded portion of the screw. The screw strength used is the bending yield strength of the steel fastener. Side plate strength is the dowel-bearing strength of the member that is being attached to the wood. The thickness of that member influences its bearing strength as well as the penetration depth of the screw. Penetration depth does not include the length of the screw's tapered tip. Edge distance is the distance from the edge of a member to the center of the nearest screw, measured perpendicular to the grain. End distance is the distance measured parallel to the grain from the end of a member to the center of the nearest screw. Spacing is the distance between screws measured along a line joining their centers. Minimum lag screw penetration for full capacity is eight times the lag screw shank diameter (8D). Penetration can be reduced to four times the shank diameter (4D) with a 50% reduction in capacity. As a worst case, minimum lag screw spacing must be five times the shank diameter (5D), minimum end distance must be seven times the shank diameter (7D), and minimum edge distance must be four times the shank diameter (4D).

A specific gravity of 0.35 covers a wide variety of standard construction woods including Douglas fir, pine, and redwood. Allowable values have been increased by a factor of 1.6 for wind and seismic loads. The lag screw allowable values into wood are listed in Table 5-6.

In most seismic connections, lag screws are subjected to combined tension and shear loads and the following equations must be used to check the connection. Calculations must be done to find the actual tension T_b and shear V_b on the lag screw. These calculations must include increases in tension caused by prying action. Using the following equation, the maximum load P_α on the lag screw can be found.

$$P_\alpha = \sqrt{T_b^2 + V_b^2}$$

The following equation can be used to determine the angle of loading (α):

$$\alpha = \tan^{-1}\left(\frac{T_b}{V_b}\right)$$

The withdrawal allowable W, shear allowable Z, and angle of loading α can be used in the following equation to determine the maximum allowable load at the angle of loading $P_\alpha allow$.

$$P_\alpha allow = \frac{WZ}{W\cos^2\alpha + Z\sin^2\alpha}$$

Then the maximum load is compared to the allowable load as shown below.

If $P_\alpha\ allow > P_\alpha$, the connection is satisfactory.

Table 5-7 lists connection examples and the chapter where they can be found.

Table 5-6 Lag-Screw Allowable Values in Wood for Wind or Seismic Loads (Minimum specific gravity = 0.35)*

Lag Screw Dia. × Length		Withdrawal Allowable		Lateral Load Allowable	
in. × in.	mm × mm	lb	kN	lb	kN
3/8 × 3	10 × 76	509	2.26	189	0.84
1/2 × 4	13 × 102	776	3.45	358	1.59
5/8 × 5	16 × 127	1087	4.84	487	2.17
3/4 × 6	19 × 152	1442	6.41	658	2.93
7/8 × 7	22 × 178	1838	8.18	844	3.75
1 × 8	25 × 203	2274	10.12	1045	4.65

* Allowable values are based on a 1/4 in. thick side plate, minimum edge distance of 7D, minimum end distance of 4D, and a minimum spacing of 5D.

Table 5-7 Connection Examples

Chapter	Items
7	Cable-braced ductwork connected to a concrete slab with expansion anchors
8	Cable-braced piping connected to a concrete slab with expansion anchors Rigid-braced piping connected to a concrete deck with expansion anchors
10	Cable-braced equipment from above the center of gravity to a concrete slab with expansion anchors
11	Combination vibration isolator/restraint connected to a steel beam Vibration isolator with a separate seismic snubber connected to a concrete slab
12	Wind load force calculation
13	Rooftop unit connected to a standard sheet metal curb with lag screws Standard sheet metal curb connected to a concrete deck
14	Cooling tower bolted to a steel beam
15	Wood beams lag screwed to the building steel

BIBLIOGRAPHY

AWC. 2005. *National Design Specification.* Washington, DC: American Forest and Paper Council, American Wood Council.

AISC. 1991. *Manual of Steel Construction*, 9th ed. Chicago, IL: American Institute of Steel Construction.

ACI. 2008. *318-08 Building Code Requirements for Structural Concrete.* Farmington Hills, MI: American Concrete Institute.

ICC. 2006. *2006 International Building Code®* (IBC®). Washington, DC: International Code Council.

ICC. 2009. *2009 International Building Code®* (IBC®). Washington, DC: International Code Council.

CBSC. 2007. *2007 California Building Code.* Sacramento, CA: California Building Standards Commission.

CBSC. 2010. *2010 California Building Code.* Sacramento, CA: California Building Standards Commission.

6 Housekeeping Pads

One of the most overlooked portions of mechanical and electrical equipment installations are the housekeeping pads, which are called plinths in many countries. They provide load transfer from the equipment anchorage to the main structure of the building. Current building codes do not address load transfer elements. Additional load transfer elements include concrete piers and pedestals, wood sleepers, and built-up wood platforms or curbs. This chapter addresses the design of housekeeping pads. Other load transfer elements should be designed on a case-by-case basis with the consultation of the structural engineer of record.

Most housekeeping pads are poured after the structural slab is poured. This happens because the exact location and size of the pads is not known until the exact piece of equipment has been selected and approved by the mechanical or electrical engineer. Nondesigned housekeeping pads usually consist of concrete poured inside a wooden formwork with or without reinforcement and are rarely positively connected to the structural slab.

In Chapter 18, photographs show that a standard nondesigned housekeeping pad can shatter like a pane of glass during a seismic event. A correctly designed housekeeping pad should be designed to handle shear, tension, and compression forces with proper reinforcement and attachments connecting the pad to the structural slab.

Tables 6-1 to 6-4 show the minimum reinforcing and doweling requirements for either lightweight or normal weight pads using IP units and Tables 6-5 through 6-8 for SI. The tables are based on F_p and are independent of seismic zones. Doweling can consist of stirrups, bent Z rebars cast in place when the structural slab is poured, bent L rebars drilled into the slab with adhesive anchors. A more positive method utilizes specifically designed anchors that screw onto male wedge anchor extensions. The tables are based on the allowable loads of expansion anchors. The nonexpansion anchor details have higher safety factors. The maximum load is the total of the equipment, vibration isolation system (if any), and the housekeeping pad. Figures 6-1 through 6-4 are details for attaching the housekeeping pad to the structural slab. The perimeter spacing of the doweling is smaller, as seismic snubbers or combination isolator/restraints are generally anchored along the perimeter.

Table 6-1 $F_p = 0.15$

Area (ft²)	Reinforcing	Interior Doweling	Perimeter Doweling	Max. Load (lb)
Up to 40	6 × 6 W1.4	1/2 in. diameter 36 in. o.c.e.w.*	1/2 in. diameter 24 in. o.c.e.w.*	8000
41 to 100	# 3 rebar 12 in. o.c.e.w.*	1/2 in. diameter 36 in. o.c.e.w.*	1/2 in. diameter 24 in. o.c.e.w.*	15,000
101 to 200	# 4 rebar 12 in. o.c.e.w.*	1/2 in. diameter 36 in. o.c.e.w.*	1/2 in. diameter 24 in. o.c.e.w.*	25,000
201 to 400	# 4 rebar 12 in. o.c.e.w.*	1/2 in. diameter 36 in. o.c.e.w.*	1/2 in. diameter 24 in. o.c.e.w.*	50,000

Notes:
1. This table applies to systems where the height of the center of gravity of the combined maximum load is less than or equal to the width of the housekeeping pad. For other conditions, the pad should be designed by the seismic restraint vendor or by other qualified design professionals.
2. Housekeeping pads that are over 400 ft^2 (37 m^2) should be designed by the seismic restraint vendor or by other qualified design professionals.
3. Reinforcing is to be installed at the centerline of the housekeeping slab height.
*On center each way (o.c.e.w.).

Table 6-2 $F_p = 0.3$

Area (ft²)	Reinforcing	Interior Doweling	Perimeter Doweling	Max. Load (lb)
Up to 40	6 × 6 W1.4	1/2 in. diameter 36 in. o.c.e.w.*	1/2 in. diameter 24 in. o.c.e.w.*	8000
41 to 100	# 3 rebar 12 in. o.c.e.w.*	1/2 in. diameter 36 in. o.c.e.w.*	5/8 in. diameter 24 in. o.c.e.w.*	15,000
101 to 200	# 4 rebar 12 in. o.c.e.w.*	1/2 in. diameter 36 in. o.c.e.w.*	5/8 in. diameter 24 in. o.c.e.w.*	25,000
201 to 400	# 4 rebar 12 in. o.c.e.w.*	1/2 in. diameter 36 in. o.c.e.w.*	5/8 in. diameter 24 in. o.c.e.w.*	50,000

Notes:
1. This table applies to systems where the height of the center of gravity of the combined maximum load is less than or equal to the width of the housekeeping pad. For other conditions, the pad should be designed by the seismic restraint vendor or by other qualified design professionals.
2. Housekeeping pads that are over 400 ft^2 (37 m^2) should be designed by the seismic restraint vendor or by other qualified design professionals.
3. Reinforcing is to be installed at the centerline of the housekeeping slab height.
*On center each way (o.c.e.w.).

Table 6-3 $F_p = 0.5$

Area (ft^2)	Reinforcing	Interior Doweling	Perimeter Doweling	Max. Load (lb)
Up to 40	#3 rebar 12 in. o.c.e.w.*	5/8 in. diameter 36 in. o.c.e.w.*	5/8 in. diameter 24 in. o.c.e.w.*	8000
41 to 100	#3 rebar 12 in. o.c.e.w.*	5/8 in. diameter 36 in. o.c.e.w.*	3/4 in. diameter 24 in. o.c.e.w.*	15,000
101 to 200	#4 rebar 12 in. o.c.e.w.*	5/8 in. diameter 36 in. o.c.e.w.*	3/4 in. diameter 24 in. o.c.e.w.*	25,000
201 to 400	#4 rebar 12 in. o.c.e.w.*	5/8 in. diameter 36 in. o.c.e.w.*	1 in. diameter 24 in. o.c.e.w.*	50,000

Notes:
1. This table applies to systems where the height of the center of gravity of the combined maximum load is less than or equal to the width of the housekeeping pad. For other conditions, the pad should be designed by the seismic restraint vendor or by other qualified design professionals.
2. Housekeeping pads that are over 400 ft^2 (37 m^2) should be designed by the seismic restraint vendor or by other qualified design professionals.
3. Reinforcing is to be installed at the centerline of the housekeeping slab height.
*On center each way (o.c.e.w.).

Table 6-4 $F_p = 0.75$

Area (ft^2)	Reinforcing	Interior Doweling	Perimeter Doweling	Max. Load (lb)
Up to 40	#3 rebar 12 in. o.c.e.w.*	5/8 in. diameter 36 in. o.c.e.w.*	3/4 in. diameter 24 in. o.c.e.w.*	8000
41 to 100	#4 rebar 12 in. o.c.e.w.*	5/8 in. diameter 36 in. o.c.e.w.*	3/4 in. diameter 24 in. o.c.e.w.*	15,000
101 to 200	#5 rebar 12 in. o.c.e.w.*	3/4 in. diameter 36 in. o.c.e.w.*	3/4 in. diameter 24 in. o.c.e.w.*	25,000
201 to 400	#5 rebar 12 in. o.c.e.w.*	3/4 in. diameter 36 in. o.c.e.w.*	1 in. diameter 24 in. o.c.e.w.*	50,000

Notes:
1. This table applies to systems where the height of the center of gravity of the combined maximum load is less than or equal to the width of the housekeeping pad. For other conditions, the pad should be designed by the seismic restraint vendor or by other qualified design professionals.
2. Housekeeping pads that are over 400 ft^2 (37 m^2) should be designed by the seismic restraint vendor or by other qualified design professionals.
3. Reinforcing is to be installed at the centerline of the housekeeping slab height.
*On center each way (o.c.e.w.).

Table 6-5 $F_p = 0.15$

Area (m^2)	Reinforcing	Interior Doweling	Perimeter Doweling	Max. Load (kg)
Up to 3.7	A 193 mesh	10 mm diameter 1 m o.c.e.w.*	12 mm diameter 500 mm o.c.e.w.*	3636
3.8 to 9.3	T8 rebar 300 mm o.c.e.w.*	10 mm diameter 1 m o.c.e.w.*	12 mm diameter 500 mm o.c.e.w.*	6818
9.4 to 18.6	T12 rebar 300 mm o.c.e.w.*	12 mm diameter 1 m o.c.e.w.*	14 mm diameter 500 mm o.c.e.w.*	13 636
18.7 to 37	T12 rebar 300 mm o.c.e.w.*	12 mm diameter 1 m o.c.e.w.*	14 mm diameter 500 mm o.c.e.w.*	22 730

Notes:
1. This table applies to systems where the height of the center of gravity of the combined maximum load is less than or equal to the width of the housekeeping pad. For other conditions, the pad should be designed by the seismic restraint vendor or by other qualified design professionals.
2. Housekeeping pads that are over 400 ft^2 (37 m^2) should be designed by the seismic restraint vendor or by other qualified design professionals.
3. Reinforcing is to be installed at the centerline of the housekeeping slab height.
*On center each way (o.c.e.w.).

Table 6-6 $F_p = 0.3$

Area (m^2)	Reinforcing	Interior Doweling	Perimeter Doweling	Max. Load (kg)
Up to 3.7	A 193 mesh	12 mm diameter 1 m o.c.e.w.*	12 mm diameter 500 mm o.c.e.w.*	3636
3.8 to 9.3	T8 rebar 300 mm o.c.e.w.*	12 mm diameter 1 m o.c.e.w.*	14 mm diameter 500 mm o.c.e.w.*	6818
9.4 to 18.6	T12 rebar 300 mm o.c.e.w.*	14 mm diameter 1 m o.c.e.w.*	16 mm diameter 500 mm o.c.e.w.*	13 636
18.7 to 37	T12 rebar 300 mm o.c.e.w.*	14 mm diameter 1 m o.c.e.w.*	16 mm diameter 500 mm o.c.e.w.*	22 730

Notes:
1. This table applies to systems where the height of the center of gravity of the combined maximum load is less than or equal to the width of the housekeeping pad. For other conditions, the pad should be designed by the seismic restraint vendor or by other qualified design professionals.
2. Housekeeping pads that are over 400 ft^2 (37 m^2) should be designed by the seismic restraint vendor or by other qualified design professionals.
3. Reinforcing is to be installed at the centerline of the housekeeping slab height.
*On center each way (o.c.e.w.).

Table 6-7 $F_p = 0.5$

Area (m^2)	Reinforcing	Interior Doweling	Perimeter Doweling	Max. Load (kg)
Up to 3.7	T8 rebar 300 mm o.c.e.w*	14 mm diameter 1 m o.c.e.w.*	16 mm diameter 500 mm o.c.e.w.*	3636
3.8 to 9.3	T12 rebar 300 mm o.c.e.w.*	16 mm diameter 1 m o.c.e.w.*	20 mm diameter 500 mm o.c.e.w.*	6818
9.4 to 18.6	T12 rebar 300 mm o.c.e.w.*	16 mm diameter 1 m o.c.e.w.*	20 mm diameter 500 mm o.c.e.w.*	13 636
18.7 to 37	T12 rebar 300 mm o.c.e.w.*	20 mm diameter 1 m o.c.e.w.*	24 mm diameter 500 mm o.c.e.w.*	22 730

Notes:
1. This table applies to systems where the height of the center of gravity of the combined maximum load is less than or equal to the width of the housekeeping pad. For other conditions, the pad should be designed by the seismic restraint vendor or by other qualified design professionals.
2. Housekeeping pads that are over 400 ft^2 (37 m^2) should be designed by the seismic restraint vendor or by other qualified design professionals.
3. Reinforcing is to be installed at the centerline of the housekeeping slab height.
*On center each way (o.c.e.w.).

Table 6-8 $F_p = 0.75$

Area (m^2)	Reinforcing	Interior Doweling	Perimeter Doweling	Max. Load (kg)
Up to 3.7	T8 rebar 300 mm o.c.e.w.*	16 mm diameter 1 m o.c.e.w.*	20 mm diameter 500 mm o.c.e.w.*	3636
3.8 to 9.3	T12 rebar 300 mm o.c.e.w.*	20 mm diameter 1 m o.c.e.w.*	20 mm diameter 500 mm o.c.e.w.*	6818
9.4 to 18.6	T16 rebar 300 mm o.c.e.w.*	20 mm diameter 1 m o.c.e.w.*	24 mm diameter 500 mm o.c.e.w.*	13 636
18.7 to 37	T16 rebar 300 mm o.c.e.w.*	24 mm diameter 1 m o.c.e.w.*	24 mm diameter 500 mm o.c.e.w.*	22 730

Notes:
1. This table applies to systems where the height of the center of gravity of the combined maximum load is less than or equal to the width of the housekeeping pad. For other conditions, the pad should be designed by the seismic restraint vendor or by other qualified design professionals.
2. Housekeeping pads that are over 400 ft^2 (37 m^2) should be designed by the seismic restraint vendor or by other qualified design professionals.
3. Reinforcing is to be installed at the centerline of the housekeeping slab height.
*On center each way (o.c.e.w.).

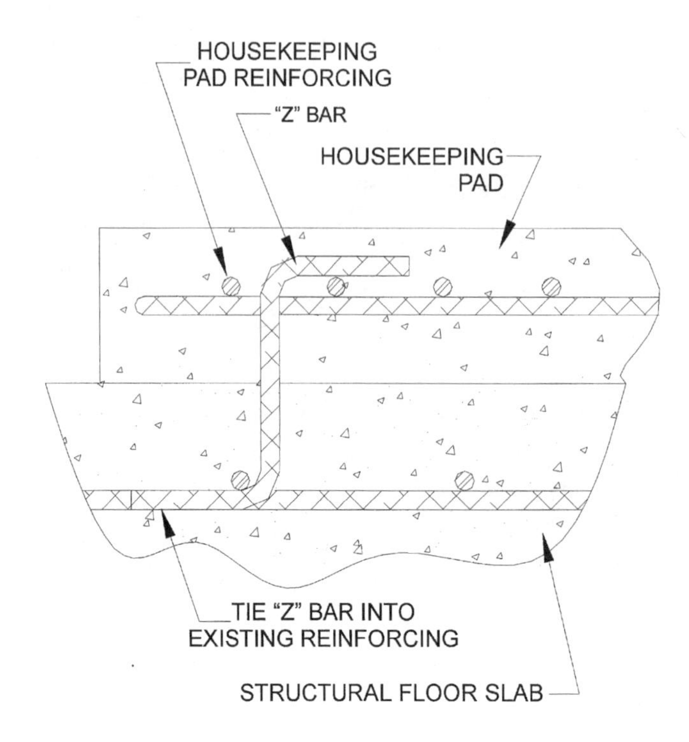

Figure 6-1 Details for attaching the housekeeping pad to the structural slab.

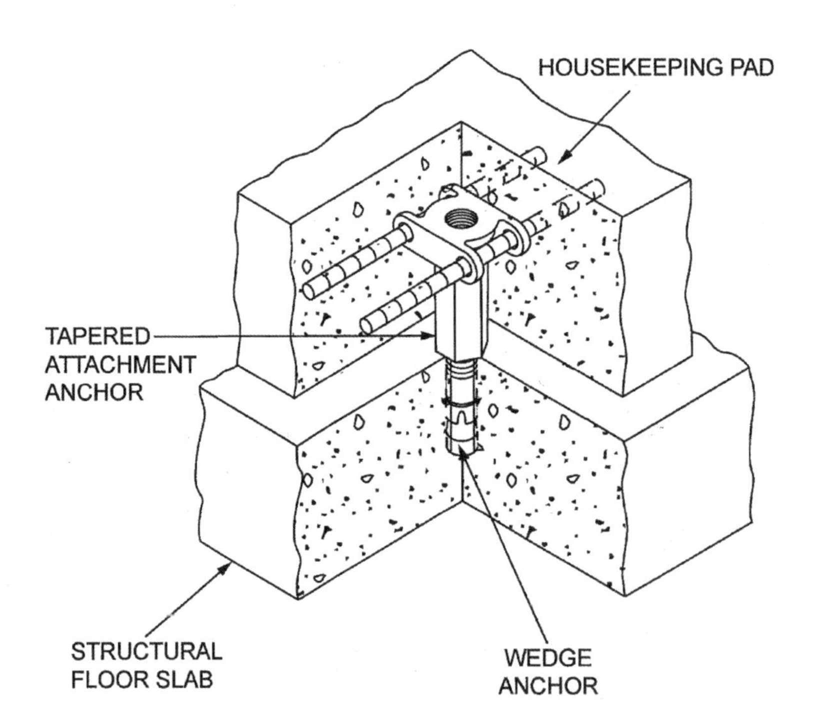

Figure 6-2 Details for attaching the housekeeping pad to the structural slab.

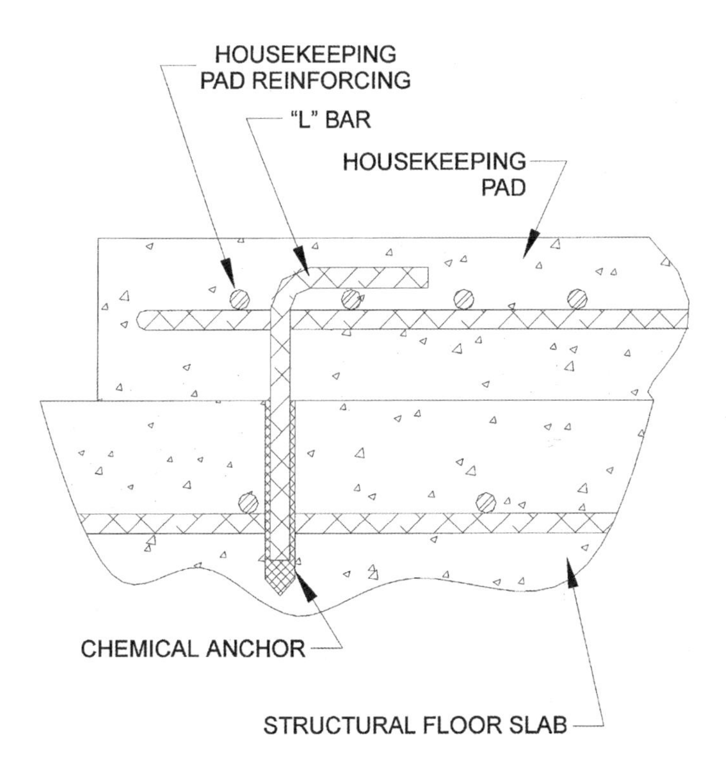

Figure 6-3 Details for attaching the housekeeping pad to the structural slab.

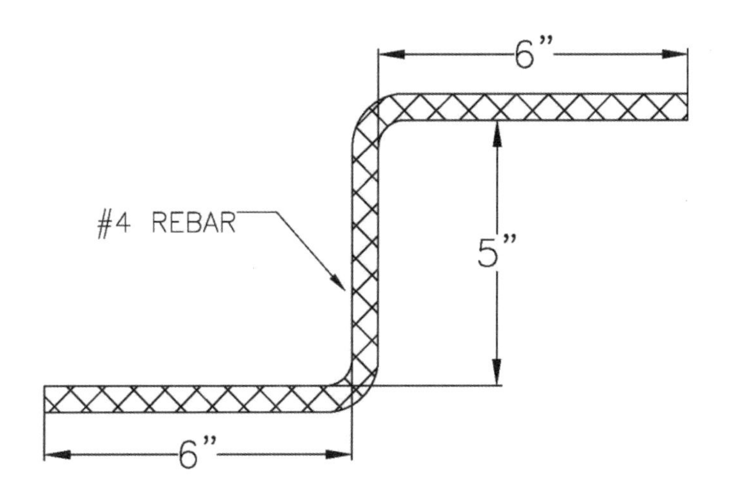

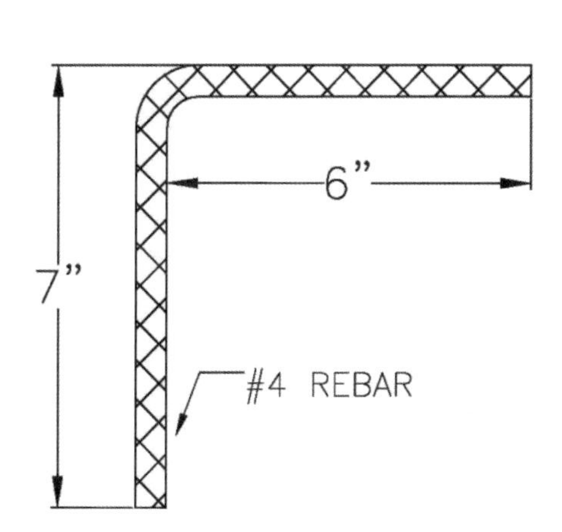

Figure 6-4 Typical rebar bends for Figures 6-1 and 6-3.

7 Suspended Ductwork

Bracing of ductwork is necessary to prevent swaying and protect surrounding piping, conduit, and equipment. Bracing is also necessary to protect duct connections to equipment and piping connections to in-line coils.

The sheet metal ductwork covered in this chapter must be constructed in accordance with the *Sheet Metal and Air Conditioning Contractors National Association (SMACNA) HVAC Duct Construction Standards*. This type of ductwork is often suspended by sheet metal straps screwed directly to the ductwork. Threaded hanger rods supporting a trapeze under the ductwork is another common support method.

Ductwork constructed of plastic or fiberglass piping, such as polyvinyl chloride (PVC), polyvinylidene fluoride (PVDF), or fiberglass-reinforced pipe (FRP), is covered in Chapter 8.

Sway bracing of suspended ductwork requires the following:

- Each individual or trapeze-supported system of ductwork should be sway braced to prevent motion in all directions.
- Maximum sway brace spacing should consider the strength of the ductwork, the duct connection locations, the seismic capacity of the sway brace, and the limitations of the building structure.
- Vertical hangers that can accept compressive loads as well as tensile loads should be used at the sway brace locations.
- The ductwork should be stiffened at sway brace locations to limit duct deformation due to localized seismic loads.
- Ductwork supported by vibration isolation hangers should be detailed and installed with isolation hangers attached directly to the structure, limit stops located above and below the isolation hanger, and cable braces to limit sway.

SWAY BRACING

There are two types of sway braces for suspended systems—solid and cable. Solid bracing consists of a single steel member, generally an angle or 12 gage (2.7 mm) channel strut, installed at the hanger rod connection to the system and extending up to the structure at an angle between 30° and 60° from horizontal. The obvious advantage of this system is that it only requires access to the structure on one side. However, there are two significant disadvantages to this system:

- Solid braces should be designed to resist compression loads in addition to tension loads, limiting the length of the installed steel brace.
- When the single solid brace is designed to resist a compression force, the hanger rod

connection will be subjected to a tension force, which, when added to the existing gravity load, may exceed the capacity of the connection to the structure. This is especially critical when brace locations are spaced at every four or more hanger locations because the ratio of seismic load to gravity load increases dramatically. As discussed throughout this guide, the hanger rod connection is critical when bracing suspended systems with solid braces.

Cable bracing consists of galvanized steel cable in tension, connected at the hanger rod location and extending up to the structure at an angle between 30° and 60° from horizontal. Steel cable can be provided with or without prestretching. Prestretched cable, combined with tested cable end connections, allows the designer to use a safety factor of 2 for seismic applications. Non-prestretched cable increases possible elongation under load. To reduce the amount of elongation non-prestretched cable should be selected with a safety factor of 5.

There are three important advantages of a cable system compared to solid bracing.

- There is no limit to the installed length of the cable.
- Because cables resist tension loads only, they do not apply any additional tension loads to the support rods.
- Cables can be cut to the approximate length and adjusted easily during installation.

The tensile load on the support rods is not affected by cable sway bracing. Bracing the duct with cables keeps the maximum tension on the support rod to a minimum. Therefore, cable bracing is recommended for retrofit projects where the hanger rod connection to the structure, especially in concrete structures, cannot be verified.

Some designers have attempted to use solid bracing on two sides of a suspended system in an attempt to design for tension loads only. This has little merit, because a solid brace is by nature a tension and compression member, and a complete analysis of member flexibility is required to determine the tension and compression loads on each member. It is impossible for this system to eliminate compression loads from brace members.

Although cable systems require access to the structure from both sides of the support connection, the lack of length limitations increases the potential angular adjustment capabilities of cable systems and makes them much easier than solid braces to install. For these reasons, cable systems are the first choice of many installing contractors.

Wall and floor penetrations can be used to control ductwork motion and work in unison with sway bracing. A wall penetration can act as a typical transverse brace. A floor penetration can act as both a transverse and longitudinal brace location for horizontal ductwork, provided the distance from the floor to the horizontal turn is less than two duct widths. However, walls or floors that cannot accept the seismic loads applied at duct penetrations cannot be used as braces.

Duct penetrations with breakaway fire dampers should not be used as typical brace locations. The ductwork connection to a wall with a fire damper should be ignored and the duct braced accordingly. This will limit any potential seismic loads on the damper connections.

A suggested layout procedure for duct bracing is as follows:

1. Each straight run of the ductwork system should be braced with a minimum of two transverse braces installed perpendicular to the ductwork and one longitudinal brace installed parallel to the ductwork, as shown in Figure 7-1.
2. Transverse braces should be located at the final support point of each run of duct with two or more supports. If the distance between the braces exceeds the maximum transverse brace spacing in Table 7-1, additional transverse braces should be located to limit the brace spacing to the maximum transverse brace spacing, as shown in Figure 7-2.
3. A longitudinal brace should be located on each straight run of duct. If the length of the duct run exceeds the maximum longitudinal brace spacing, additional longitudinal

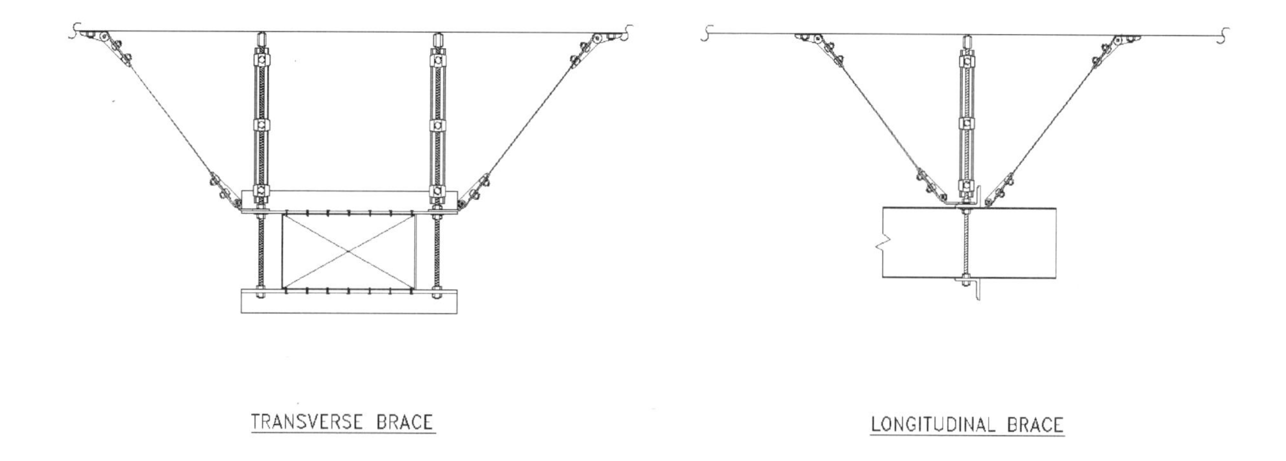

Figure 7-1 Bracing of suspended ductwork.

Table 7-1 Maximum Brace Spacing of Sheet Metal Ductwork*

Maximum Seismic Acceleration Input (g)	Maximum Transverse Brace Spacing, ft (m)	Maximum Longitudinal Brace Spacing, ft (m)
0.25	40 (12.2)	80 (24.4)
0.50	30 (9.1)	60 (18.2)
1.0	30 (9.1)	60 (18.2)
2.0	20 (6.1)	40 (12.2)

*Sheet metal should be constructed according to SMACNA HVAC duct construction standards.

braces should be located on the duct run while limiting the brace spacing to the maximum longitudinal brace spacing, as shown in Figure 7-3.

4. A transverse brace located within two duct widths of a 90° turn can provide limited longitudinal bracing for the straight run of duct around the 90° turn. The length of ductwork longitudinally braced by this transverse brace is equal to one-half the maximum transverse brace spacing minus the distance from the transverse brace to the 90° turn, as shown in Figure 7-4.
5. If a straight run of ductwork has less than two support points, is connected to a braced straight run of ductwork at each end, and its total length is less than two duct widths, brace "across" the run by adding its length to the transverse and longitudinal brace design of the connected runs, as shown in Figure 7-5. If its length is greater than two duct widths, a support point with a transverse brace should be provided.
6. Vertical drops from horizontal runs of ductwork to equipment require a transverse brace at the final support location before the drop, as shown in Figure 7-6. The total length of ductwork from the support point to the equipment connection or flexible connector should be less than one-half the maximum sway brace spacing of the transverse brace, and the length of ductwork from the support point to the drop should be less than two duct widths.
7. Avoid bracing a duct to separate portions of the structure that may act differently in response to an earthquake. For example, do not connect a transverse brace to a wall and a longitudinal brace to a floor or roof at the same brace location if the floor and wall will act differently in an earthquake. This is a special concern in a lightly framed penthouse mechanical room. The wall and floor tend to move in unison in a basement mechanical room, so attachment of braces to the wall and floor is acceptable.

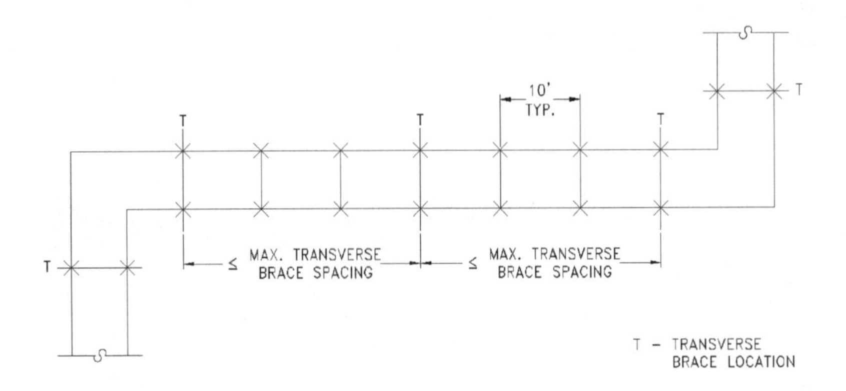

Figure 7-2 Transverse brace layout.

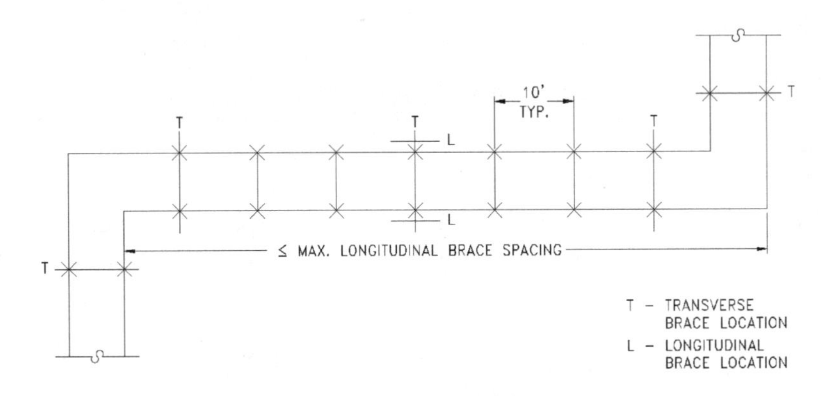

Figure 7-3 Transverse and longitudinal brace layout.

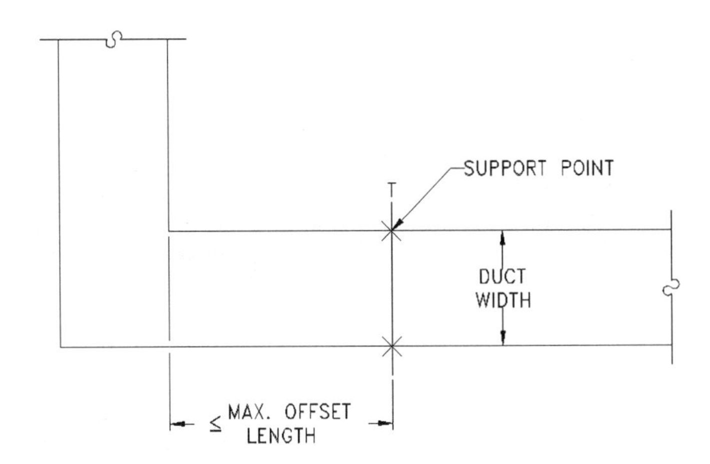

Figure 7-4 Transverse brace with offset length.

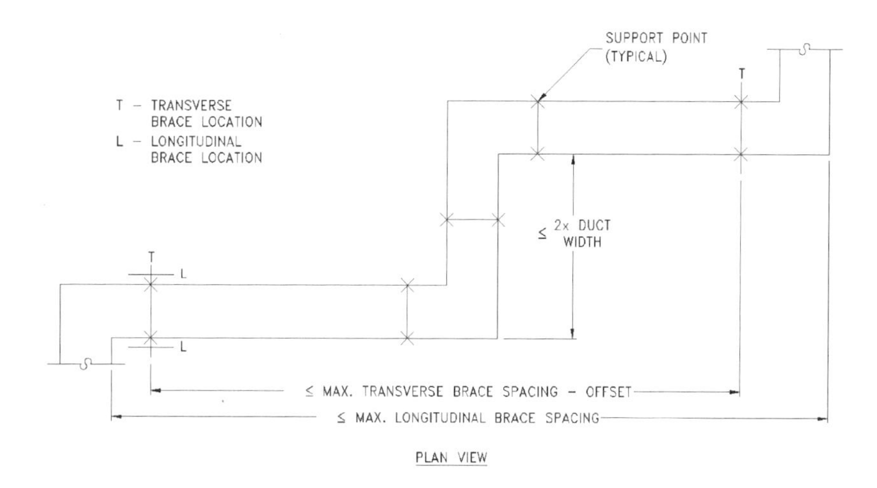

Figure 7-5 Bracing of duct with an offset.

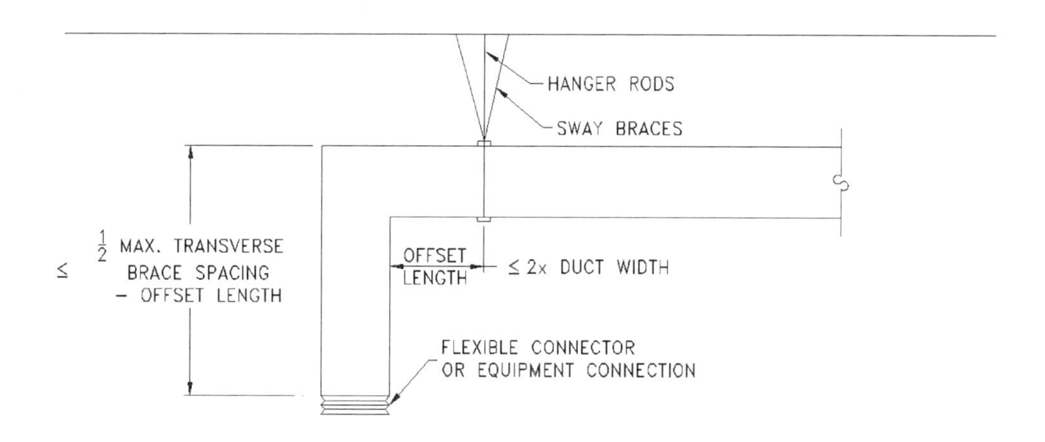

Figure 7-6 Bracing of duct drop to equipment supported on floor.

8. Do not mix solid bracing with cable bracing in the same direction on any duct run. Solid bracing is likely more rigid than cable bracing, so they do not share the loads evenly. This is very important for large, rigid ducts, but less important on smaller-diameter, less-rigid ducts, which are more flexible than either solid or cable bracing.
9. Multiple or stacked trapezes that share hanger rods should be braced independently from one another. Otherwise, the trapezes can be connected with a rigid frame at the seismic brace locations and the frames braced with standard methods and details.

Maximum sway brace spacing for sheet metal ductwork is indicated in Table 7-1. Maximum brace spacings tabulated are referenced in part from the *Seismic Evaluation Procedure for Equipment in U.S. Department of Energy Facilities*, the *SMACNA Seismic Restraint Manual—Guidelines for Mechanical Systems*, and the *Mason Industries Seismic Restraint Guidelines for Suspended Piping, Ductwork, and Electrical Systems.* Maximum brace spacing of all other ductwork materials and connections should be individually designed.

HANGER ROD REQUIREMENTS

The motion of suspended systems installed without sway bracing both bends and adds tension to the hanger rods as the system swings to resist the horizontal seismic force. However, when sway bracing is connected from the hanger rod attachment to the ductwork support and up to the structure, a vertical seismic load will be applied to the hanger rod at the seismic brace location. This load is a compression load when the sway brace is in tension and is a tension load when the sway brace is in compression. Because there is a tension load on the hanger rod due to the system weight, the vertical seismic load should be added to or subtracted from the existing gravity load.

The dynamics of vertical and horizontal seismic loads eliminates the use of simple connections of hanger rods to the structure that are not tested or certified for dynamic loads, such as the c-clamp connected to one side of a steel beam flange as shown in Figure 7-7. In this case, a safety hook or strap should be provided to prevent the hanger from falling.

The resultant load on the hanger rods is also dependent on the connection point of the sway brace to the hanger relative to the ductwork center of gravity. The duct will rotate about its center of gravity creating additional tension and compression loads on the hanger rods. The magnitude of the force is dependent on the vertical distance from the connection point of the sway brace to the center of gravity and the minimum horizontal distance between hanger rods. Refer to Figures 7-8 through 7-11 for determining resultant tensile or compressive loads on the hanger rods. Each figure indicates the external seismic and gravity loads and the reactions of both the sway brace at the hanger rod connection and the hanger rods at their connection points to the structure.

As shown in Figure 7-8, for solid and cable brace systems where the brace connection is above the ductwork center of gravity:

Hanger Rod 1

$$\text{If } (T_w + T_{over} + T_{Fpv}) > C_{Fph},$$

$$\text{then } T_{(Rod\ 1)max} = (T_w + T_{over} + T_{Fpv}) - C_{Fph}, \tag{7-1}$$

$$\text{else } C_{(Rod\ 1)max} = (C_{Fph} + C_{Fpv}) - (T_w + T_{over}). \tag{7-2}$$

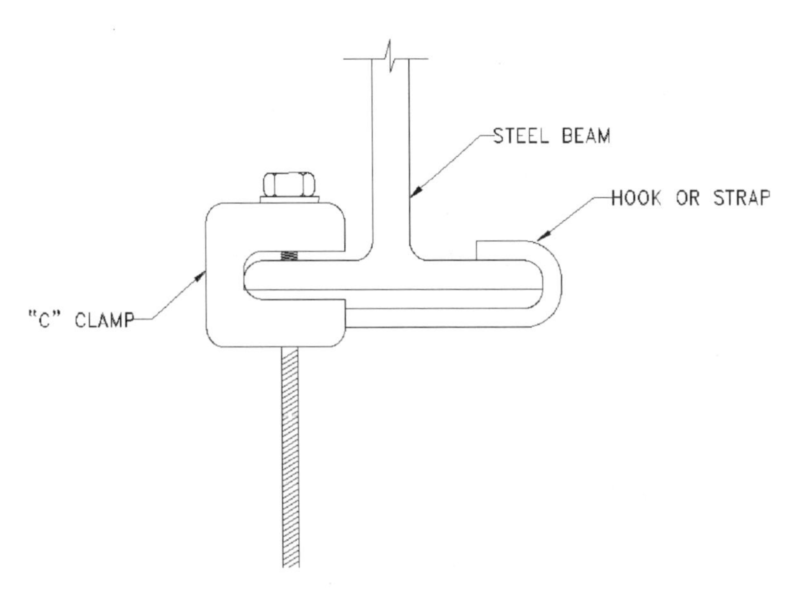

Figure 7-7 "C" clamp hanger installation.

Hanger Rod 2

$$\text{If } (T_w + T_{Fpv}) > C_{over},$$

$$\text{then } T_{(Rod\ 2)max} = (T_w + T_{Fpv}) - C_{over}, \quad (7\text{-}3)$$

$$\text{else } C_{(Rod\ 2)max} = (C_{over} + C_{Fpv}) - T_w. \quad (7\text{-}4)$$

As shown in Figure 7-9, for solid brace systems where the brace connection is above the ductwork center of gravity:

Hanger Rod 1

$$\text{If } (T_w + T_{Fph} + T_{Fpv}) > C_{over},$$

$$\text{then } T_{(Rod\ 1)max} = (T_w + T_{Fph} + T_{Fpv}) - C_{over}, \quad (7\text{-}5)$$

$$\text{else } C_{(Rod\ 1)max} = (C_{over} + C_{Fpv}) - (T_w + T_{Fph}). \quad (7\text{-}6)$$

Hanger Rod 2

$$\text{If } (T_w + T_{over}) > C_{Fpv},$$

$$\text{then } T_{(Rod\ 2)max} = T_w + T_{Fpv} + T_{over}, \quad (7\text{-}7)$$

$$\text{else } C_{(Rod\ 2)max} = C_{Fpv} - (T_w + T_{over}). \quad (7\text{-}8)$$

As shown in Figure 7-10, for solid and cable brace systems where the brace connection is below the ductwork center of gravity:

Hanger Rod 1:

$$\text{If } (T_w + T_{Fpv}) > (C_{Fph} + C_{over}),$$

$$\text{then } T_{(Rod\ 1)max} = (T_w + T_{Fpv}) - (C_{Fph} + C_{over}), \quad (7\text{-}9)$$

$$\text{else } C_{(Rod\ 1)max} = (C_{Fph} + C_{over} + C_{Fpv}) - T_w. \quad (7\text{-}10)$$

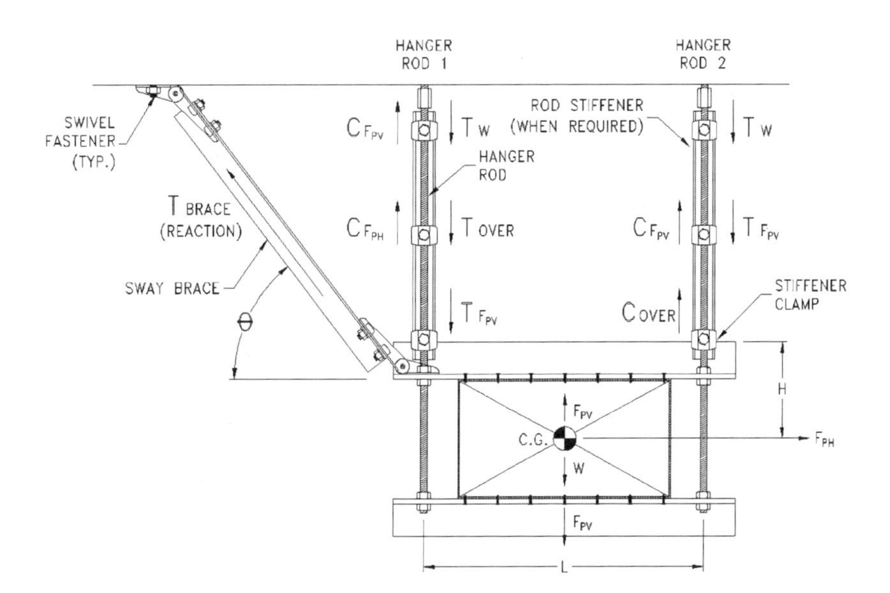

Figure 7-8 Analysis of hanger rod loads for solid or cable bracing connected above the center of gravity.

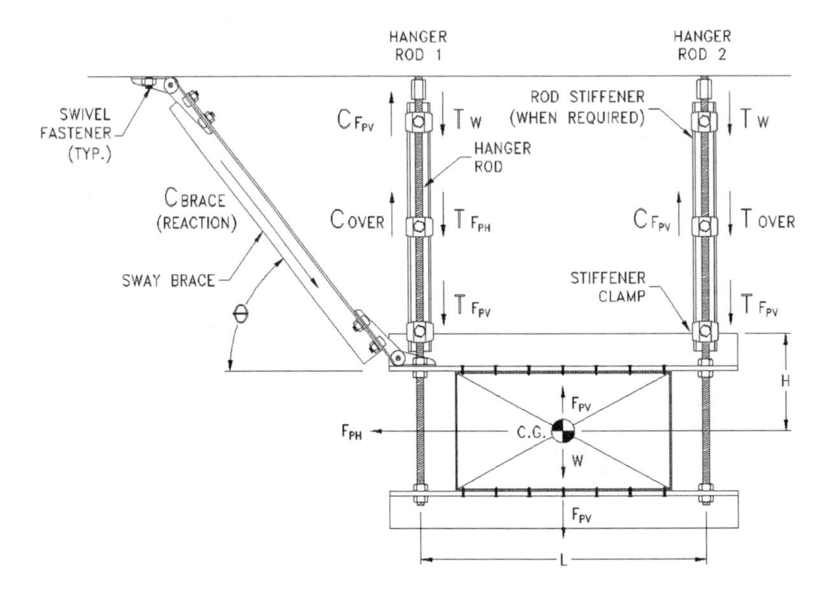

Figure 7-9 Analysis of hanger rod loads for solid bracing connected above the center of gravity.

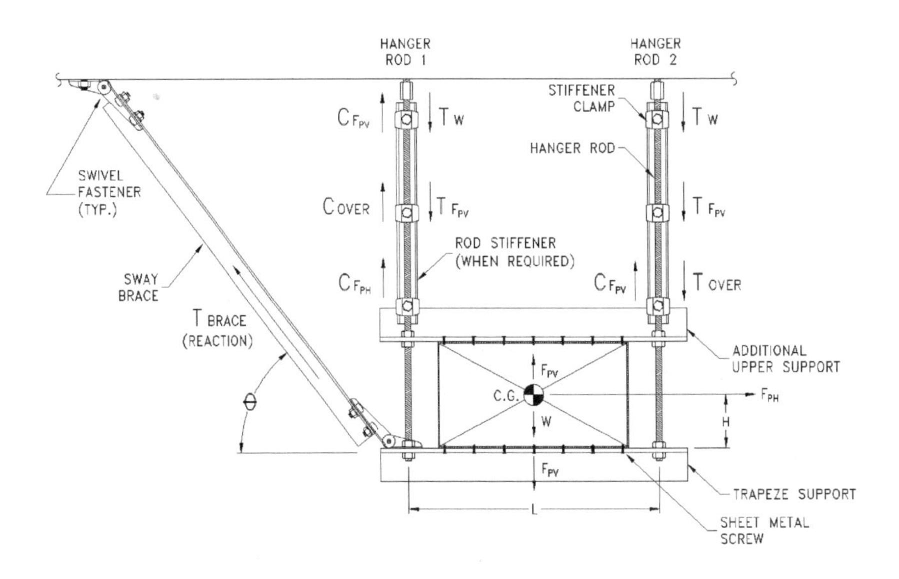

Figure 7-10 Analysis of hanger rod loads for solid or cable bracing below the center of gravity.

Hanger Rod 2

$$\text{If } (T_w + T_{over}) > C_{Fpv},$$

$$\text{then } T_{(Rod\ 2)max} = T_w + T_{over} + T_{Fpv}, \tag{7-11}$$

$$\text{else } C_{(Rod\ 2)max} = C_{Fpv} - (T_w + T_{over}) \tag{7-12}$$

As shown in Figure 7-11, for solid brace systems where the brace connection is below the ductwork center of gravity:

Hanger Rod 1

$$\text{If } (T_w + T_{Fph} + T_{over}) > C_{Fpv},$$

$$\text{then } T_{(Rod\ 1)max} = T_w + T_{Fpv} + T_{Fph} + T_{over}, \tag{7-13}$$

$$\text{else } C_{(Rod\ 1)max} = C_{Fpv} - (T_w + T_{Fph} + T_{over}). \tag{7-14}$$

Hanger Rod 2

$$\text{If } (T_w + T_{Fpv}) > C_{over},$$

$$\text{then } T_{(Rod\ 2)max} = (T_w + T_{Fpv}) - C_{over}, \tag{7-15}$$

$$\text{else }_{(Rod\ 2)max} = (C_{over} + C_{Fpv}) - T_w. \tag{7-16}$$

where

W = total weight of ductwork = (duct wt/ft) × (maximum transverse brace spacing),
L = span between hanger rods along width of ductwork,
H = vertical distance from ductwork center of gravity to brace connection,
Fph = seismic horizontal force,
Fpv = seismic vertical force,
N = number of hangers carrying duct over the transverse brace spacing span,
T_w = tensile force due to $W = (W / N)$,

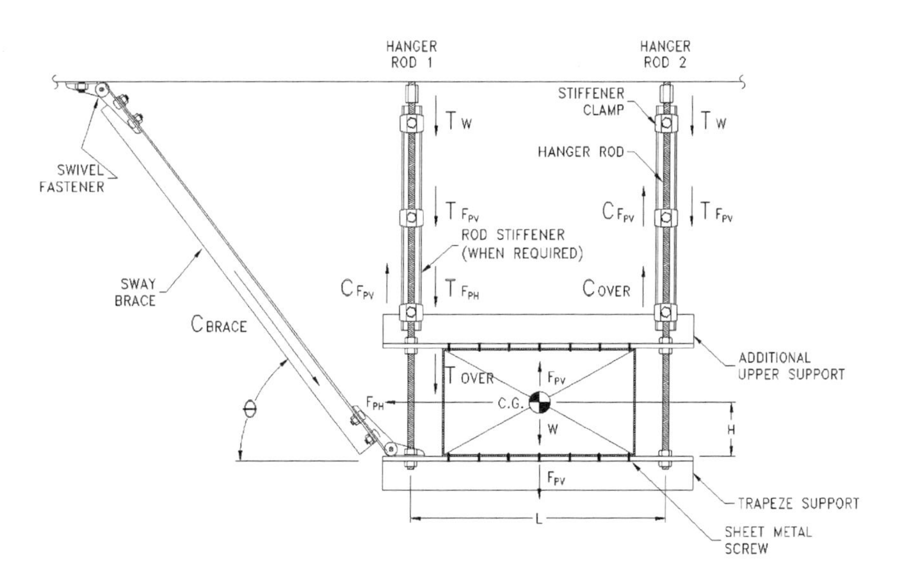

Figure 7-11 Analysis of hanger rod loads for solid bracing below the center of gravity.

$T_{Fph} = C_{Fph}$ = tensile or compressive force due to $F_{ph} = F_{ph} \tan \theta$,

$T_{Fpv} = C_{Fpv}$ = tensile or compressive force due to $F_{pv} = F_{pv} / N$,

$T_{over} = C_{over}$ = tensile or compressive force due to ductwork overturning $= (Fph \times H) / [(L / 2)(N / 2)]$,

$T_{brace} = C_{brace}$ = tensile or compressive force on the brace due to $Fph = Fph/\cos \theta$

Design loads are the resolution of forces, and the resolution is either a tension or compression load on the vertical hanger. For tension loads, check both the hanger rod and its connection to the structure. The allowable value is based on the tensile strength of the threaded rod shown in Table 7-2. Connection of the hanger rod to the structure and its allowable value is dependent on the structure and method of connection, as shown in this manual in Chapter 5. Hanger rod connections to structural steel may use the allowable values in Table 7-2 but are limited to the maximum loads outlined by the structural engineer of record for the project.

Threaded rods can accept a limited amount of compressive force. The maximum length of the hanger rod that can accept the compressive force is called the maximum unbraced rod length. If the installed length of the hanger rod is greater than the maximum unbraced rod length, the hanger rod should be braced with steel angle or 12 gage (2.7 mm) strut channel. The brace is attached to the rod with clamps installed at specified locations and spacings, as shown in Figure 7-12 and Table 7-3. The length of the braced hanger rod is limited as well. Bracing of hanger rods is only required at sway brace connection points. Welding of angles or strut channels to a hanger rod is not generally acceptable. Refer to Tables 7-4 and 7-5 for maximum unbraced rod lengths.

Table 7-2 Threaded Rod Allowable Tension Loads

Threaded Rod Diameter, in. (mm)	Allowable Working Load, lb (kN)	Allowable Combined Working and Seismic Load, lb (kN)
3/8 (10)	610 (2.7)	810 (3.6)
1/2 (13)	1130 (5.0)	1500 (6.6)
5/8 (16)	1810 (8.0)	2410 (10.7)
3/4 (19)	2710 (12.0)	3610 (16.0)
7/8 (22)	3770 (16.7)	5030 (22.3)
1 (25)	4960 (22.1)	6610 (29.4)
1 1/4 (32)	8000 (35.5)	10,660 (47.4)

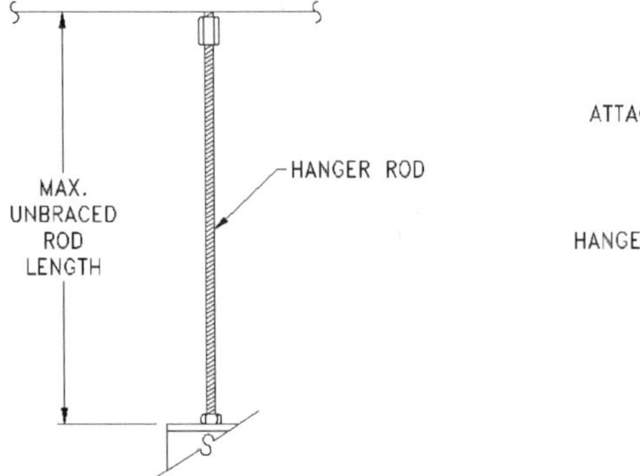

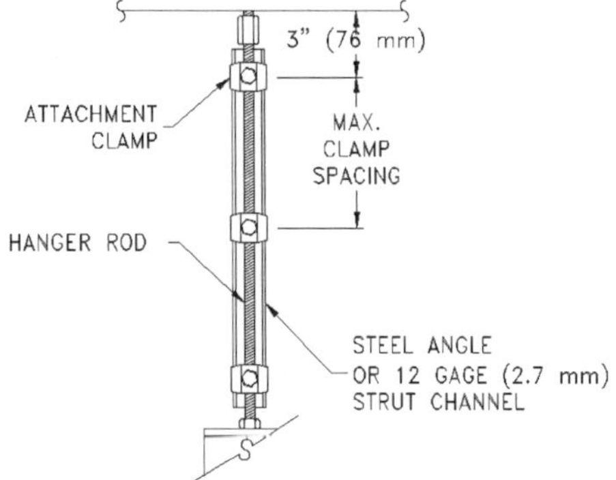

Figure 7-12 Vertical hanger rod installation.

Table 7-3 Maximum Clamp Bracing for Rod Stiffeners

Support Rod Diameter, in. (mm)	Maximum Clamp Spacing, in. (mm)
3/8 (10)	12 (305)
1/2 (13)	16 (406)
5/8 (16)	21 (533)
3/4 (19)	25 (635)
7/8 (22)	30 (762)
1 (25)	38 (965)
1-1/4 (32)	48 (1219)

Table 7-4 Ductwork Braced Above Its Center of Gravity

Maximum Duct Weight per Foot *m*, lb (kg)				Seismic Horizontal Force, *Fph*, lb (kN)	Vertical Hanger Rod Diameter, in. (mm)	Connection of Hanger Rod to Structure		Maximum Unbraced Rod Length, in. (mm)	Brace Member Size		Connection to Structure
0.25 *g*	0.5 *g*	1.0 *g*	2.0 *g*			Solid	Cable		Solid	Cable	
12 (17)	8 (12)	4 (6)	3 (4)	120 (0.6)	3/8 (10)	A	A	24 (610)	A	A	B
20 (29)	13 (19)	6 (9)	5 (7)	200 (0.9)	1/2 (13)	B	B	34 (864)	A	B	C
48 (71)	32 (47)	16 (24)	12 (17)	480 (2.1)	1/2 (13)	E	B	22 (559)	B	C	D
62 (92)	41 (61)	20 (29)	15 (22)	620 (2.8)	5/8 (16)	F	C	32 (813)	C	C	E
124 (184)	82 (122)	41 (61)	31 (46)	1240 (5.6)	3/4 (19)	H	E	33 (838)	D	D	F
200 (297)	140 (208)	70 (104)	50 (74)	2100 (9.4)	7/8 (22)	H	F	36 (914)	D	D	H

Table 7-5 Ductwork Braced Below Its Center of Gravity

Maximum Duct Weight per Foot *m*, lb (kg)				Seismic Horizontal Force, *Fph*, lb (kN)	Vertical Hanger Rod Diameter, in. (mm)	Connection of Hanger Rod to Structure		Maximum Unbraced Rod Length, in. (mm)	Brace Member Size		Connection to Structure
0.25 *g*	0.5 *g*	1.0 *g*	2.0 *g*			Solid	Cable		Solid	Cable	
12 (17)	8 (12)	4 (6)	3 (4)	120 (0.6)	3/8 (10)	B	A	18 (457)	A	A	B
20 (29)	13 (19)	6 (9)	5 (7)	200 (0.9)	1/2 (13)	C	B	25 (635)	A	B	C
48 (71)	32 (47)	16 (24)	12 (17)	480 (2.1)	5/8 (16)	F	C	26 (660)	B	C	D
62 (92)	41 (61)	20 (29)	15 (22)	620 (2.8)	5/8 (16)	F	E	23 (584)	C	C	E
124 (184)	82 (122)	41 (61)	31 (46)	1240 (5.6)	3/4 (19)	H	F	24 (610)	D	D	F
200 (297)	140 (208)	70 (104)	50 (74)	2100 (9.4)	7/8 (22)	H	G	26 (660)	D	D	H

It may be practical to add a completely separate sway brace location for ductwork supported by hanger straps. The vertical hanger can be designed as an added trapeze with threaded rods or using steel angle or 12 gage (2.7 mm) strut channel without a threaded rod, as shown in Figure 7-13.

STIFFENING OF THE DUCTWORK AT SWAY BRACE LOCATIONS

Sheet metal ductwork is generally supported at or near joint connections where the rigidity is greater rather than at a midpoint between joints. When sway bracing is connected to these points, it is recommended that horizontal braces sandwich the duct on top and bottom using threaded vertical hanger rods, as shown in Figure 7-14. If the sway brace is more than two feet away from a joint, a complete frame around the duct should be provided, as shown in Figure 7-15. The stiffener member sizes should be equal to the trapeze support

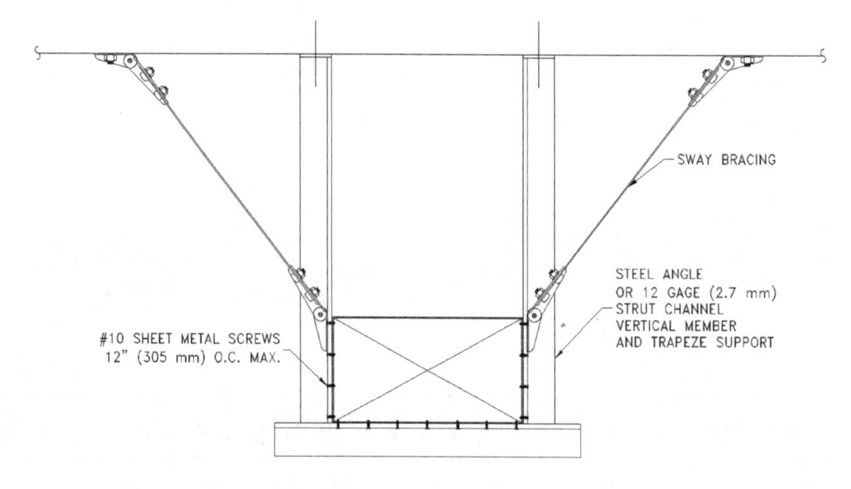

Figure 7-13 Optional vertical hanger installation.

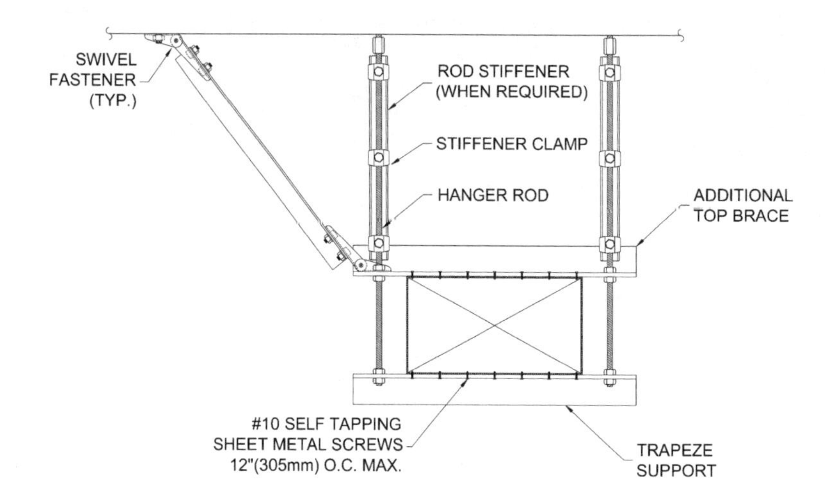

Figure 7-14 Stiffening of ductwork near joint connection.

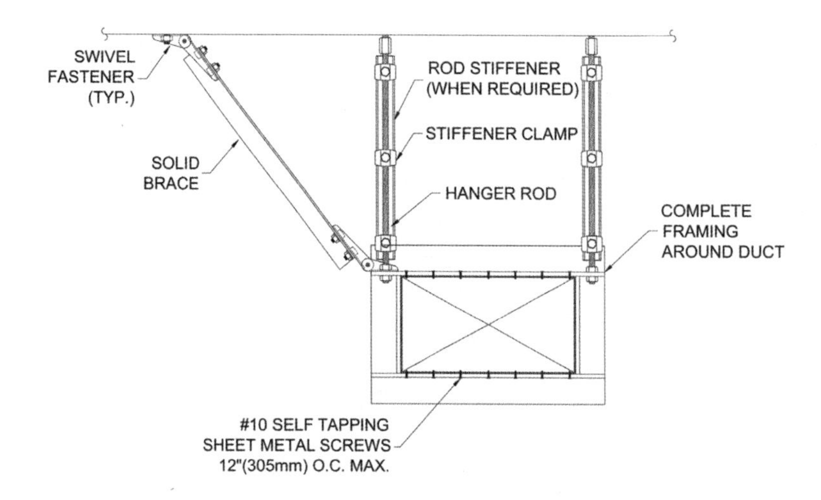

Figure 7-15 Stiffening of ductwork away from joint connection.

member size designed to support the ductwork. Round ductwork should be stiffened with a 2 1/2 in. × 12 gage (64 mm × 2.7 mm) sheet metal strap around the ductwork at all sway brace locations, as shown in Figure 7-16. Stiffening members should be connected to the ductwork with #10 sheet metal screws at 12 in. (305 mm) on center, as shown in Figures 7-14, 7-15, and 7-16.

VIBRATION ISOLATED DUCTWORK

Ductwork suspended by spring and/or neoprene vibration-isolation hangers should only be braced with cable bracing. Steel cables should be installed so that they do not carry any duct loads. Cables should be installed with a minor amount of sag due to the weight of the cable, but without additional slack. This prevents vibration transmission to the structure but does not allow excessive motion.

Installation of the vibration isolation hangers is critical to the performance of the sway braces. As discussed earlier, the function of a hanger rod is to resist tensile and compressive loads. The vibration isolation hanger introduces a break in the solid support of the ductwork with a flexible material designed to accept gravity loads but not compressive loads.

At seismic brace locations the vibration isolation hanger boxes must be designed to accept compressive forces without excessive bending, and the hangers should be installed to prevent excessive upward motion. Neoprene washers with steel inserts must be installed above and below the hanger. The hanger is installed with the top of the neoprene washer in contact with the supporting structure. The upper neoprene element in combination spring and neoprene vibration-isolation hangers deflects under load to create the 1/4 in. (6 mm) clearance. The lower limit neoprene washer and steel insert is installed below the hanger box with a nut to limit upward motion to 1/4 in. (6 mm), all as shown in Figure 7-17.

Contact of the top of the hanger box assembly against the structure, combined with contact of the limit stop against the bottom of the hanger box, will limit vertical motion of the equipment without compromising the performance of the vibration isolators.

THERMAL EXPANSION

All ductwork experiences some thermal expansion or contraction. In typical HVAC systems, the temperature change is small and the duct runs relatively short. These typical systems can be braced without special considerations for thermal expansion. Systems with larger temperature changes and long, straight duct runs should be sway-braced by designing a longitudinal brace for the entire straight run, even if it exceeds 80 ft, as shown in Figure 7-18. This will prevent damage to the ductwork or to conflicting longitudinal sway braces installed on a straight run. If an anchor is designed into the system, it should be capable of sway bracing the entire straight run of duct longitudinally and a portion of it transversely in addition to any thermal loads. Careful consideration is recommended for selecting and locating sway bracing ductwork when thermal expansion is a concern, especially where the contents are hazardous such as combustion, engine, or kitchen exhaust ductwork.

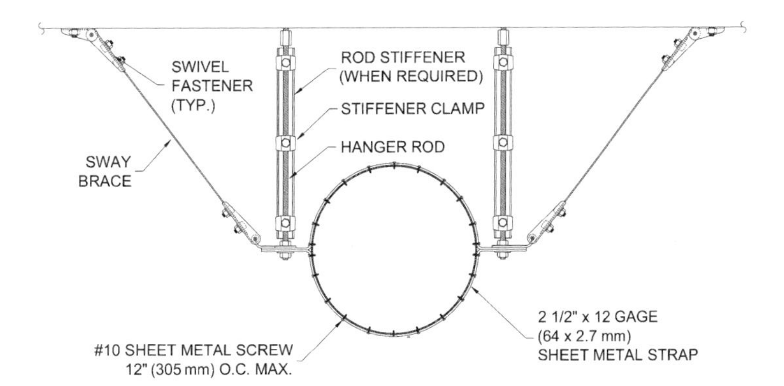

Figure 7-16 Stiffening of round duct.

SEISMIC JOINTS

Ductwork systems crossing building seismic joints, passing from building to building, or supported from different portions of the building should be designed to accept differential support displacements without damaging the duct, equipment connections, or support connections. The amount of displacement can vary, depending on acceleration input, soil conditions, and vertical distance from grade to the ductwork. To accept the motion, which is usually determined from the structural engineer of record, ductwork systems can be designed by one of the following methods:

1. Design the ductwork system with offsets and branches to develop the inherent flexibility required to accept the differential motion, as shown in Figure 7-19.
2. Localize the area at which differential motion will occur by anchoring to each building and utilize an offset in the duct to accept the motion, as shown in Figure 7-20A.
3. Localize the area at which differential motion will occur by anchoring to each building and utilize a flexible connector to accept the motion, as shown in Figure 7-20B. Size the connector to accept the predicted motion from data provided by the manufacturer.

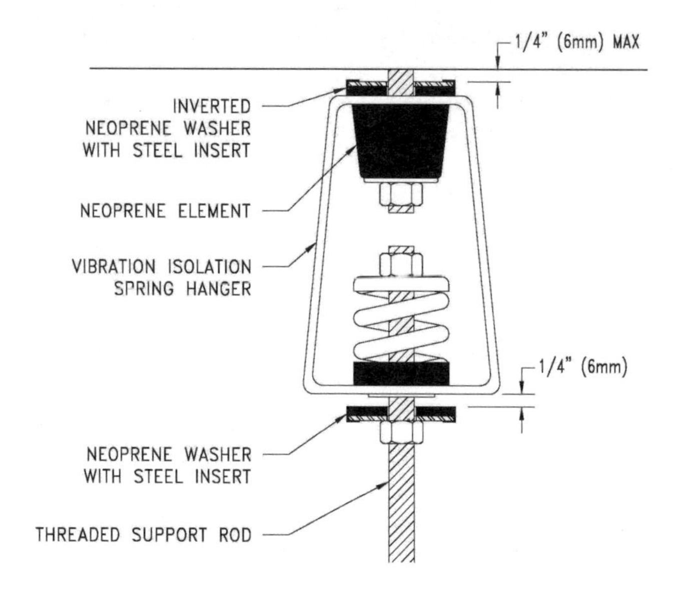

Figure 7-17 Vibration isolation hanger installation.

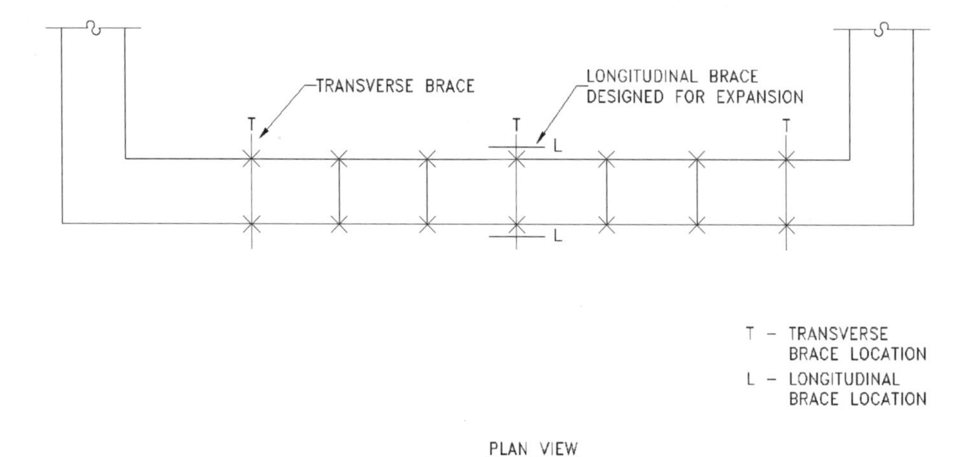

Figure 7-18 Longitudinal bracing with thermal movement.

4. Where the differential motion is more than a few inches (centimeters), a single duct flexible connector may not have the required motion capability. In these situations, a double flexible connector arrangement may be required, as shown in Figure 7-21. The amount of acceptable transverse motion is a function of the angular offset capability of the duct flexible connectors and the length of the duct section between the flexible connectors. If the flexible connectors have a low stiffness, it may be possible to use seismic bracing to localize the motion.

DESIGN TABLES

Tables 7-4 to 7-8 can be used to determine the vertical hanger rod diameter, maximum unbraced rod length, hanger rod connections to the structure, and solid or cable brace member size for bracing installed up to a 60° angle from horizontal. Separate tables are included for brace connections to the duct above and below the center of gravity. Refer to Chapter 5 for allowable values for connections used to formulate Table 7-6.

The charts assume the following:

$$Fpv \leq Fph$$

$$N \geq 4$$

$$H \leq L/2$$

$$\theta \leq 60°$$

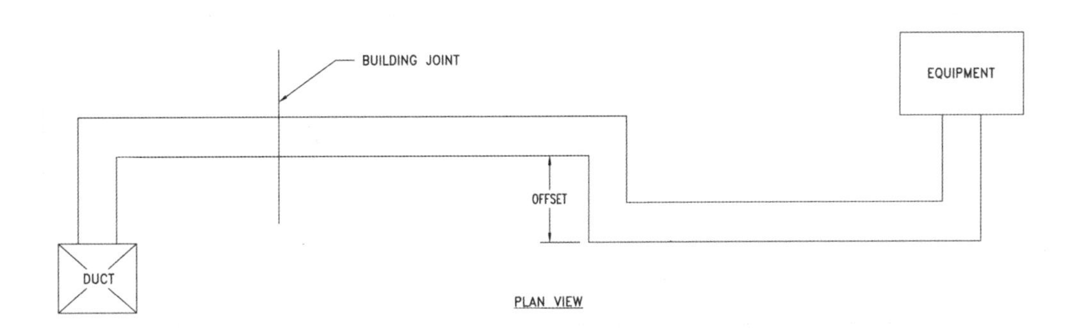

Figure 7-19 System design.

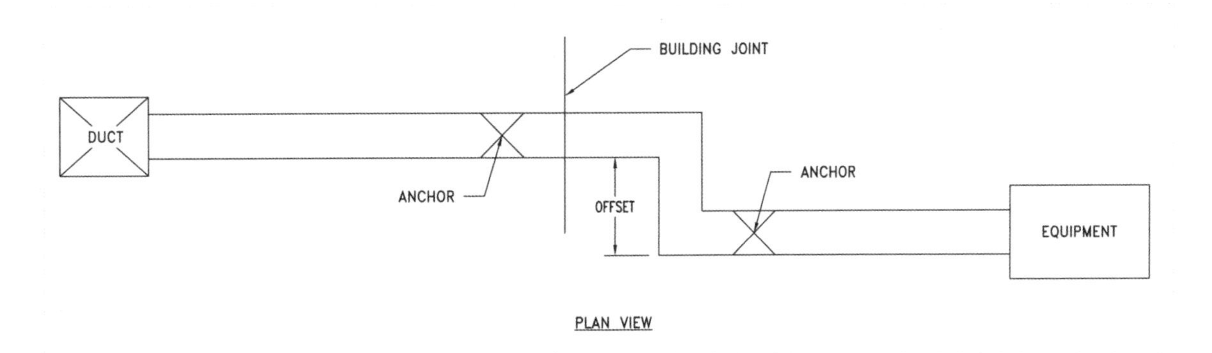

Figure 7-20A Localize displacement.

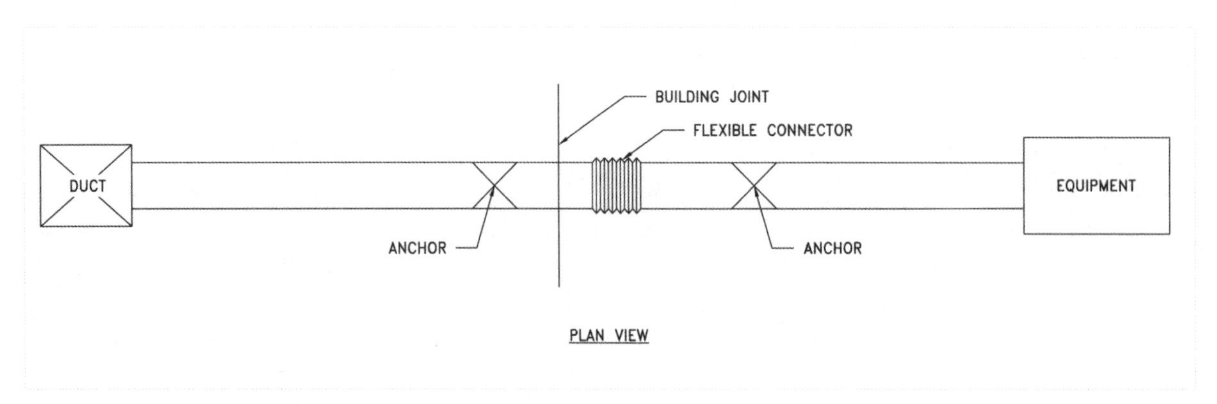

Figure 7-20B Localize displacement with a flexible connector.

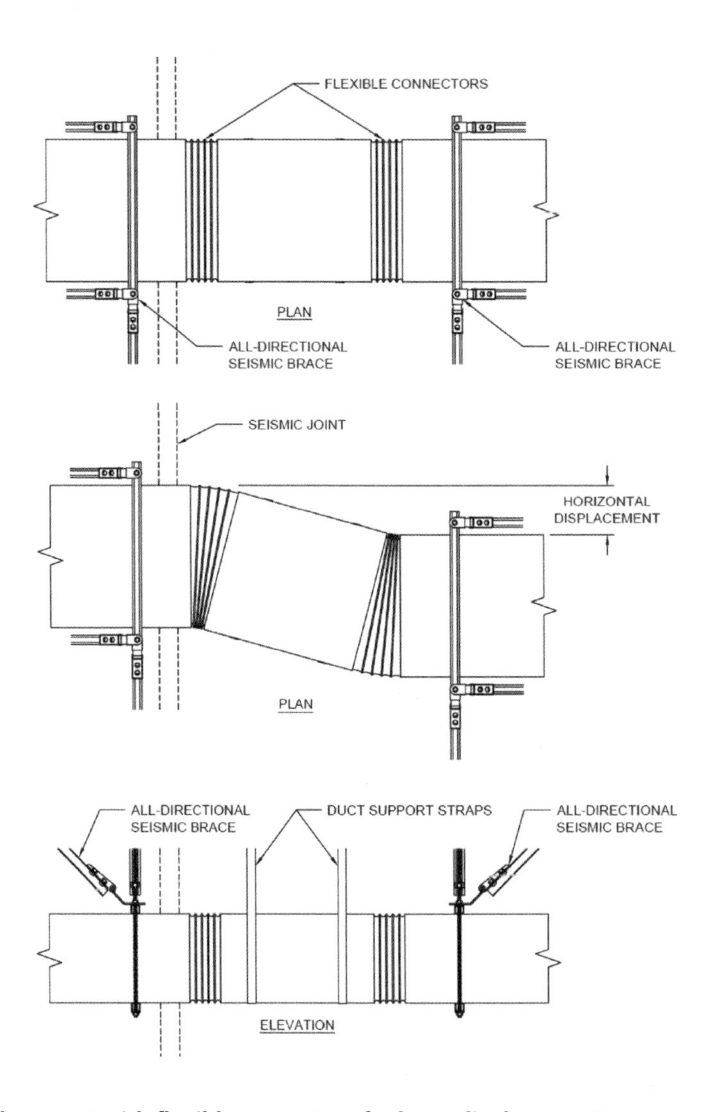

Figure 7-21 Localize displacement with flexible connectors for large displacements.

Table 7-6 Solid Brace Members

	Structural Steel Angle—Max. Length of 9 ft, 6 in. (2.9 m), in. (mm)	12 Gauge Channel Strut—Max. Length of 9 ft, 6 in. (2.9 m), in. (mm)
A	2 × 2 × 1/8 (51 × 51 × 3)	1 5/8 × 1 5/8 (41 × 41 × 2.7)
B	2 × 2 × 1/4 (51 × 51 × 6)	1 5/8 × 1 5/8 (41 × 41 × 2.7)
C	3 × 3 × 1/4 (76 × 76 × 6)	1 5/8 × 3 1/4 (41 × 83 × 2.7)
D	4 × 4 × 1/4 (102 × 102 × 6)	1 5/8 × 3 1/4 (41 × 83 × 2.7)

Table 7-7 Cable Brace Members (Minimum Breaking Strength Required)

	Prestretched Steel Cable—Safety Factor of 2, lb (kN)	Standard Steel Cable—Safety Factor of 5, lb (kN)
A	640 (2.9)	1600 (7.1)
B	1600 (7.1)	4000 (17.8)
C	4000 (17.8)	10,000 (44.5)
D	10,000 (44.5)	25,000 (111.2)

Table 7-8 Connections to Structure

	Expansion Anchors into a Concrete Slab, Dia. × Embed., in. (mm)	Expansion Anchors into a Concrete Deck, Dia. × Embed., in. (mm)	Steel Bolts into Structural Steel, Diameter, in. (mm)	Lag Bolts into a Wood Structure, Dia. × Embed., in. (mm)
A	3/8 × 2 1/2 (10 × 64)	3/8 × 3 (10 × 76)	3/8 (10)	3/8 × 3 (10 × 76)
B	1/2 × 3 (13 × 76)	1/2 × 3 (13 × 76)	1/2 (13)	1/2 × 4 (13 × 102)
C	5/8 × 3 1/2 (16 × 89)	3/4 × 5 1/4 (19 × 83)	1/2 (13)	two 1/2 × 4 (two 13 × 102)
D	two 1/2 × 3 (two 13 × 76)	two 1/2 × 3 (two 13 × 76)	5/8 (16)	two 5/8 × 5 (two 16 x 127)
E	two 5/8 × 3 1/2 (two 16 × 89)	two 5/8 × 5 (two 16 × 127)	5/8 (16)	two 5/8 × 5 (two 16 × 127)
F	four 5/8 × 3 1/2 (four 16 × 89)	four 5/8 × 5 (four 16 × 127)	3/4 (19)	four 5/8 × 5 (four 16 × 127)
G	four 3/4 × 4 1/2 (four 19 × 114)	—	7/8 (22)	four 5/8 × 5 (four 16 × 127)
H	—	—	1 (25)	—

EXCEPTIONS FROM SWAY BRACING

Some ductwork may not require sway bracing. The applicable code for the project may list some minimum requirements, as shown in Chapter 2. Some building codes allow for exceptions based on ductwork size. In all cases, life safety issues should be considered when determining bracing exceptions. The following is a list of suspended ductwork installations that the engineer of record should consider excluding from bracing unless specifically addressed in local codes.

1. All rectangular/oval ductwork with a cross-sectional area of less than 6 ft^2 (0.56 mm^2).
2. Any ductwork run supported by duct straps where the distance as measured from the top of the ductwork to the structure to which it is attached is less than 12 in. (305 mm) for the entire

length of the run *and* the connection point of each duct strap support to the duct is within 2 in. (51 mm) from the top of the duct. This rule also applies to duct supported by threaded rods however the hanger rod connection to the structure should be free to pivot so as to not develop a moment such as a swivel, eye bolt, or vibration isolation hanger connection.

Note: A single support point, which meets the requirements of exclusion 2, does not constitute a seismic sway brace location. The purpose of the exception is to allow the ductwork to swing over a short distance of 12 in. (305 mm), which, in most cases, will rule out the possibility of the ductwork striking another system or piece of equipment.

PIPING CONNECTIONS TO INLINE COILS

Many small diameter pipe connections to coils in variable-air-volume (VAV) terminal units and fan-coil units installed in suspended ductwork failed in recent earthquakes, causing extensive water damage. Special considerations should be made to select a flexible connector for these piping connections that is designed to accept the potential differential motion between the duct and the connected piping. This is especially true if the piping is unbraced, the duct is unbraced, or both are unbraced. Because suspended systems can have very large potential differential displacement, braided-hose V loops or other flexible connectors with large motion capacities should be used. Refer to Chapter 11 for additional information on flexible connector attachments to in-line coils and other components.

EXAMPLE

The following example illustrates how to use the tables outlined in this chapter.

Support type: trapeze
Duct material: 60 in. × 42 in. × 18 gage (1524 mm × 1067 mm × 1.21 mm) sheet metal duct
Structure type: concrete deck
Brace type and orientation: cable bracing above duct center of gravity
Structural connection: expansion anchors
Seismic input: 1.0 *g*

Combined duct weight: 32 lb/ft (48 kg/m)

From Table 7-1, at 1.0 *g*:

Maximum transverse brace spacing = 30 ft (9.1 m).

From Table 7-4, at 1.0 *g* and maximum duct weight of 41 lb/ft (61 kg/m), the transverse brace requirements are as follows:

Vertical hanger rod diameter = 3/4 in. (19 mm)

Hanger rod connection = *E*, two 5/8 in. (16 mm) diameter expansion anchors

Maximum unbraced rod length = 33 in. (838 mm)

Cable brace member = *D*, prestretched cable with a breaking strength of 10,000 lb (44.5 kN)

Brace member connection to structure = *F*, four 5/8 in. (16 mm) diameter expansion anchors

BIBLIOGRAPHY

MSS. 2009. ANSI/MSS SP-58-Pipe Hangers and Supports. Falls Church, VA: Manufacturers Standardization Society.

Mason Industries. 2009. Seismic Restraint of Suspended Piping, Ductwork and Electrical Systems. Hauppauge, NY.

SMACNA. 2005. HVAC Duct Construction Standard, Metal and Flexible. Chantilly, VA: Sheet Metal and Contractors National Association.

SMACNA. 2008. Seismic Restraint Manual-Guidelines for Mechanical Systems. Chantilly, VA: Sheet Metal and Contractors National Association.

DOE. 1997. Seismic Evaluation Procedure for Equipment in U.S. Department of Energy Facilities. Springfield, VA: U.S. Department of Energy.

8 Suspended Piping

Seismic bracing is required on all types of pipe materials, including steel, copper, cast iron, and specialty plastic or fiberglass piping. The performance of piping in earthquakes varies based upon the ductility of the pipe material itself and the flexibility of the individual pipe connections. The most common weak link is at pipe connections to valves and equipment. Pipe sections with screwed fittings can fail at threads when subjected to repeated flexural movements. Proper sway bracing of suspended piping prevents damage to adjacent systems and also protects critical connections of pipe sections and to equipment.

Steel and copper pipe with welded, brazed, or groove-fitted joints is extremely ductile and perform well in earthquakes. Although calculations indicate large pipe deflections between sway braces, the motion is typically not great enough to strike adjacent systems or overstress the pipe. Piping itself simply doesn't fail in earthquakes. However, motion can be too great at the end of pipe runs, specifically at connections to equipment. Flexible connectors at the equipment can add to the motion capability of the pipe and protect the equipment connections. However, the flexible connectors can fail if the pipe is allowed to sway excessively. The combination of properly designed flexible connectors and properly located sway bracing near the end of the pipe run will tend to prevent damage to equipment connections.

Although specialty plastic pipe materials, such as polyvinyl chloride (PVC) or polyvinylidene fluoride (PVDF), are flexible and allow large deflections, similar to steel and copper pipe, the solvent welded joints for these pipe materials tend to introduce areas of stress concentration, reducing overall ductility and requiring a reduction in brace spacing.

Screwed joints for steel and copper pipe also tend to introduce areas of stress concentration and may be subject to reduced brace spacing. The ends of pipe runs at equipment connections are especially critical and require proper sway bracing.

No-hub pipe, such as cast iron or glass pipe, connected with flexible, no-hub rubber couplings, shields, and two-band clamp assemblies, has some flexibility and system ductility, but only because of the couplings. Sway brace spacing of this pipe should be reduced to limit the loads on the nonductile cast iron or glass pipe and the joint connectors. A shield and four-band clamp assembly, as shown in Figure 8-1, is available for joint connections requiring a higher safety factor without compromising the required flexibility provided by the couplings. This assembly is recommended on no-hub pipe installed over critical life-safety areas of the building. The shield and four-band clamps are specifically required in surgery and other critical areas on all OSHPD projects. Because the flexibility of the no-hub pipe is greatly reduced with the addition of clamps and steel bar joint reinforcement brackets, these brackets should be limited to use on unsupported piping or special conditions.

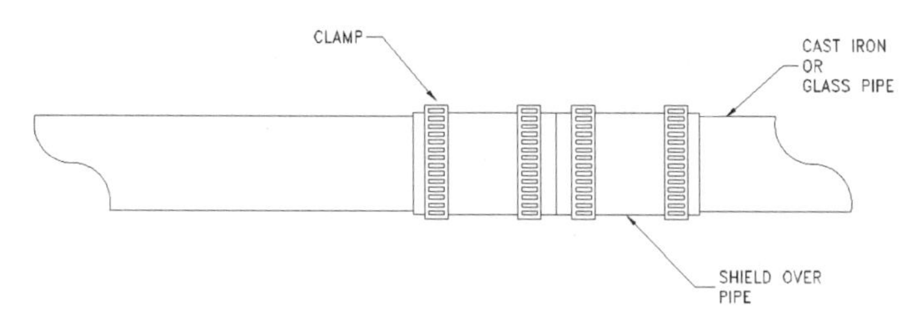

Figure 8-1 Shield and 4-band clamp assembly.

Fiberglass-reinforced pipe (FRP) does not possess the flexibility inherent in other pipe materials. Additionally, the connection points tend to introduce stress concentrations, creating weak points at midspan that could overstress and crack if sway brace spacing is too large.

Most piping is supported by individual clevis hangers or trapeze supports. Large-diameter FRP may be supported by two hangers. In all cases, the pipe is hung from threaded hanger rods.

Sway bracing of suspended piping requires the following:

1. Each individual or trapeze-supported piping run should be sway braced to limit motion in all directions.
2. Maximum sway brace spacing should consider the strength of the piping, the pipe connections, the seismic capacity of the sway brace, and the limitations of the building structure.
3. Vertical hangers that can accept a compressive load as well as a tensile load should be used.
4. Existing individual pipe hangers should be used or pipe clamps added to connect sway bracing to the pipe.
5. Hold down devices, such as U-bolts or clamps for pipes supported on a trapeze, should be used.
6. Piping hung by vibration isolators should be detailed and installed with the isolators attached directly to the structure, limit stops located directly below the isolation housings, and cable braces arranged to limit swaying.
7. Anchors and guides used to control thermal expansion and contraction should also be designed to act as sway braces.

SWAY BRACING

There are two types of sway braces, solid and cable, each with advantages and disadvantages as discussed in Chapter 7.

Figure 8-2A shows a typical transverse solid brace installation for individually supported pipe with arrows indicating the external seismic and gravity loads and the reactions of both the sway brace at the hanger rod connection and the hanger rod at its connection point to the structure. Figures 8-2B and 8-2C outline specific examples of loads and ratios between dead load and combined dead and seismic load when the solid brace is installed at 45° and 60° from horizontal.

Figures 8-3A, 8-3B, and 8-3C show a typical transverse solid brace installation for trapeze-supported pipe with external loads and reactions identified and specific examples. The ratio between dead load and combined dead and seismic load is considerably larger than when compared to the individually supported pipe examples. In most cases, existing hanger rod connections to the structure for trapeze-supported pipe cannot accept the additional seismic loads resulting when the solid brace resists a compression load. Solid sway bracing is not recommended on trapeze installations unless proper consideration is given to the maximum tension loads, as shown in the hanger rod requirements later in this chapter.

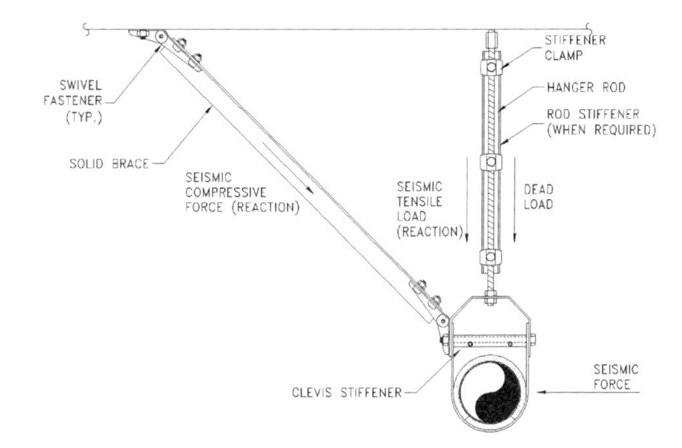

Figure 8-2A Analysis of solid brace compression for individually supported pipe.

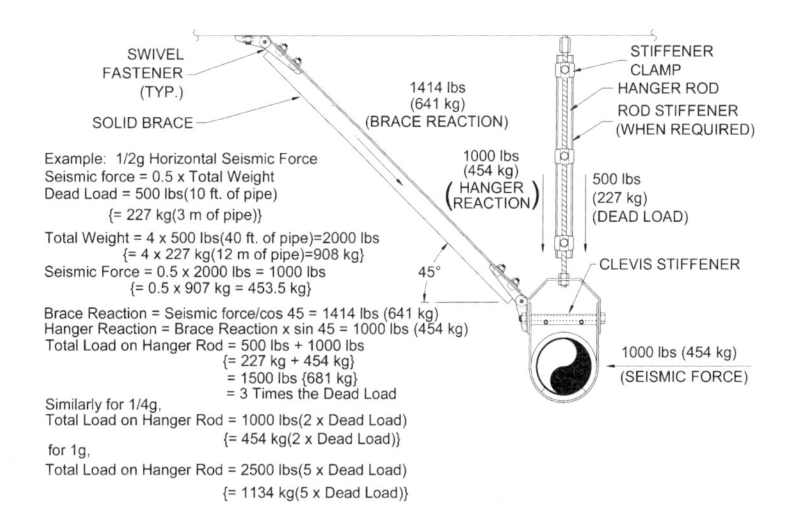

Figure 8-2B Example of hanger rod loads at 45° brace angle.

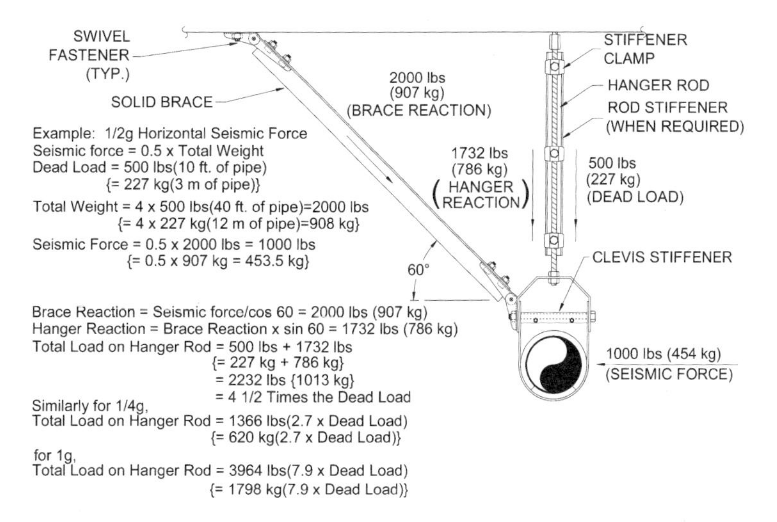

Figure 8-2C Example of rod loads at 60° brace angle.

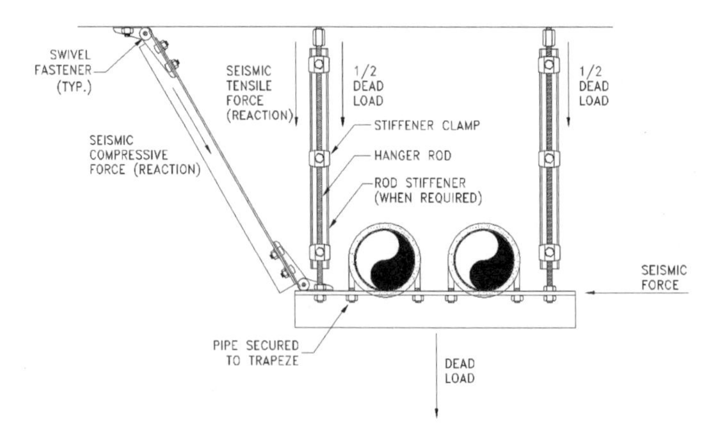

Figure 8-3A Analysis of solid brace in compression for trapeze pipe.

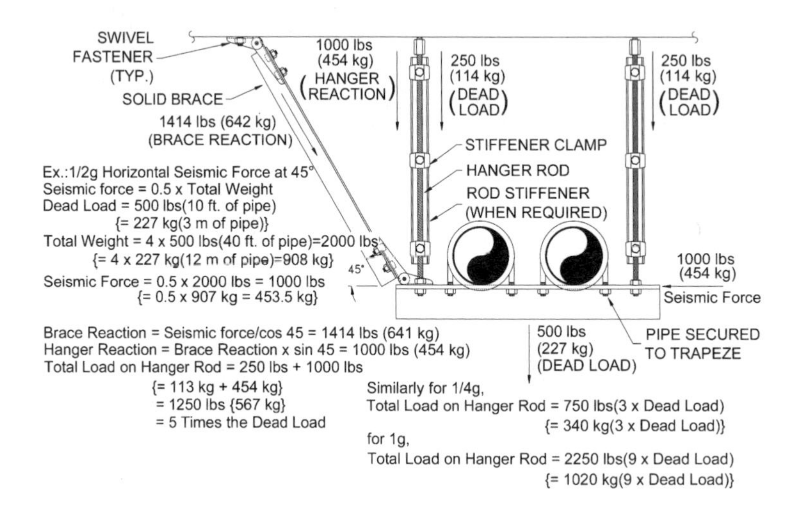

Figure 8-3B Example of hanger rod loads at 45° brace angle.

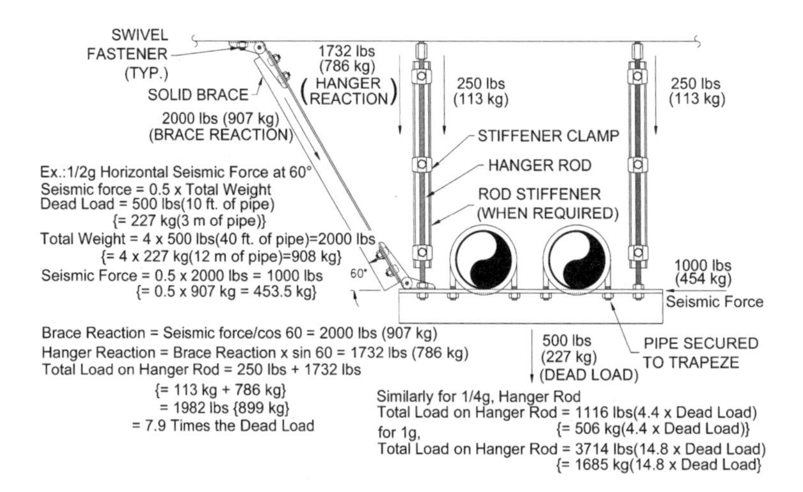

Figure 8-3C Example of hanger rod loads at 60° brace angle.

Figures 8-4 and 8-5 show typical transverse cable brace installations with external loads and reactions for individually supported and trapeze-supported pipe.

Wall and floor penetrations seem like natural sway brace locations. However, when compared to ductwork, piping generates much higher loads in a more concentrated area, which may damage walls. In addition, walls required for area separation, smoke barriers, or other fire or life safety related functions should not be used to brace piping. In all other cases, the structural engineer of record should determine whether to allow specific floor and wall penetrations to act as sway bracing. A wall penetration can act as a typical transverse brace location and a floor penetration can act as both a transverse and longitudinal brace location. A floor penetration can only be considered a longitudinal brace if the distance from the floor to the horizontal run of pipe is less than the maximum offset length, which is described later.

A suggested layout procedure for pipe bracing is as follows:

1. Each straight run of the piping system should be braced with a minimum of two transverse braces installed perpendicular to the piping and one longitudinal brace installed parallel to the piping, as shown in Figure 8-6.

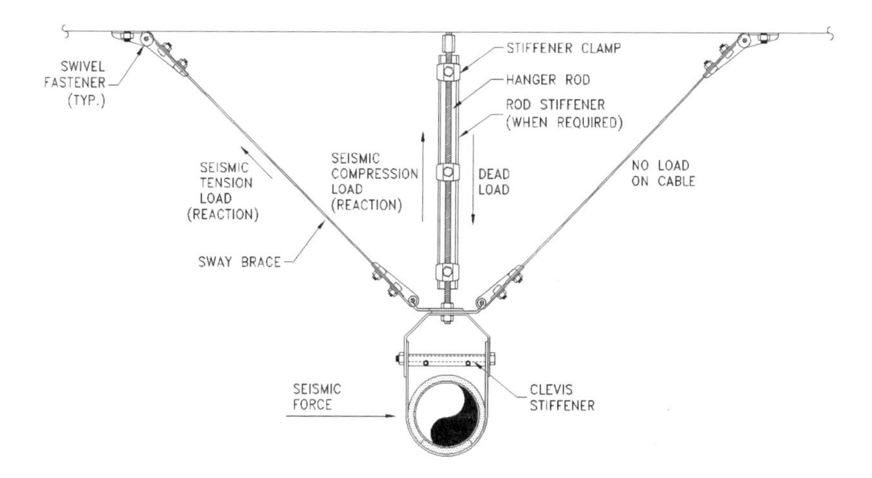

Figure 8-4 Analysis of cable brace for horizontal suspended piping.

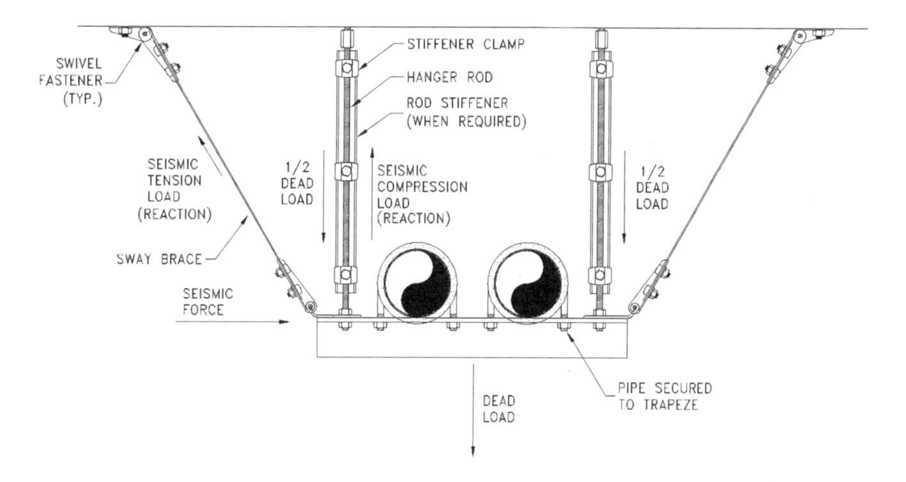

Figure 8-5 Analysis of cable brace for trapeze suspended piping.

2. Transverse braces should be located at the final support point of each run of pipe with two supports. If the distance between the braces exceeds the maximum transverse brace spacing, additional transverse braces should be located to limit the brace spacing to the maximum transverse brace spacing, as shown in Figure 8-7.

3. A longitudinal brace should be located on each straight run of pipe. If the length of the pipe run exceeds the maximum longitudinal brace spacing, additional longitudinal braces should be located on the pipe run while limiting the brace spacing to the maximum longitudinal brace spacing, as shown in Figure 8-8.

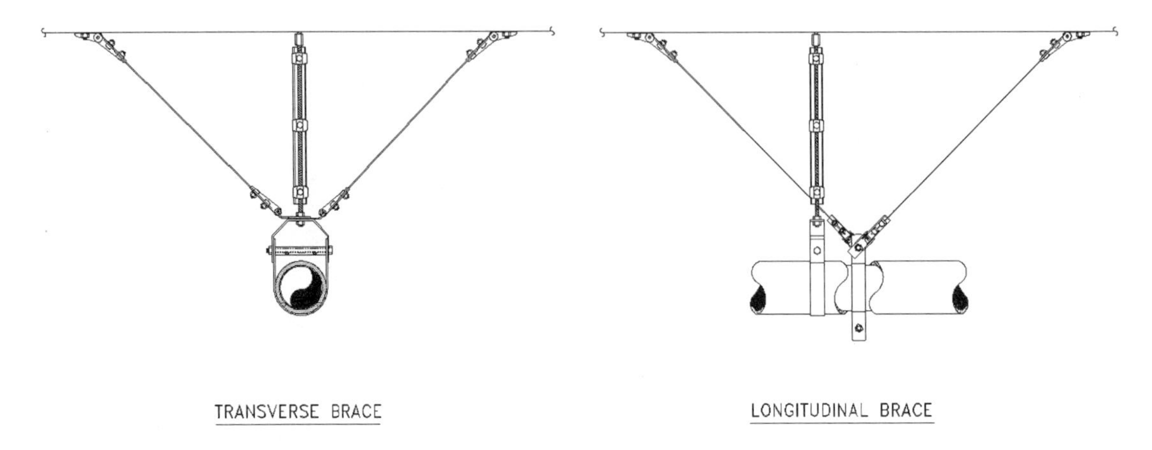

Figure 8-6 Braces of piping system.

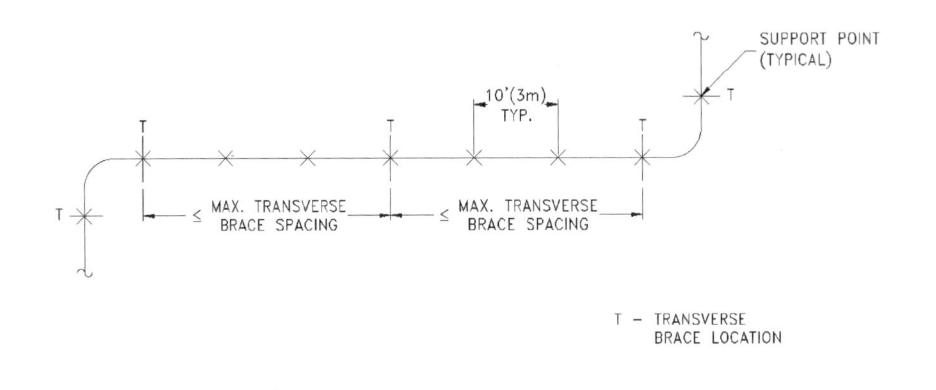

Figure 8-7 Transverse brace layout.

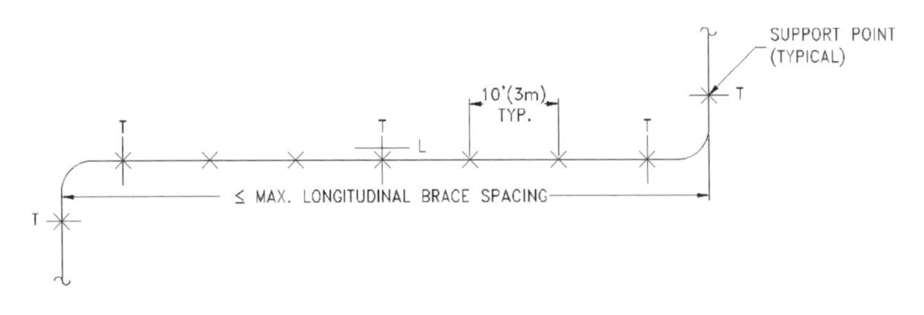

Figure 8-8 Transverse and longitudinal brace layout.

4. A transverse brace located within the maximum offset length of the pipe can provide limited longitudinal bracing for the straight run of pipe around a 90° turn, or elbow. The maximum offset length is based on the maximum stress of the pipe and a maximum 1/ in. (6 mm) deflection of the offset of the pipe with welded, brazed, or groove-fitted joints. The maximum offset lengths of different pipe materials are listed in Tables 8-1 and 8-2. As illustrated in Figure 8-9, the longitudinal length of pipe that can be braced by this transverse brace is equal to one-half the maximum transverse brace spacing minus the actual offset length.
5. If a straight run of pipe has less than two support points, is connected to a straight run of pipe sway braced at each end, and its total length is less than the maximum offset length, brace across the run by adding its length to the transverse and longitudinal brace design of the connected runs, as shown in Figure 8-10. If its length is greater than the maximum offset length, a support point with a transverse brace is required.
6. Vertical drops from horizontal runs of piping to the equipment require a transverse brace at the final support location before the pipe drop. This is the most critical brace on the entire pipe run. As illustrated in Figure 8-11, the total length of pipe from the support point to the equipment connection or flexible connector should be less than one-half the maximum sway brace spacing of the transverse brace *and* the length of pipe from the support point to the drop should be less than maximum offset length. Proper location of the brace, as close as reasonably possible to the pipe drop, and proper size of the brace to handle the loads, including the pipe drop, are critical.

Table 8-1 Maximum Offset Length for Steel Pipe with Welded or Grooved Connections (Maximum Length of Pipe Braced = 80 ft [24.4 m], 40 ft [12.2 m] for 2.0 g)

Maximum Pipe Diameter, in. (mm)	0.25 g Seismic Acceleration Input, ft (m)	0.5 g Seismic Acceleration Input, ft (m)	1.0 g Seismic Acceleration Input, ft (m)	2.0 g Seismic Acceleration Input, ft (m)
1 (25)	3 (0.9)	1 (0.3)	0	0
2 (51)	4 (1.2)	2 (0.6)	1 (0.3)	1 (0.3)
3 (76)	8 (2.4)	4 (1.2)	2 (0.6)	2 (0.6)
4 (102)	10 (3.0)	6 (1.8)	3 (0.9)	3 (0.9)
6 (152)	10 (3.0)	10 (3.0)	5 (1.5)	5 (1.5)
8 (203)	10 (3.0)	10 (3.0)	7 (2.1)	7 (2.1)
10 (254)	10 (3.0)	10 (3.0)	9 (2.7)	9 (2.7)
12 (305)	10 (3.0)	10 (3.0)	9 (2.7)	9 (2.7)
14 (356)	10 (3.0)	10 (3.0)	10 (3.0)	10 (3.0)

Table 8-2 Maximum Offset Length for Copper Pipe with Brazed Connections (Maximum Length of Pipe Braced = 80 ft [24.4 m], 40 ft [12.2 m] for 1.0 g, 20 ft [6.1 m] for 2.0 g)

Maximum Pipe Diameter, in. (mm)	0.25 g Seismic Acceleration Input, ft (m)	0.5 g Seismic Acceleration Input, ft (m)	1.0 g Seismic Acceleration Input, ft (m)	2.0 g Seismic Acceleration Input, ft (m)
2 1/2 (64)	2 (0.6)	1 (0.3)	0	0
3 (76)	2 (0.6)	1 (0.3)	0	0
4 (102)	4 (1.2)	2 (0.6)	1 (0.3)	1 (0.3)
6 (152)	8 (2.4)	4 (1.2)	2 (0.6)	2 (0.6)
8 (203)	10 (3.0)	8 (2.4)	4 (1.2)	4 (1.2)
10 (254)	10 (3.0)	10 (3.0)	5 (1.5)	5 (1.5)
12 (305)	10 (3.0)	10 (3.0)	6 (1.8)	6 (1.8)

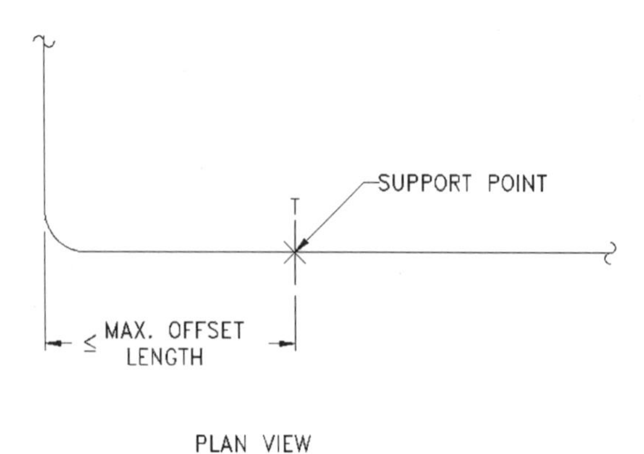

Figure 8-9 Transverse brace within offset length.

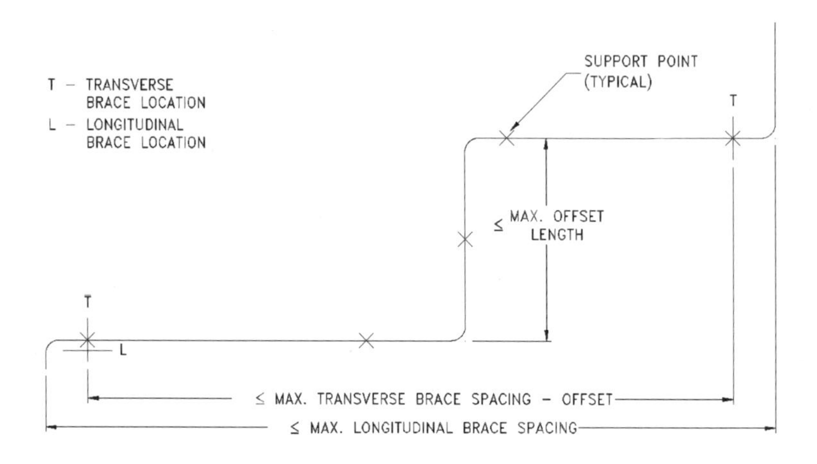

Figure 8-10 Bracing of pipe with an offset.

7. Avoid bracing a pipe to separate portions of the structure that may act differently in response to an earthquake. For example, do not connect a transverse brace to a wall and a longitudinal brace to a floor or roof at the same brace location. This is special concern in a lightly framed penthouse mechanical room. The wall and floor tend to move in unison in a basement mechanical room, so attachment of braces to the wall and floor is acceptable.
8. Do not mix solid bracing with cable bracing in the same direction on any pipe run. Solid bracing is likely more rigid than cable bracing, so they do not share the loads evenly. This is very important for large, rigid piping, but less important for smaller-diameter, less-rigid piping, which is more flexible than either solid or cable bracing.
9. Multiple or stacked trapezes that share hanger rods should be braced independently from one another. Otherwise, the trapezes can be connected with a rigid frame at the seismic brace locations and the frames braced with standard methods and details.

Maximum sway brace spacing for different pipe materials is indicated in Tables 8-3 through 8-5. Brace spacings, offset lengths, and connections for other pipe materials should be individually designed.

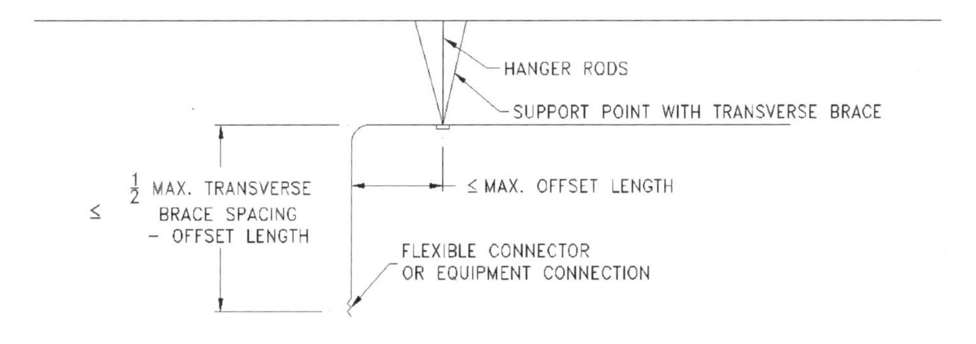

Figure 8-11 Bracing of pipe drop to equipment supported on floor.

Table 8-3 Steel and Copper Pipe with Welded, Brazed, Grooved, or Screwed Connections

Maximum Seismic Acceleration Input (*g*)	Maximum Transverse Brace Spacing, ft (m)	Maximum Longitudinal Brace Spacing, ft (m)
0.25	50 (15.2)	80 (24.4)
0.50	40 (12.2)	80 (24.4)
1.0	40 (12.2)	80 (24.4)
2.0	20 (6.1)	40 (12.2)

Table 8-4 PVC or PVDF Pipe with Solvent-Welded Connections

Maximum Seismic Acceleration Input (*g*)	Maximum Transverse Brace Spacing, ft (m)	Maximum Longitudinal Brace Spacing, ft (m)
0.25	25 (7.6)	40 (12.2)
0.50	20 (6.1)	40 (12.2)
1.0	20 (6.1)	40 (12.2)
2.0	10 (3.0)	20 (6.1)

Table 8-5 No-Hub Pipe with Shield and Clamp Connections

Maximum Seismic Acceleration Input (*g*)	Maximum Transverse Brace Spacing, ft (m)	Maximum Longitudinal Brace Spacing, ft (m)
0.25	25 (7.6)	40 (12.2)
0.50	20 (6.1)	40 (12.2)
1.0	20 (6.1)	40 (12.2)
2.0	10 (3.0)	20 (6.1)

HANGER ROD REQUIREMENTS

The effects of sway bracing on the hanger rod are discussed in more detail in Chapter 7. Following is a discussion of how loads are applied to hanger rods for suspended piping.

Figures 8-12 to 8-15 include calculation formats used to resolve tensile and compressive loads of hanger rods. Each figure indicates the external seismic and gravity loads and the reactions of both the sway brace at the hanger rod connection and the hanger rods at their connection points to the structure.

As shown in Figure 8-12, for individually supported pipe with solid bracing:

$$T_{(Rod)max} = (T_w + T_{Fph} + T_{Fpv}) \quad (8\text{-}1)$$

$$C_{(Rod)max} = (C_{Fph} + C_{Fpv}) - T_w \quad (8\text{-}2)$$

As shown in Figure 8-13, for individually supported pipe with cable bracing:

$$T_{(Rod)max} = (T_w + T_{Fpv}) \quad (8\text{-}3)$$

$$C_{(Rod)max} = (C_{Fph} + C_{Fpv}) - T_w \quad (8\text{-}4)$$

As shown in Figure 8-14,for trapeze-supported pipe with solid bracing:

Hanger Rod 1

$$T_{(Rod\ 1)max} = (T_w + T_{Fph} + T_{Fpv}) \quad (8\text{-}5)$$

$$C_{(Rod\ 1)max} = (C_{Fph} + C_{Fpv}) - T_w \quad (8\text{-}6)$$

Hanger Rod 2

$$T_{(Rod\ 2)max} = T_w + T_{Fpv} \quad (8\text{-}7)$$

$$C_{(Rod\ 2)max} = \text{No Compression}$$

As shown in Figure 8-15, for trapeze-supported pipe with cable bracing:

Hanger Rod 1

$$T_{(Rod\ 1)max} = T_w + T_{Fpv} \quad (8\text{-}8)$$

$$C_{(Rod\ 1)max} = (C_{Fph} + C_{Fpv}) - T_w \quad (8\text{-}9)$$

Hanger Rod 2

$$T_{(Rod\ 2)max} = T_w + T_{Fpv} \quad (8\text{-}10)$$

$$C_{(Rod\ 2)max} = \text{No Compression}$$

where

w = total weight of pipe = pipe wt/ft (M) × (maximum transverse brace spacing)

Fph = seismic horizontal force

Fpv = seismic vertical force

N = number of vertical hangers carrying pipe over transverse brace spacing span

T_w = tensile force due to $W = (W/N)$

$T_{Fph} = C_{Fph}$ = tensile or compressive force due to $Fph = Fph \tan \theta$

$T_{Fpv} = C_{Fpv}$ = tensile or compressive force due to $Fpv = Fpv/N$

$T_{brace} = C_{brace}$ = tensile or compressive force on the brace due to $Fph = Fph/\cos \theta$

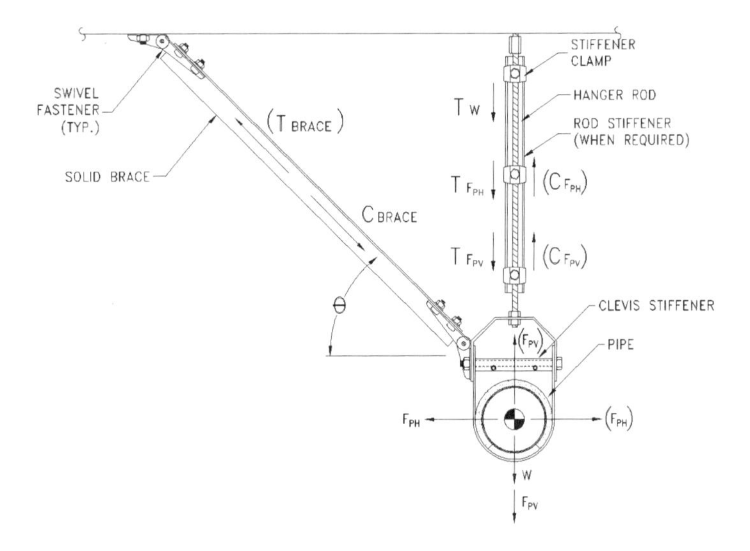

Figure 8-12 Analysis of hanger rod loads for individually supported pipe with cable bracing.

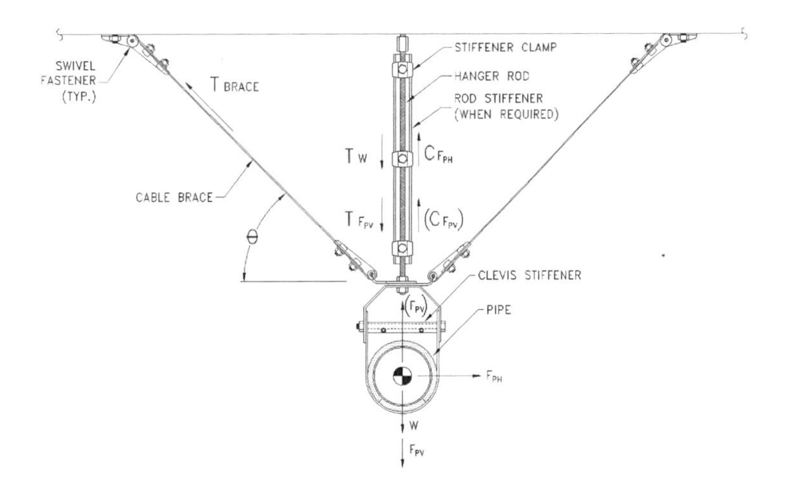

Figure 8-13 Analysis of hanger rod loads for individually supported pipe with solid bracing.

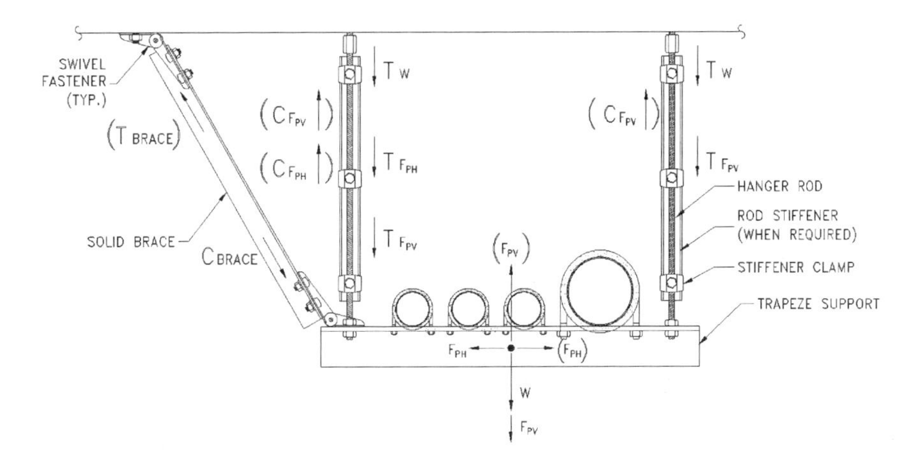

Figure 8-14 Analysis of hanger rod loads for trapeze-supported pipe with solid bracing.

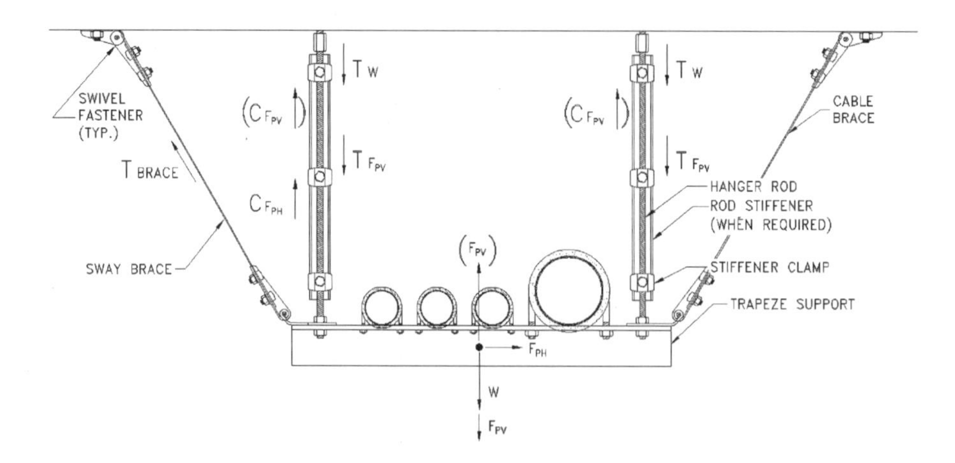

Figure 8-15 Analysis of hanger rod loads for trapeze-supported pipe with cable bracing.

For resultant tension loads, check both the hanger rod and its connection to the structure. Threaded hanger rod allowable values are tabulated in Table 8-6. For compressive forces, determine if the hanger rod requires a stiffener, as discussed in Chapter 7 and shown in Figure 7-12. Refer to Table 7-3 for maximum clamp spacing for attaching the stiffener to the rod and Table 8-7 for maximum unbraced rod lengths.

TRANSVERSE SWAY BRACING OF INDIVIDUALLY SUPPORTED PIPE

Braces can be attached to individual clevis, roller-clevis, or J-type clevis pipe hangers. Figures 8-16 and 8-17 show two options for connecting the sway brace to a pipe hanger. The most versatile position is shown in Figure 8-16, connection to the top of the hanger where the hanger rod intersects the hanger. This is certainly preferred for hardware coordination because the rod size stays the same for all pipe of the same diameter, but the clevis size and cross-bolt size will change with different insulation thickness requiring different hardware. In either option, the cross-bolt in clevis hangers should be reinforced as shown in Figure 8-18 to prevent the clevis from slipping and potentially failing.

A pipe clamp can also be used as a transverse sway brace connection point, as shown in Figure 8-19. The pipe clamp is located within 4 in. (102 mm) of a hanger rod by convention. But this dimension can vary, especially on larger pipe where clearance from the clevis is necessary for larger restraint brackets. The vertical hanger is subject to the requirements outlined above. For insulated pipe, clamps for transverse braces can be installed over hard inserts.

LONGITUDINAL SWAY BRACING OF INDIVIDUALLY SUPPORTED PIPE

Longitudinal braces should never be attached to pipe clevis supports or other supports that do not restrain the pipe longitudinally. Longitudinal sway braces should be connected directly to the pipe with the use of a pipe clamp, as shown in Figure 8-20. The pipe clamp should be located within 4 in. (102 mm) of a hanger rod by convention, but this dimension can vary, especially on larger pipe where clearance from the clevis is necessary for larger restraint brackets. The hanger rod is braced for potential compressive loads. If a transverse sway brace is located at the same hanger rod location, the seismic loads based on the longitudinal sway brace may be larger and should be used for design of the hanger. For insulated pipe, the pipe clamp should be attached directly to the pipe and the vapor seal installed over the clamp or sealed if required. Sway braces connected directly to any type of clevis hanger cannot prevent the pipe from moving longitudinally and are not acceptable.

Table 8-6 Threaded Rod Allowable Tension Loads

Threaded Rod Diameter, in. (mm)	Allowable Working Load, lb (kN)	Allowable Combined Working and Seismic Load, lb (kN)
3/8 (10)	610 (2.7)	810 (3.6)
1/2 (13)	1130 (5.0)	1500 (6.6)
5/8 (16)	1810 (8.0)	2410 (10.7)
3/4 (19)	2710 (12.0)	3610 (16.0)
7/8 (22)	3770 (16.7)	5030 (22.3)
1 (25)	4960 (22.1)	6610 (29.4)
1 1/4 (32)	8000 (35.5)	10,660 (47.4)

Table 8-7 Individually or Trapeze-Supported Pipe Transverse Brace Requirements

Maximum Pipe Weight per Foot *m*, lb (kg)				Seismic Horizontal Force, *Fph*, lb (kN)	Hanger Rod Diameter, in. (mm)	Connection of Hanger Rod to Structure		Maximum Unbraced Rod Length, in. (mm)	Brace Member Size		Connection to Structure
0.25 g	0.5 g	1.0 g	2.0 g			Solid	Cable		Solid	Cable	
10 (14)	6 (9)	3 (4)	3 (4)	125 (0.6)	3/8 (10)	A	A	20 (508)	A	A	B
16 (23)	10 (14)	5 (7)	5 (7)	200 (0.9)	1/2 (13)	B	B	29 (736)	A	B	C
36 (53)	24 (35)	12 (17)	12 (17)	480 (2.1)	1/2 (13)	E	B	18 (457)	B	C	D
50 (74)	31 (46)	15 (22)	15 (22)	625 (2.8)	5/8 (16)	F	C	26 (660)	C	C	E
100 (148)	62 (92)	31 (46)	31 (46)	1250 (5.6)	3/4 (19)	H	E	27 (685)	D	D	F
178 (264)	111 (165)	55 (81)	55 (81)	2225 (9.9)	7/8 (22)	H	E	29 (736)	D	D	H

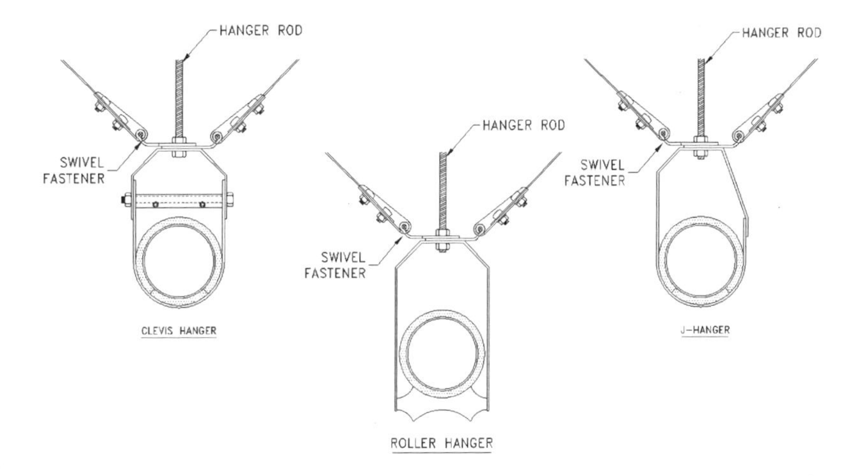

Figure 8-16 Transverse braces connected to support rod.

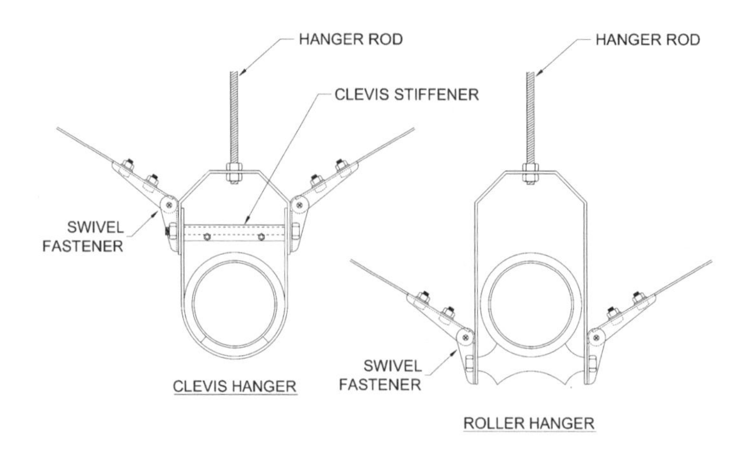

Figure 8-17 Transverse cross-bolt to pipe hangers.

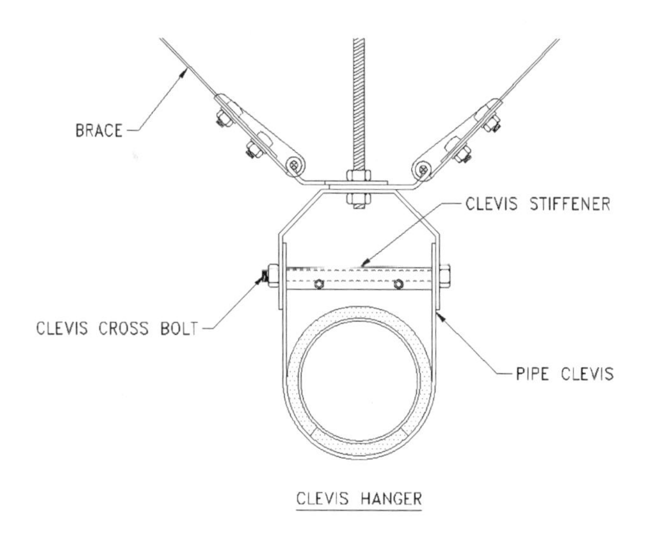

Figure 8-18 Clevis cross-bolt stiffener.

SWAY BRACING OF TRAPEZE SUPPORTED PIPE

The connection of the hanger rods to the trapeze is also the best connection point of the sway braces. All pipe supported by the trapeze should be held down to the trapeze with U-bolts or clamps, as shown in Figures 8-21 to 8-24.

There are two methods for sway bracing trapeze-supported pipe in all directions. One method is to use separate transverse and longitudinal braces and follow the layout procedure outlined above and shown in Figure 8-21 for a transverse brace and Figure 8-22 for a longitudinal brace. Refer to Figure 8-23 for an all-directional brace, which is a transverse and longitudinal brace located at the same trapeze support location. The second method is to provide all-directional braces at each sway brace location, as shown in Figure 8-24A. However, this method requires access in all directions at every sway brace location, does not take advantage of bracing capability around a 90° turn, and, in most cases, requires more sway brace locations. In most cases, the same number of braces is required for either method.

VIBRATION ISOLATED PIPING

Piping suspended by spring and/or neoprene vibration isolation hangers should only be braced with cable bracing. Steel cables should be installed so that they do not carry any dead loads. Cables should be installed with a minor amount of sag due to the weight of the cable,

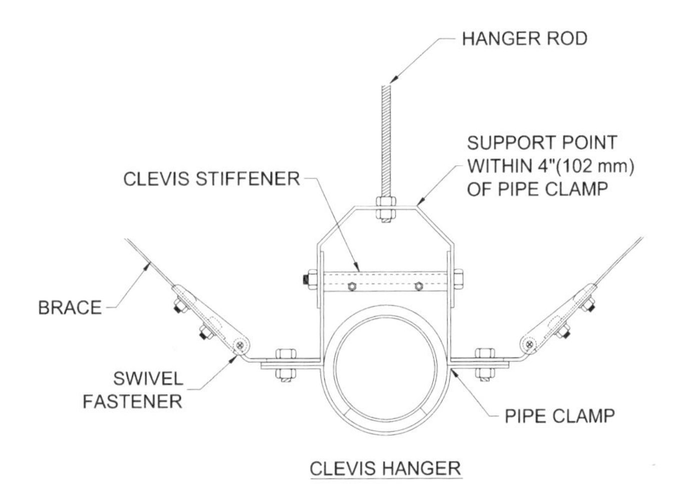

Figure 8-19 Optional transverse brace at pipe clamp.

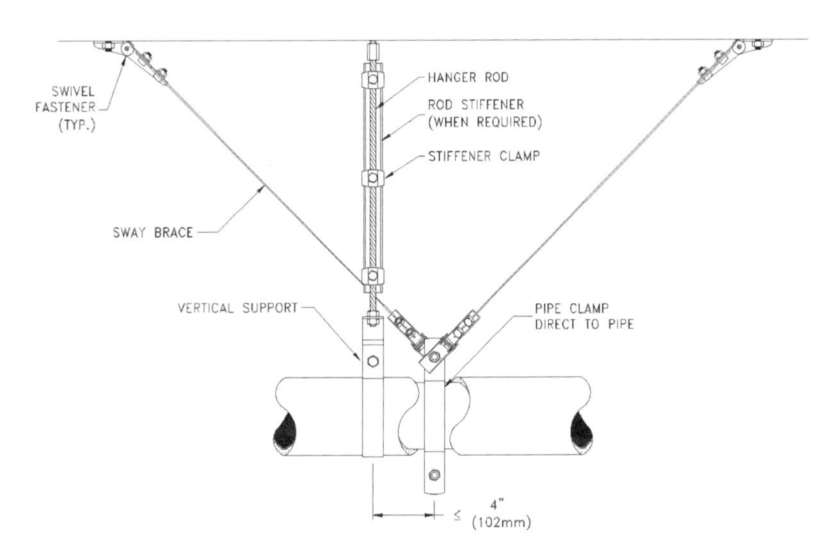

Figure 8-20 Longitudinal brace for individually supported pipe.

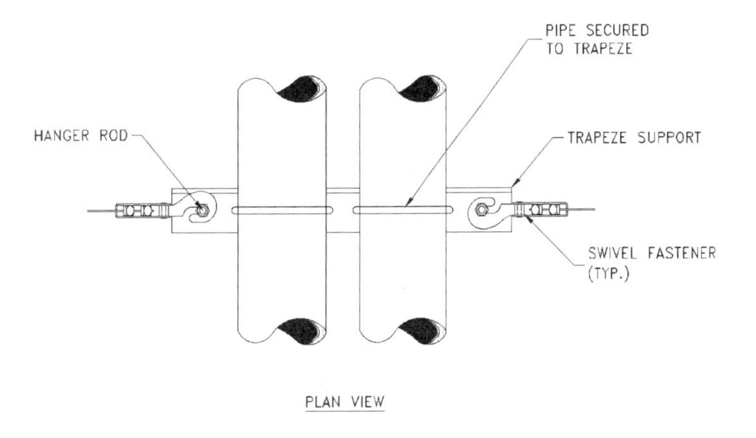

Figure 8-21 Transverse brace for trapeze-supported pipe.

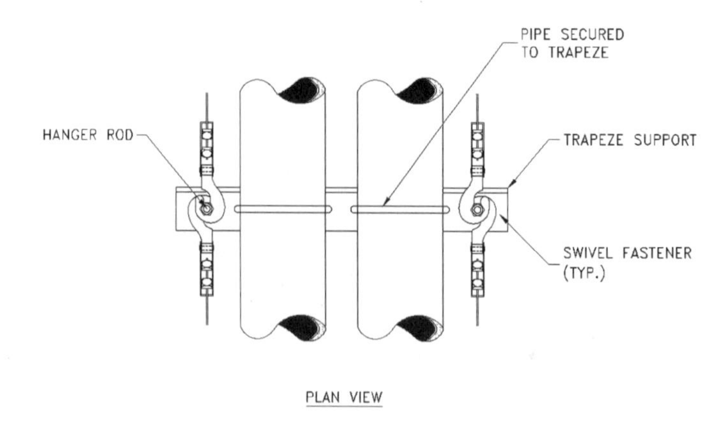

Figure 8-22 Longitudinal brace for trapeze-supported pipe.

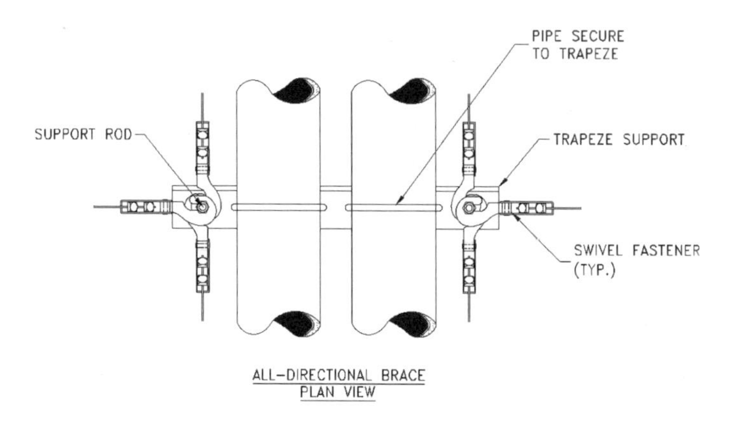

Figure 8-23 All-directional brace for trapeze-supported pipe.

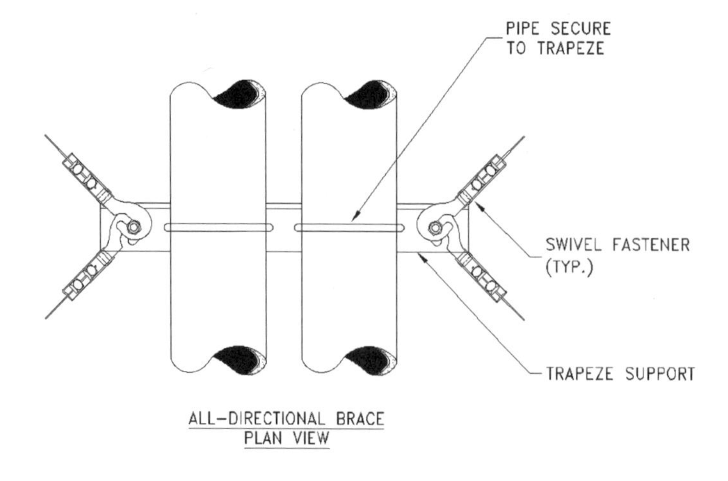

Figure 8-24A All-directional brace for trapeze-supported pipe.

but without additional slack. This prevents vibration transmission to the structure but does not allow excessive motion.

Installation of the vibration isolation hangers is critical to the performance of the sway braces. As discussed earlier, the function of a hanger rod is to resist tensile and compressive loads. The vibration isolation hanger introduces a break in the solid support of the piping with a flexible material designed to accept gravity loads but not compressive loads.

At seismic brace locations, the vibration isolation hanger boxes must be designed to accept compressive forces without excessive bending, and the hangers should be installed to prevent excessive upward motion. Neoprene washers with steel inserts must be installed above and below the hanger. The hanger is installed with the top of the neoprene washer in contact with the supporting structure. The upper neoprene element in combination spring and neoprene vibration isolation hangers deflects under load to create the 1/4 in. (6 mm) clearance. The lower limit neoprene washer and steel insert is installed 1/4 in. (6 mm) below the hanger box, all as shown in Figure 8-24B.

Contact between the top of the hanger box and the structure, combined with contact between the limit stop and the bottom of the hanger box, will limit vertical motion of the equipment without compromising the performance of the vibration isolators.

THERMAL EXPANSION

All piping experiences some degree of thermal expansion or contraction. When the motion is large enough, pipe loops or expansion joints should be designed into the system. Anchors and guides as shown in Figure 8-25 are usually necessary to force the pipe loops or expansion joints to accept the thermal motion.

Anchors and guides should be designed to accept the combination of loads due to both thermal and seismic motion. As indicated in Tables 8-3 to 8-5 and shown in Figure 8-26, anchors should be designed to restrain up to 50 ft (15.2 m) of pipe in the transverse direction and the entire straight run of pipe in the longitudinal direction, even if it is greater than 80 ft (24.4 m). This will prevent seismic damage to the anchors. To prevent excessive stress on the pipe, guides should be located at the maximum transverse brace spacing or less. Guides should be designed to restrain up to 50 ft (15.2 m) of pipe.

Additional sway bracing in lieu of guides is not recommended. The flexibility of the braces may prevent any effective restraint of the pipe, overloading and damaging the more rigid guides and anchors. Or the added sway braces may create inadvertent anchor locations, disrupting the thermal design of the system and creating excessive pipe bending and failure.

SEISMIC JOINTS

Piping systems that cross building seismic joints, pass from building to building, or are supported from different portions of the building should be designed to accept differential

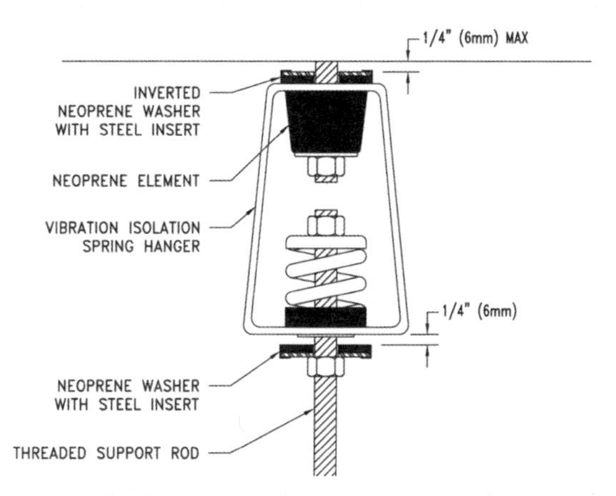

Figure 8-24B Vibration isolation hanger installation.

support displacements without damaging the pipe, equipment connections, or support connections. The amount of displacement can vary, depending on acceleration input, soil conditions, and vertical distance from grade to the piping and is usually determined by the structural engineer of record. To accept these differential displacements, the entire piping system can be designed to accept the displacement or the displacement can be localized between anchors as illustrated by Figures 8-27 to 8-31 using one of the following methods:

1. Design the piping system to accept the differential motion as shown in Figure 8-27. This may require special piping offsets, loops, supports, and connections to equipment to accept the motion without damage.
2. Localize the area at which differential motion will occur by anchoring to each building and design the piping between the anchors to accept the motion, as shown in Figure 8-28. This may result in large anchor loads and pipe stresses.

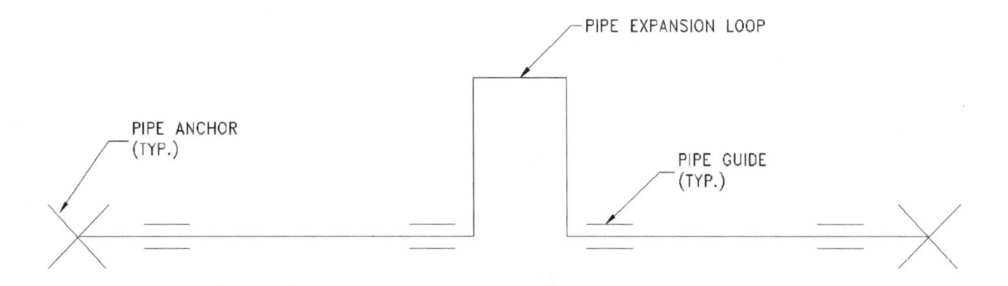

Figure 8-25 Thermal expansion design.

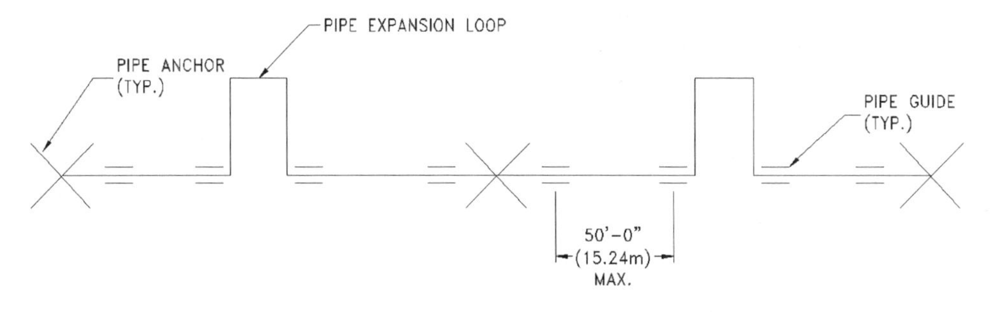

Figure 8-26 Thermal expansion and seismic design.

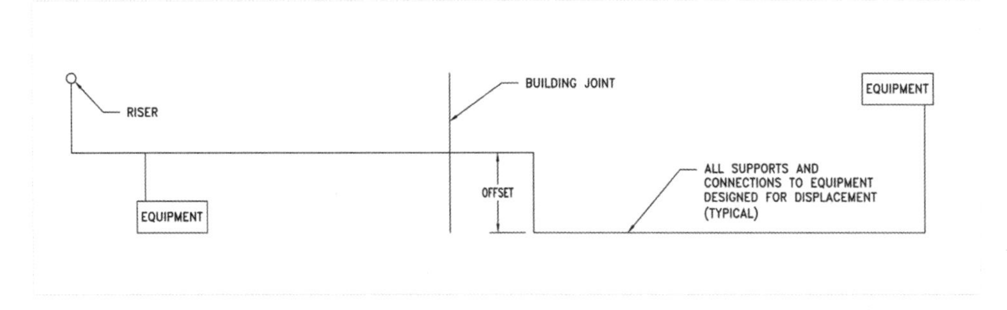

Figure 8-27 System design to accept displacement.

3. Localize the area at which differential motion will occur by anchoring to each building and provide one of the following type of manufactured pipe expansion products in the systems:
    a. A braided-flexible-hose dogleg assembly, installed at the pipe offset as shown in Figure 8-29. This will tend to reduce the anchor loads and pipe stress on the system.
    b. A braided-flexible-hose V loop, installed in the straight pipe line as shown in Figure 8-30. This will also tend to reduce the anchor loads and pipe stress on the system and does not require an offset in the piping for installation. Additionally, the V loop can usually be nested or rotated to fit in the available space. Braided-hose V loops may be the best choice for standard HVAC piping systems. V loops are also available with CSA approval for gas lines and UL approval for refrigeration piping.
    c. A pipe loop or knuckle assembly with ball joints, as shown in Figure 8-31. This assembly may require very carefully designed supports to prevent sagging but allow full unrestrained motion. Care should be taken in the use of traditional (3) ball joint arrangements that do not readily accept all-directional motion without developing large anchor loads and pipe stress. Ball joints may be the best choice for high-pressure steam and high-pressure hot-water systems.

Fire sprinkler piping crossing building seismic joints must adhere to National Fire Protection Association (NFPA) document NFPA-13 using grooved connections with an Underwriter's Laboratory (UL) listing or Factory Mutual (FM) rating.

In most cases, flexible connectors designed to accept the displacements transversely provide the optimum performance. Connectors designed to accept motion in compression

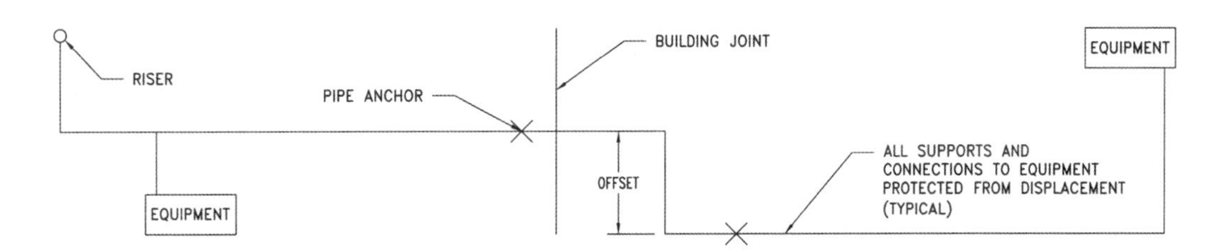

Figure 8-28 Localize displacement between anchors.

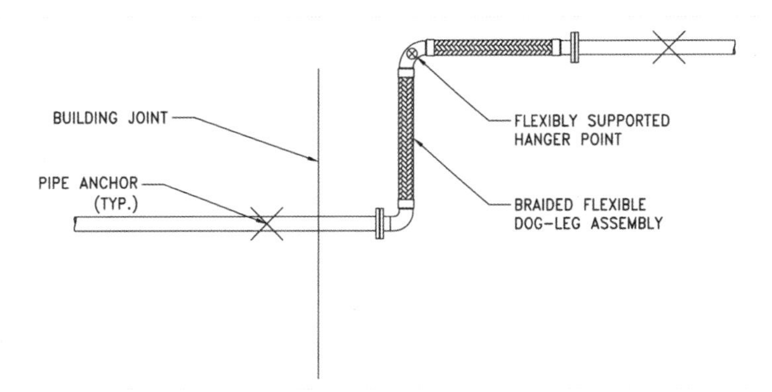

Figure 8-29 Braided flexible hose dogleg assembly.

or extension require anchors designed to restrain pipeline pressure thrust forces and should only be used where other designs are not possible. Where possible, avoid locating flexible connectors within finished areas of the building to reduce the possibility and extent of damage to ceilings, walls, and furnishings and injury to building occupants. This is especially true with elastomeric connectors or other connectors requiring flange retightening and access for service. When using flexible connectors, follow manufacturers' recommendations regarding allowable connector displacement, proper support, shutoff valve locations, access to allow service, and required guide and anchor locations.

The system anchors are the components that allow each design method to work. Similar to pipe systems designed for thermal motion, anchors for seismic joints are used to control and define where motion will occur to avoid overstressing the pipe, supports or equipment connections. In some cases, the anchor is the most critical portion of the overall design. Unfortunately, unlike systems designed for thermal motion, the seismic piping design is not tested until an earthquake occurs and improperly designed and/or installed anchors will go unnoticed until they fail in an earthquake. Therefore, a qualified design professional should design and inspect the anchors.

Engineers should consider the effects of building displacements where utility systems enter the building. Refer to the *Tri-Services Technical Manual,* Chapter 12, Utility Systems, for additional discussion and details.

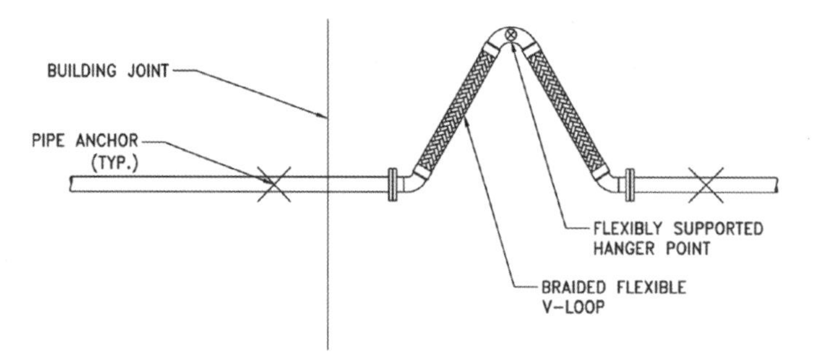

Figure 8-30 Braided flexible hose V loop.

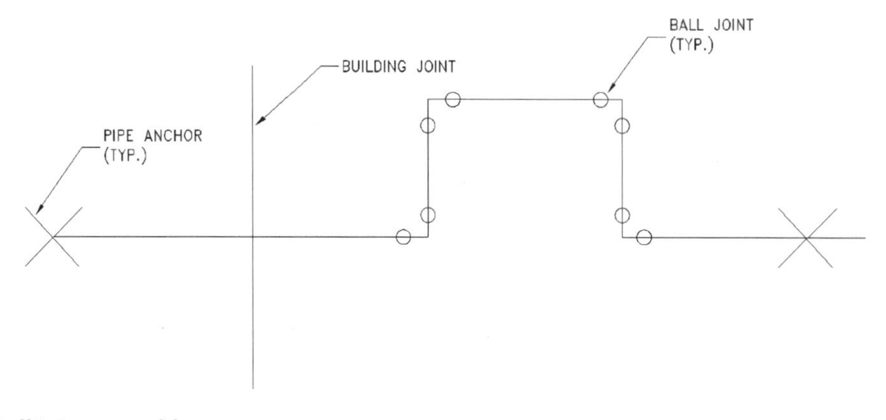

Figure 8-31 Ball joint assembly.

DESIGN TABLES

Tables 8-7 to 8-10 can be used to determine the vertical hanger rod diameter, maximum unbraced rod length, hanger rod connections to the structure, and solid or cable brace member size installed up to a 60° brace angle from horizontal.

The charts are based on the following parameters:

$$Fpv \leq 1/3\ Fph$$

$$N \geq 4$$

$$\theta \leq 60^\circ$$

Table 8-8 Solid Brace Members

	Structural Steel Angle—Max. Length of 9 ft, 6 in. (2.9 m), in. (mm)	12 Gauge Channel Strut—Max. Length of 9 ft, 6 in. (2.9 m), in. (mm)
A	2 × 2 × 1/8 (51 × 51 × 3)	1 5/8 × 1 5/8 (41 × 41 × 2.7)
B	2 × 2 × 1/4 (51 × 51 × 6)	1 5/8 × 1 5/8 (41 × 41 × 2.7)
C	3 × 3 × 1/4 (76 × 76 × 6)	1 5/8 × 3 1/4 (41 × 83 × 2.7)
D	4 × 4 × 1/4 (102 × 102 × 6)	1 5/8 × 3 1/4 (41 × 83 × 2.7)

Table 8-9 Cable Brace Members (Minimum Breaking Strength Required)

	Prestretched Steel Cable—Safety Factor of 2, lb (kN)	Standard Steel Cable—Safety Factor of 5, lb (kN)
A	640 (2.9)	1600 (7.1)
B	1600 (7.1)	4000 (17.8)
C	4000 (17.8)	10,000 (44.5)
D	10,000 (44.5)	25,000 (111.2)

Table 8-10 Connections to Structure

	Expansion Anchors into a Concrete Slab, Dia. × Embed., in. (mm)	Expansion Anchors into a Concrete Deck, Dia. × Embed., in. (mm)	Steel Bolts into a Structural Steel Diameter, in. (mm)	Lag Bolts into a Wood Structure, Dia. × Embed., in. (mm)
A	3/8 × 2 1/2 (10 × 64)	3/8 × 3 (10 × 76)	3/8 (10)	3/8 × 3 (10 × 76)
B	1/2 × 3 (13 × 76)	1/2 × 3 (13 × 76)	1/2 (13)	1/2 × 4 (13 × 102)
C	5/8 × 3 1/2 (16 × 89)	3/4 × 5 1/4 (19 × 83)	1/2 (13)	two 1/2 × 4 (two 13 × 102)
D	two 1/2 × 3 two 13 × 76)	two 1/2 × 3 (two 13 × 76)	5/8 (16)	two 5/8 × 5 two 16 × 127)
E	two 5/8 × 3 1/2 (two 16 × 89)	two 5/8 × 5 (two 16 × 127)	5/8 (16)	two 5/8 × 5 two 16 × 127)
F	four 5/8 × 3 1/2 (four 16 × 89)	four 5/8 × 5 (four 16 × 127)	3/4 (19)	four 5/8 × 5 (four 16 × 127)
G	four 3/4 × 4 1/2 (four 19 × 114)	—	7/8 (22)	four 5/8 × 5 (four 16 × 127)
H	—	—	1 (25)	—

EXCEPTIONS FROM SWAY BRACING

Some piping may not require sway bracing. The applicable code for the project may list some minimum requirements, as shown in Chapter 2. Some building codes allow for exceptions based on pipe diameter. In all cases, the life safety implications should be considered when determining bracing exceptions. Following is a list of suspended piping installations that the engineer of record may consider excluding from bracing unless specifically addressed in local codes:

1. All gas piping with a nominal diameter less than 1 in. (25 mm).
2. All other steel and copper piping throughout the building with a nominal diameter less than 3 in. (75 mm).
3. Any individually supported pipe run where the distance, as measured from the top of the pipe to the structure to which it is attached, is less than 12 in. (305 mm) for the entire length of the run. Trapeze-supported pipe runs may also be excluded from bracing, provided the distance as measured from the top of the trapeze to the structure to which it is attached is less than 12 in. (305 mm) for the entire length of the run. In addition, the hanger connection to the structure should be free to pivot so as to not develop a moment. This can be done with the addition of a swivel, eye bolt, or vibration isolation hanger connection.

   Note: A single support location that meets the 12 in. (305 mm) rule above does not constitute a seismic sway brace location. The purpose of this exception is to allow the piping to swing over a short distance of 12 in. (305 mm), which, in most cases, will not result in piping damage or the piping striking another system or piece of equipment.

   Two additional special conditions can be considered:

   a. The trapeze may support multiple small-diameter pipes that would be bracing exceptions according to the list above.
   b. One pipe of many on a trapeze would not be a bracing exception according to the list above.

   The following notes address these special conditions.

1. Any combination of small piping supported on a trapeze where the total weight exceeds 10 lb/ft (14.9 kg/m) should be sway braced regardless of pipe size.

   Note: If directional changes or offsets to equipment connections allow for flexibility of the trapeze system (e.g., long offsets or flexible connectors), the system can be excluded from bracing as long as all pipes supported by the trapeze are listed in exceptions 1 and 2 above.

2. If only one pipe supported on a trapeze requires sway bracing, consider the total combined weight of all pipes on the trapeze to determine sway brace components and anchorage for the entire trapeze.
3. The 12 in. (305 mm) rule can still be used in these situations.

PIPING CONNECTIONS TO INLINE COILS

Many small-diameter pipe connections to coils in variable-air-volume (VAV) terminal units and fan-coil units failed in recent earthquakes, causing extensive water damage. Special considerations should be made to select a flexible connector for these piping connections that is designed to accept the potential differential motion between the duct and the connected piping. This is especially true if the piping is unbraced, the duct is unbraced, or both are unbraced. Because suspended systems can have very large potential differential displacement, braided-hose V loops or other flexible connectors with large motion capacities supported by displacement force test data should be used. Refer to Chapter 11 for additional information on flexible connector attachments to in-line coils and other components.

EXAMPLES

The following examples illustrate how to use the tables outlined in this chapter.

Example 1

Support type: individual clevis
Pipe material: 5 in. (127 mm) diameter, schedule 40 pipe
Pipe connection type: welded
Structure type: concrete slab
Brace type: cable bracing
Structural connection: expansion anchors
Seismic input: 0.5 *g*

Pipe weight – 26 lb/ft (39 kg/m), insulated and full of water

From Table 8-3, at 0.5 *g*:
Maximum transverse brace spacing = 40 ft (12.2 m)

From Table 8-7, at 0.5 *g* and maximum pipe weight of 31 lb/ft (46 kg/m), the transverse brace requirements are as follows:
Hanger rod diameter = 5/8 in. (16 mm)
Hanger rod connection = *C*, one 5/8 in. (16 mm) diameter expansion anchor
Maximum unbraced rod length = 26 in. (660 mm)
Cable brace member = *C*, prestretched steel cable with a breaking strength of 4000 lb (17.8 kN)
Brace member connection to structure = *E*, two 5/8 in. (16 mm) diameter expansion anchors

Example 2

Support type: trapeze
Pipe material: two 2 1/2 in. (64 mm) and two 3 in. (76 mm) diameter copper pipes
Pipe connection type: brazed
Structure type: concrete deck
Brace type: solid bracing
Structural connection: expansion anchors
Seismic input: 0.25 *g*

Combined pipe weight: 32 lb/ft (48 kg/m), insulated and full of water.

From Table 8-3, at 0.25 *g*:
Maximum transverse brace spacing = 50 ft (15.2 m).

From Table 8-7, at 0.25 *g* and maximum pipe weight of 36 lb/ft (54 kg/m), the transverse brace requirements are as follows:
Hanger rod diameter = 1/2 in. (13 mm)
Hanger rod connection = *E*, two 5/8 in. (16 mm) diameter expansion anchors
Maximum unbraced rod length = 18 in. (457 mm)
Solid brace member = *B*, 2 × 2 × 1/4 in. (51 × 51 × 6 mm) steel angle
Brace member connection to structure = *D*, two 1/2 in. (13 mm) diameter expansion anchors

BIBLIOGRAPHY

Departments of the Army, Navy, and Air Force. 1982. *Seismic design for buildings—Technical manual* or *Tri-services technical manual*. Springfield, VA.

DOE. 1997. *Seismic Evaluation Procedure for Equipment in U.S. Department of Energy Facilities*. Springfield, VA: U.S. Department of Energy.

Manufacturers Standardization Society of the Valve and Fittings Industry, Inc. 2003. *ANSI/MSS SP-58-Pipe Hangers and Supports*. Falls Church, VA: Manufacturers Standardization Society.

Mason Industries, Inc. 2009. *Seismic Restraint of Suspended Piping, Ductwork and Electrical Systems*. Hauppauge, NY.

SMACNA. 2008. *Seismic Restraint Manual—Guidelines for Mechanical Systems*. Chantilly, VA: Sheet Metal and Contractors National Association.

9 Piping Risers

Riser supports should be designed to carry the weight of the riser and allow thermal expansion or contraction. Anchors are used where needed to direct axial motion. Guides are used to maintain pipe alignment and restrict transverse motion due to seismic forces.

DESIGN STANDARDS

In addition to the seismic requirements for piping in local building codes, requirements for the support and restraint of pipe risers are included in the following:

- The American National Standards Institute/American Society of Mechanical Engineers (ANSI/ASME) B31.9, *Building Services Piping* (Section 920.1.1).
- The ANSI/ASME B31.1, *Power Piping* (Section 101.5.3).
- Other standards require earthquake forces to be considered in the design of pipe supports including the following:
- The Manufacturers' Standardization Society's MSS-SP-69 includes a statement that calculations for pipe risers shall give consideration to the loads due to seismic forces.
- The Sheet Metal and Air-Conditioning Contractors National Association's (SMACNA's) *Seismic Restraint Manual/Guidelines for Mechanical Systems* includes a statement that risers must be individually engineered. It also includes a detail of joint restraint for seismic loads on hubless pipe risers.

RISER DESIGN FOR THERMAL EXPANSION AND CONTRACTION

All risers are subject to movement caused by thermal changes. During construction, riser supports are installed at ambient temperature, which can vary widely. Movements caused by these temperature variations can be greater than those caused by thermal changes during daily operation. Risers can also see unusual thermal changes during cleaning and start-up.

Straight solid risers can be rigidly supported at any one point. In buildings where the risers begin in a basement, the riser supports can be located at the bottom of the risers, with special slab reinforcement or footings for large risers. The thermal growth or contraction is a function of the distance from this support and is greatest at the highest floor.

In situations where maximum expansion and contraction must be reduced, the support should be located at the center of the riser. This cuts the maximum thermal motion in half at each end. Unfortunately, supporting large risers at the midpoint of the building means there will be concentrated loads on one floor, often requiring modifications to the building structure.

Risers can be designed with piping offsets or loops to take up expansion or contraction. Each straight riser section should be independently supported and anchored to force the offsets or loops to take up the expansion or contraction. Offsets and loops can be quite large and require a great deal of space planning to fit into the structure.

The use of expansion joints can eliminate the need for pipe loops, but anchors must be designed for pressure thrust forces. The failure of an expansion joint means not only the loss of heating or cooling but can cause water damage. Adding valves above and below the expansion joint will allow shutdown and replacement but will not eliminate initial water damage.

Each straight riser section must be independently supported and anchored. Since a riser with expansion joints is not a rigid element, the water load occurring at the bottom of the riser cannot be shared by the supports; all of the water load must be carried by the supports on the bottom pipe section of the riser.

Anchors must also be designed to handle thrust forces during hydrostatic pressure tests. Under no circumstances should anchors be eliminated or design forces reduced due to the theoretical balancing of thrust forces from adjacent expansion joints or couplings. Once valves are closed and a joint removed, there is no possible balancing; anchors must be designed for the full pressure thrust forces as a worst case.

Grooved-pipe couplings are often used on piping to ease installation. Pressurized piping becomes a rigid element due to pressure thrust forces, and any initial pipe gaps in the couplings will only allow the piping to increase in overall length. This may not be a problem on horizontal piping, but increasing the overall length of pipe risers can result in major load shifts at the support points. Although it might be possible to readjust the supports to accommodate the new riser length, complex installation and potential coupling leaks in a riser shaft make grooved coupling risers difficult to install and risky. Because there is no real expansion and contraction capability, grooved couplings should not be used on risers. Solid risers are recommended to allow accurate load distribution on multiple floors and eliminate the chance of water leakage.

SEISMIC RESTRAINT OF RISERS

Anchors and guides used to provide pipe stability and control expansion and contraction should be designed for seismic restraint. The use of separate seismic restraints is not practical due to space limitations and the difficulties in determining load distribution between different types of restraint elements. Seismic loads on any restraint element can be estimated by adding the total weight of pipe and water for one-half the distance to adjacent restraint elements and applying the seismic force factor.

RISER SUPPORTS

Riser clamps are the most common riser supports. These two-piece clamps bolt together and have a load rating based on clamping capacity. Riser clamps reinforce the pipe and distribute forces evenly to minimize pipe wall stress concentrations that would otherwise develop with welded lugs or brackets. During construction, risers may be supported on riser clamps on each floor. This arrangement must be temporary, because it does not allow for expansion and contraction and would result in major load shifts as clamps lift off some floors, overloading other clamps and floors.

According to MSS-SP-69, only one rigid support is allowed on (straight) risers subject to thermal expansion and contraction. According to B31.1, B31.9, and MSS-SP-69, the clamp must be sized for two times the dead load, and there must be a positive means of engagement between the clamp and the riser. For risers supported at one point, the standard riser clamps in pipe hanger manufacturers' catalogs are undersized. For major riser loads and long distances to support steel, special heavy-duty clamps should be used. Welding the heavy-duty clamp to the pipe is the most effective means of meeting the positive engagement requirement. The riser clamp can also be welded to plates that are bolted to the structure. This type of assembly is shown in Figure 9-1.

Welding the riser clamp to copper pipe is not possible. In these applications, a copper sleeve or slip-on flange can be soldered or brazed to the pipe directly above the clamp to provide a positive means of engagement. A sleeve or flange can be added directly under

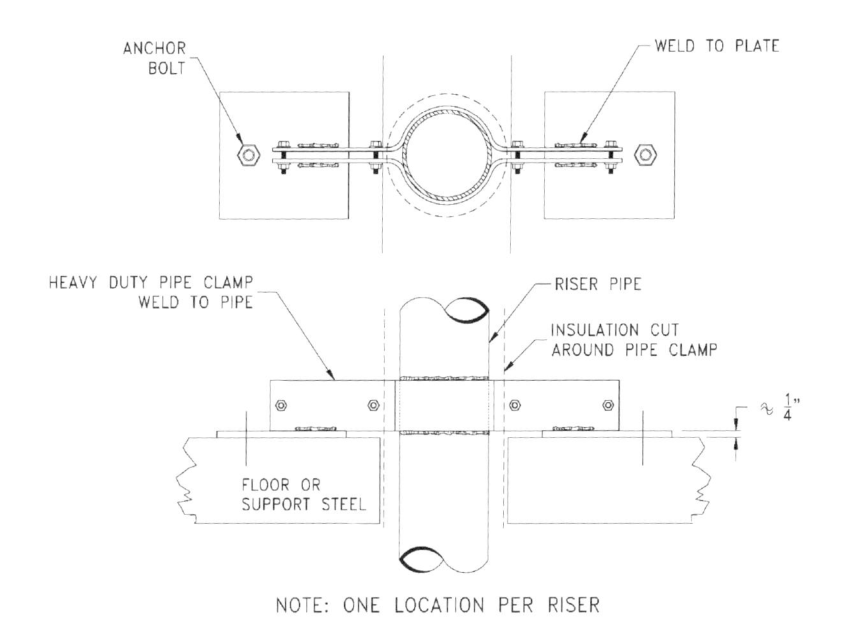

Figure 9-1 Riser clamp support.

the clamp to prevent upward motion. A detail of the sleeve attachment is shown in Figure 9-2.

Risers can be supported on neoprene pads in a limited number of situations. Because pads do not have adjustment bolts to control initial deflections and deflection of the pad is limited to 0.1 in., pads should generally be used on risers supported at only one floor. Using pad supports on multiple floors can result in unpredictable floor loads and should be avoided. This type of support is shown in Figure 9-3.

Where structural considerations require the riser loads to be spread over two or more floors, adjustable open-style spring mounts can be used to support the riser and allow both expansion and contraction. Simple load calculations can predict load shifts. Springs are bolted to welded plates on the riser clamp or custom bracket. This type of arrangement is shown in Figure 9-4.

Springs should be selected for a deflection of four times the riser expansion or contraction at the spring location to allow a maximum 25% load change between installed and operating conditions. The load on each floor is known when the riser is installed, empty, full, and at operating temperature. The riser is supported with minor load changes on each floor and is also vibration isolated.

On straight solid risers, spring supports can be designed to force the riser to expand or contract at any point on the riser, commonly called the neutral point. The riser can be free-floating or an anchor can be used at the neutral point to correct for variations in support stiffness or horizontal pipe stiffness. A properly balanced riser will put very little load on the anchor due to thermal expansion or contraction.

The only disadvantage of spring-supported risers is that the entire system is subject to upward motion when the water is drained. Unlike loads from thermal motion, loads caused by water weight removal cannot be balanced. A free-floating riser can be allowed to lift as the water is drained as long as the motion does not damage supports or piping. The anchor of an anchored riser must be designed to resist this load. Other solutions, including spring readjustment during water drainage, or individual precision spring limit stops, are impractical and should not be used.

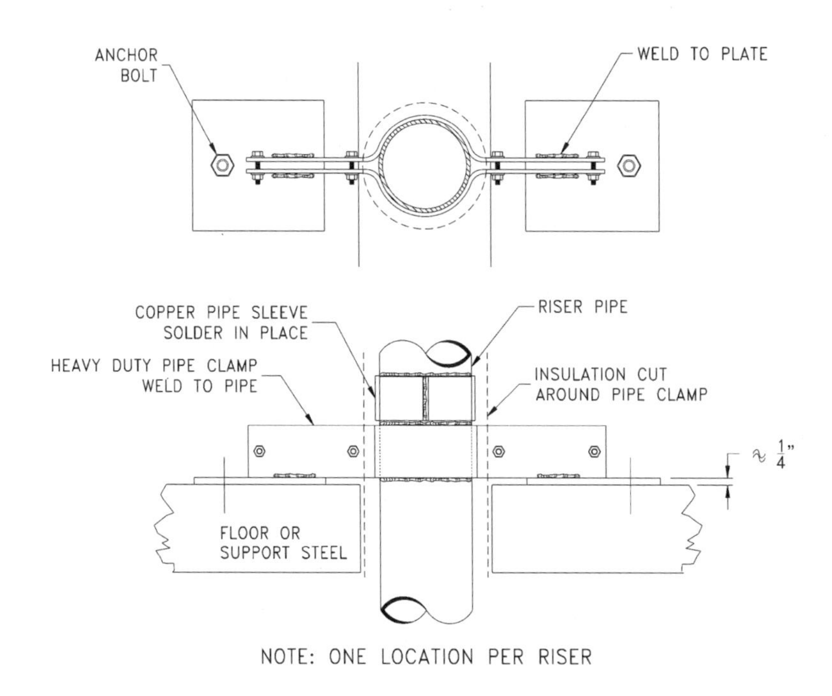

Figure 9-2 Riser clamp support on copper pipe.

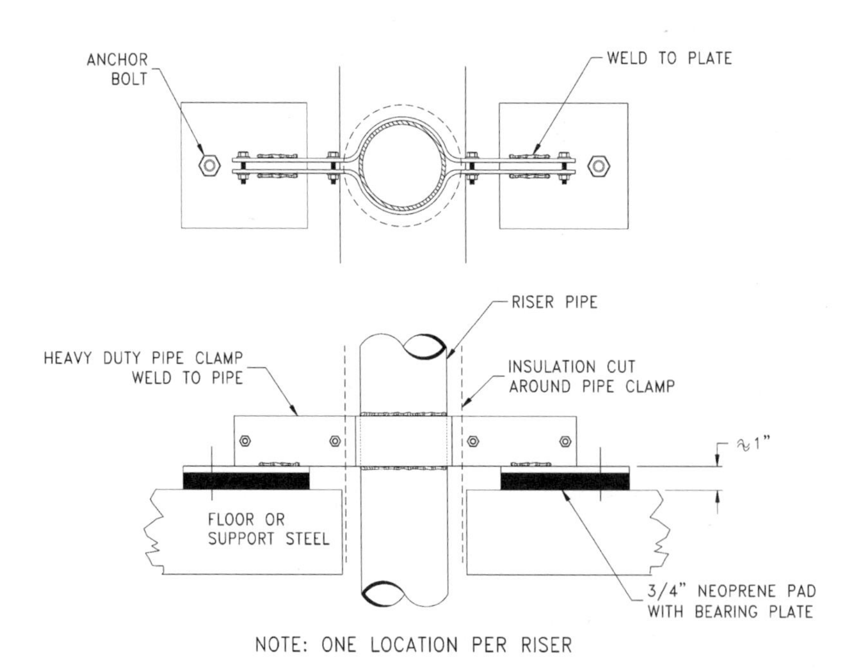

Figure 9-3 Riser clamp supported with neoprene pad.

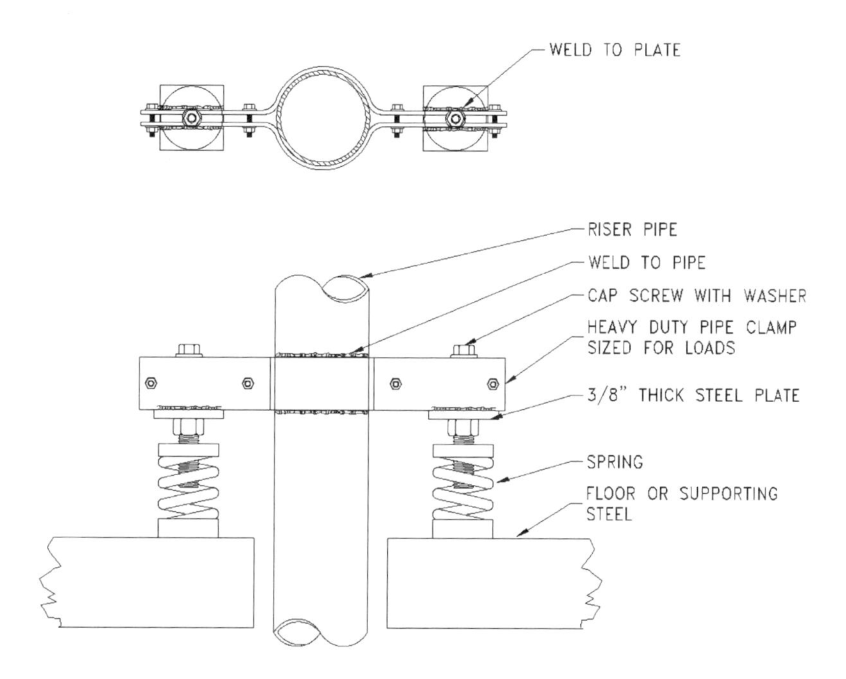

Figure 9-4 Riser spring support.

RISER ANCHORS

A riser anchor must provide vertical restraint, allow proper installation of fire packing, provide adjustment for pipe alignment, and allow proper installation of the vapor seal on chilled-water pipes. This can best be achieved with a welded riser clamp or custom bracket that is connected to external anchor assemblies located outboard of the riser, between the riser clamp or bracket and the building steel or floor. This arrangement is shown in Figure 9-5.

Resilient anchors should be used to avoid short-circuiting vibration isolators. Resilient anchors can also be used on adjacent floors to distribute anchor forces that are too large for any one floor. This type of anchor is shown in Figure 9-6.

An anchor can be used to control the uplift of the riser caused by water drainage. Anchors should also be used on risers with expansion joints, offsets, or loops to resist pipe pressure thrust and to force motion at offsets and pipe loops. Springs can distribute the dead loads but should not be used to control any of these internal or external forces, because excessive motion and problems can result.

RISER GUIDES

Pipe guides must allow axial motion of the pipe and provide lateral restraint. If cored or sleeved holes in floor are used, alignment from floor to floor must be very accurate and the gap between the pipe and the floor filled with an appropriate material. If the penetration is fire rated, the material is limited to fire packing and fire seal materials that may crush and allow misalignment. Spider guides or oversized clamps introduce vapor barrier problems and fire-rated penetration access issues.

External sliding guide assemblies located outboard of the riser, between a welded riser clamp or bracket and the building steel or floor, can provide transverse restraint, allow proper installation of fire packing, provide adjustment for pipe alignment, and allow proper installation of the vapor seal on chilled-water pipes.

The guides should include a neoprene bushing. This bushing allows some flexibility and prevents short-circuiting of vibration isolated risers. The neoprene bushing also allows seismic loads to be cushioned and distributed to several guides. Unlike oversized riser clamps, spider guides, or cored or sleeved holes, the welded clamp or bracket assembly provides dead load support in a catastrophic situation. This arrangement is illustrated in Figure 9-7.

The guide assemblies should include neoprene sleeves to provide acoustical isolation and prevent rattling. This type of guide is shown in Figure 9-8.

GUIDE SPACING

All risers require guides to maintain pipe alignment, prevent column buckling, and provide transverse restraint. The maximum distance between guides varies with the pipe diameter and the magnitude of the seismic force.

Special guide-spacing requirements apply to risers with offsets, expansion loops, expansion joints, or grooved-pipe fittings. Guide spacing for solid straight steel risers and solid steel risers with offsets or loops for seismic forces up to 1.0 *g* are shown in Table 9-1. Guide spacing for steel risers with expansion joints for seismic forces up to 1.0 *g* are shown in Table 9-2.

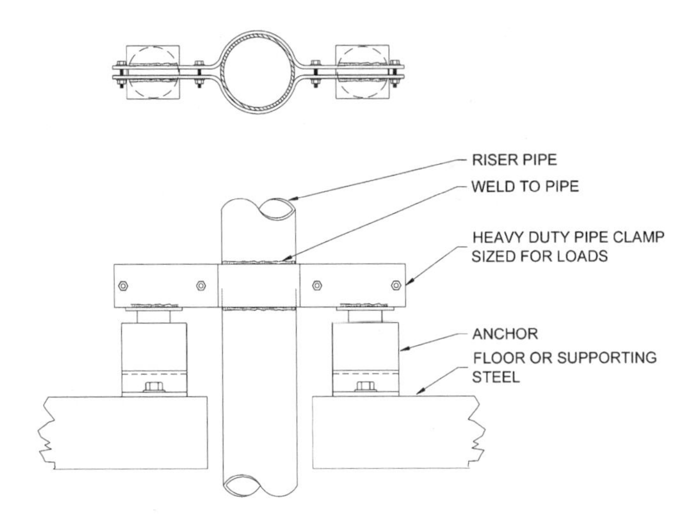

Figure 9-5 Riser anchor.

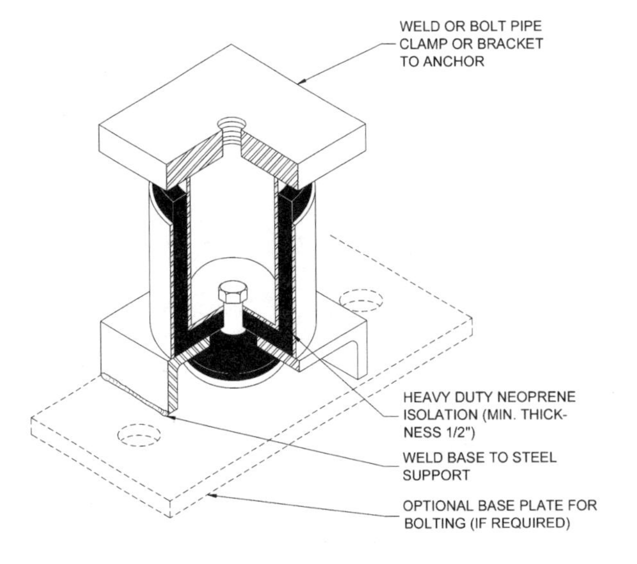

Figure 9-6 Anchor detail.

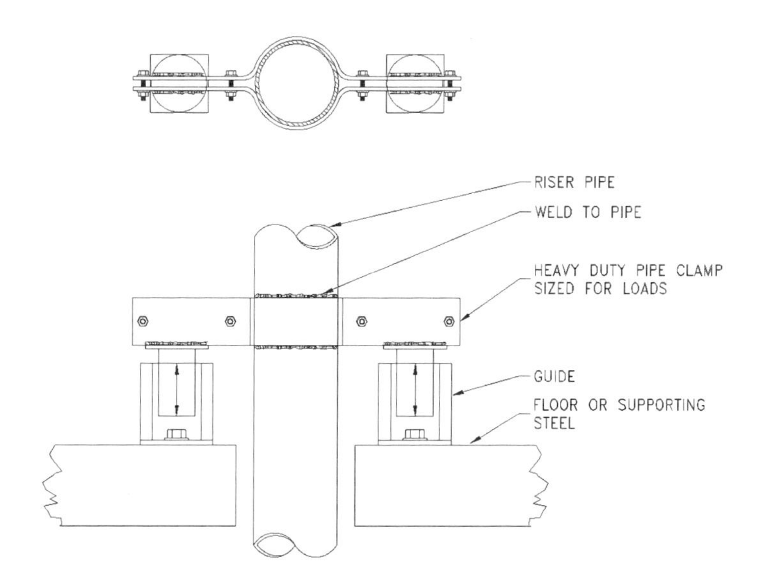

Figure 9-7 Riser guide.

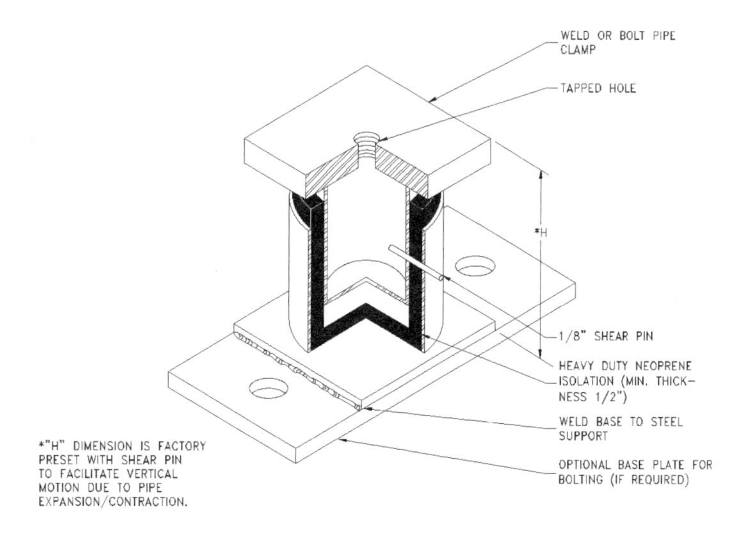

Figure 9-8 Guide detail.

Table 9-1 Maximum Recommended Vertical Guide Spacing of Solid Steel Risers for Seismic Forces up to 1.0 g*

	Straight Solid Riser	Offset Solid Riser
Pipe Size (in.)	One Guide Each End and Maximum Distance between Guides (ft)	One Anchor Each End and Maximum Distance between Guides (ft)
1	40	24
1 1/4	40	24
1 1/2	40	24
2	40	24
2 1/2	40	30
3	40	36
4	40	36
5	40	36
6	40	36
8	40	40
10	40	40
12	40	40
14	40	40
16	40	40
18	40	40
20	40	40

*Maximum guide spacing for pipes of other materials or for seismic forces greater than 1.0 *g* must be designed individually.

OTHER RISER DESIGN CONSIDERATIONS

All concrete structures shorten because of concrete curing, and buildings of all types shorten as construction adds weight. There is further compression from occupancy loads. These changes in the structure must be considered, particularly if the risers are installed as the construction progresses rather than after completion.

All new structures are designed to sway during an earthquake or in high wind conditions. The amount of sway, otherwise known as interstory drift, varies with the type of structure but is at least 1% of story height. Risers must be designed to move with the building, and riser supports must be designed to accommodate this motion. External sliding guides with neoprene bushings and free-standing spring supports will allow angular misalignment of the riser.

Horizontal pipe attached to risers must be supported to accommodate the riser movements without excessive strain. When possible, the horizontal pipe should be supported on spring hangers for the first three supports from the riser. Springs should be selected for a deflection of four times the riser expansion or contraction to allow a maximum 25% load change between installed and operating conditions. Where spring hangers are not possible, braided stainless steel flexible hoses can be used on the horizontal pipe. See Chapter 11 for more information on flexible connectors. As a precaution, valves should be used on each side of the connectors to allow removal and replacement without draining the riser and to minimize the impact on other portions of the building.

Risers in open shafts require special consideration. All supplemental support steel should be designed to support the riser dead loads and seismic loading. Access should be provided at all support, expansion joint, flexible connector, valve, and pipe loop locations. Service flooring or grating should be considered at these locations as well. Most riser shaft

Table 9-2 Maximum Recommended Vertical Guide Spacing of Steel Risers with Expansion Joints for Seismic Forces up to 1.0 g*

Pipe Size (in.)	Risers with Expansion Joints				
	Stainless Steel		Neoprene	Balance of Guides—Distance between Joint and Anchor*	
	One Anchor—Each End of Pipeline		One Anchor—Each End of Pipeline	Operating Pressures	
	Distance—Joint to First Guide†	Distance—Joint to Second Guide‡	Distance—Joint to First Guide (second not required)	Up to 150 psi (ft)	151 – 300 psi (ft)
1	8 in.	2 ft – 0 in.	8 in.	12	12
1 1/4	8 in.	2 ft – 0 in.	8 in.	12	12
1 1/2	10 in.	3 ft – 0 in.	10 in.	12	12
2	10 in.	3 ft – 0 in.	10 in.	12	12
2 1/2	12 in.	3 ft – 6 in.	12 in.	12	12
3	12 in.	3 ft – 6 in.	12 in.	17	14
4	1 ft – 4 in.	4 ft – 8 in.	1 ft – 4 in.	25	19
5	2 ft – 0 in.	7 ft – 0 in.	2 ft – 0 in.	30	23
6	2 ft – 0 in.	7 ft – 0 in.	2 ft – 0 in.	37	27
8	2 ft – 6 in.	9 ft – 4 in.	2 ft – 6 in.	40	33
10	3 ft – 4 in.	11 ft - 8 in.	3 ft – 4 in.	40	40
12	4 ft – 0 in.	14 ft – 0 in.	4 ft – 0 in.	40	40
14	4 ft – 8 in.	16 ft – 4 in.	4 ft – 8 in.	40	40
16	5 ft – 4 in.	18 ft - 8 in.	5 ft – 4 in.	40	40
18	6 ft – 0 in.	21 ft – 0 in.	6 ft – 0 in.	40	40
20	6 ft – 8 in	23 ft - 4 in.	6 ft – 8 in.	40	40
24	8 ft – 0 in.	28 ft – 0 in.	8 ft – 0 in.	40	40
26	10 ft – 0 in.	35 ft - 0 in.	10 ft – 0 in.	40	40
28	10 ft – 0 in.	35 ft - 0 in.	10 ft – 0 in.	40	40
30	10 ft – 0 in.	35 ft - 0 in.	10 ft – 0 in.	40	40

*Maximum guide spacing for pipes of other materials or for seismic forces greater than 1.0 *g* must be designed individually.
†Guides beyond the anchor may be spaced as in a straight solid run because there is no thrust force.
‡If the anchor is next to the joint on one side, no guide is needed on that side.

walls are fire rated. Wall penetrations for branch piping or takeoffs should be fire rated as well. Flexible connectors should be used on the horizontal pipe if motion exceeds the capability of the fire-rated penetration.

HUBLESS CAST IRON PIPE RISERS

Hubless cast iron pipe should be supported and guided at each floor or at a maximum spacing of 20 ft. Hubless cast iron risers can be supported on riser clamps on each floor. Refer to Figure 9-1 for a typical support detail and Figure 9-3 for support detail for risers on neoprene pads, except that welding of the clamp to the cast iron pipe is not possible or required. Where fire-rated floor penetrations are made in sleeves or cored holes, anchoring of the riser clamps may not be necessary if the fire-rated material has the compressive strength to resist damage in a seismic event.

Joints on unsupported pipe sections between floors should be braced in accordance with the details in SMACNA, as shown in Figure 9-9. It should be noted that the couplings

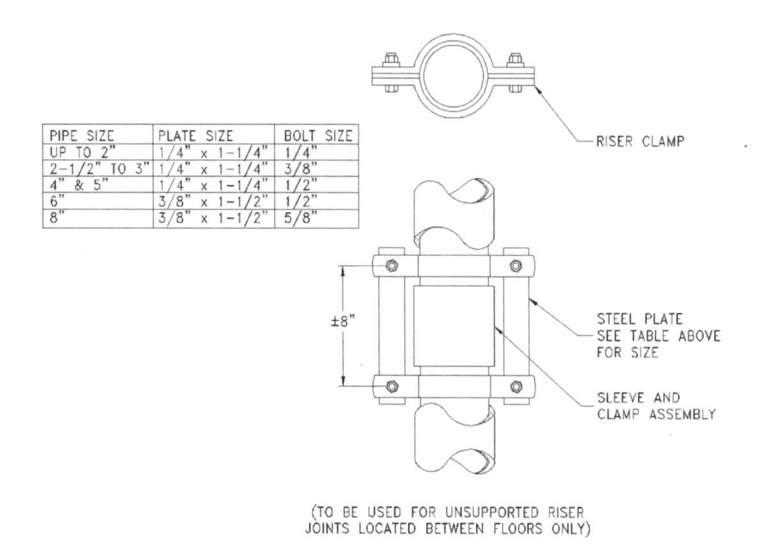

PIPE SIZE	PLATE SIZE	BOLT SIZE
UP TO 2"	1/4" x 1-1/4"	1/4"
2-1/2" TO 3"	1/4" x 1-1/4"	3/8"
4" & 5"	1/4" x 1-1/4"	1/2"
6"	3/8" x 1-1/2"	1/2"
8"	3/8" x 1-1/2"	5/8"

Figure 9-9 Riser bracing for hubless pipes.

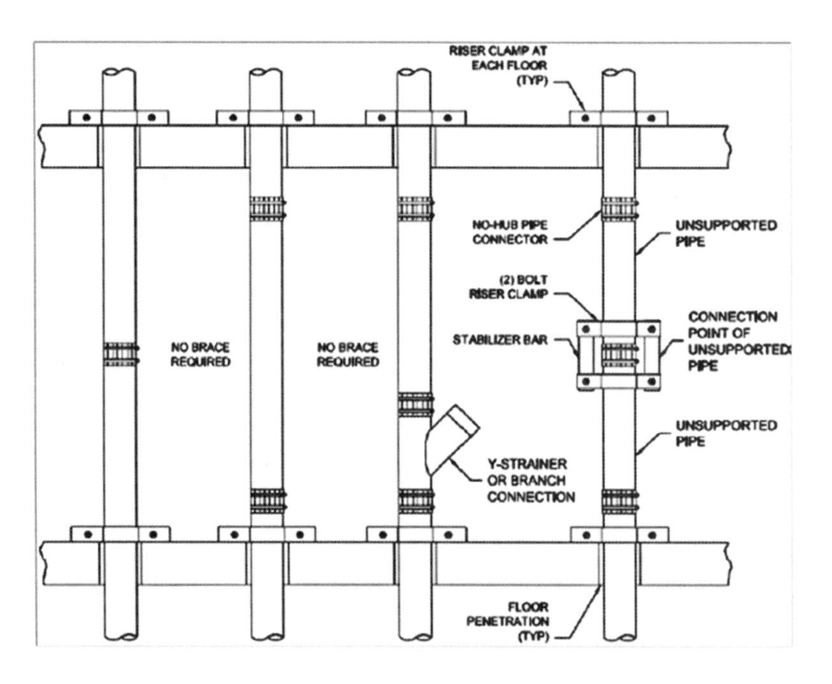

Figure 9-10 Hubless Cast Iron Risers

on hubless cast iron piping are critically important in that they allow some flexibility and prevent straining and breakage of the nonductile piping. For this reason, some unbraced couplings on risers are essential to allow the piping to move and absorb interstory drift. See Figure 9-10.

BIBLIOGRAPHY

ASME. 2008. *Building Services Piping, ANSI/ASME B31.9-2008*. New York.

MSS. 2009. *ANSI/MSS-SP-58, Pipe Hangers and Supports*. Falls Church, VA: Manufacturers Standardization Society of the Valve and Fittings Industry.

SMACNA. 2008. *Seismic Restraint Manual—Guidelines for Mechanical Systems*. Chantilly, VA: Sheet Metal and Air Conditioning Contractors' National Association.

10 Suspended Equipment

The basic design requirements for bracing of suspended equipment are as follows:

1. Sway braces should be arranged so that they limit motion of the equipment in all directions.
2. Threaded rods supporting equipment should be designed to resist vertical seismic loads and support equipment.
3. Equipment supported by vibration isolation hangers should be detailed and installed with isolation hangers close to the structure and upward limit stops located directly below the hangers.
4. Avoid bracing equipment to separate portions of the structure that may act differently in response to an earthquake. For example, do not connect a transverse brace to a wall and a longitudinal brace to a floor or roof at the same brace location.

SWAY BRACING

Sway bracing of suspended equipment differs from piping, ductwork, or other suspended systems. Equipment is braced independently of connected systems, such as ductwork and piping, and requires restraint in all horizontal and vertical directions.

There are two types of sway braces, solid and cable, each with advantages and disadvantages as discussed in Chapter 7.

Figures 10-1 and 10-2 show typical solid and cable brace arrangements for suspended equipment. In the extreme case, the unit is square in plan; it is possible the unit may rotate; and therefore an eight-cable arrangement is recommended, as shown in Figure 10-3.

HANGER ROD REQUIREMENTS

The effects of sway bracing on the hanger rod are discussed in more detail in Chapter 7. Following is a discussion of how loads are applied to hanger rods for suspended equipment.

Figures 10-4 to 10-7 show the reactive tensile and compressive loads from external seismic and dead loads where the hanger rods connect to the structure.

Figure 10-4 shows cable and solid-brace systems where the brace connection is above the equipment center of gravity.

Hanger Rod 1

$$\text{If } (T_w + T_{over} + T_{Fpv}) > C_{Fp},$$

$$\text{Then } T_{(Rod\ 1)max} = (T_w + T_{over} + T_{Fpv}) - C_{Fp} \tag{10-1}$$

$$\text{Else } C_{(Rod\ 1)max} = (C_{Fp} + C_{Fpv}) - (T_w + T_{over}). \tag{10-2}$$

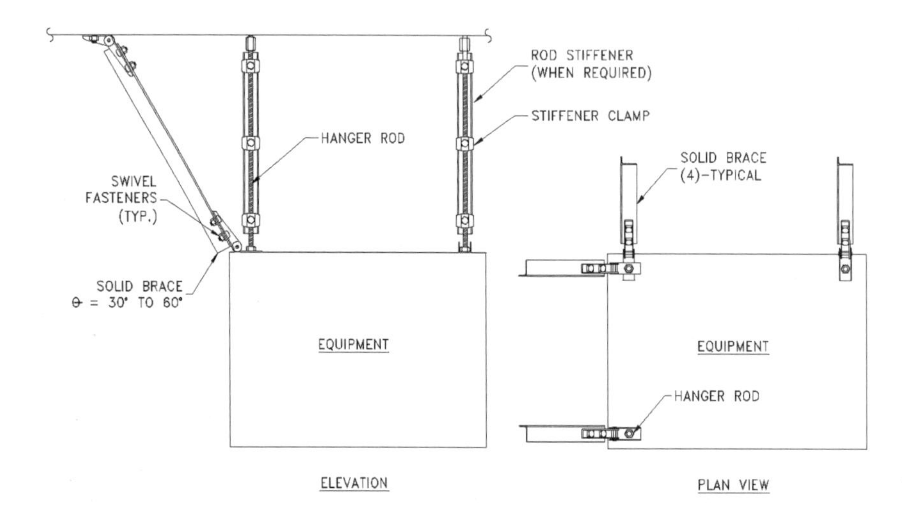

Figure 10-1 Typical solid-brace arrangement.

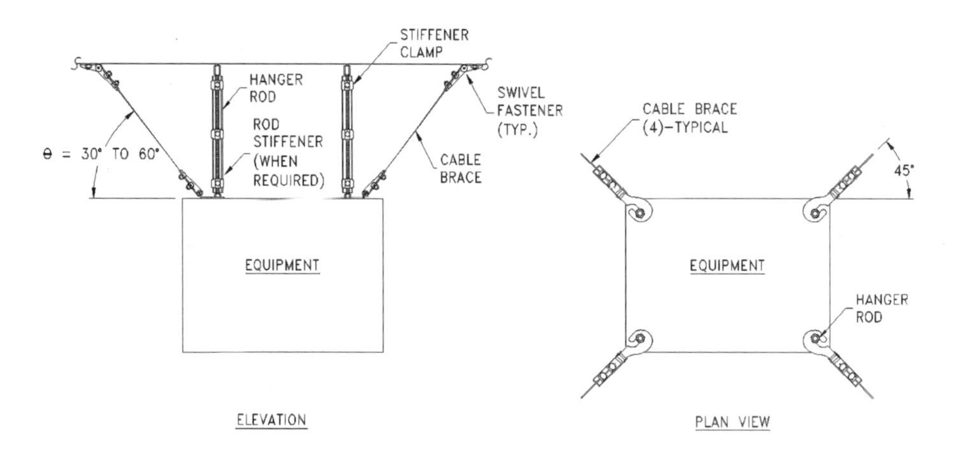

Figure 10-2 Typical solid-brace arrangement.

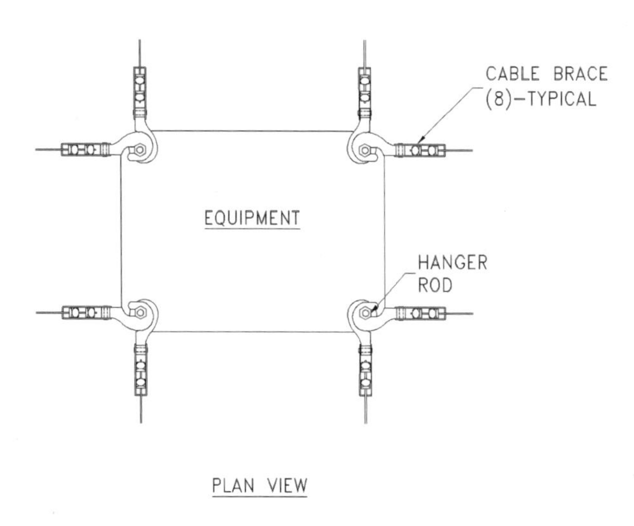

Figure 10-3 Typical eight-cable arrangement.

Hanger Rod 2

$$\text{If } (T_w + T_{Fpv}) > C_{over},$$

$$\text{Then } T_{(Rod\ 2)max} = (T_w + T_{Fpv}) - C_{over}, \tag{10-3}$$

$$\text{Else } C_{(Rod\ 2)max} = (C_{over} + C_{Fpv}) - T_w. \tag{10-4}$$

Figure 10-5 shows solid-brace systems where the brace connection is above the equipment center of gravity.

Hanger Rod 1

$$\text{If } (T_w + T_{Fp} + T_{Fpv}) > C_{over},$$

$$\text{Then } T_{(Rod\ 1)max} = (T_w + T_{Fp} + T_{Fpv}) - C_{over}, \tag{10-5}$$

$$\text{Else } C_{(Rod\ 1)max} = (C_{over} + C_{Fpv}) - (T_w + T_{Fp}). \tag{10-6}$$

Hanger Rod 2

$$\text{If } (T_w + T_{over}) > C_{Fpv},$$

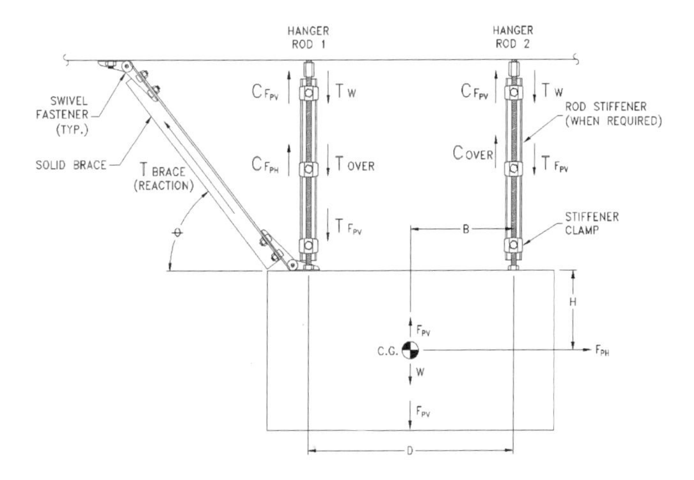

Figure 10-4 Solid-braced equipment connected above the center of gravity with brace in tension.

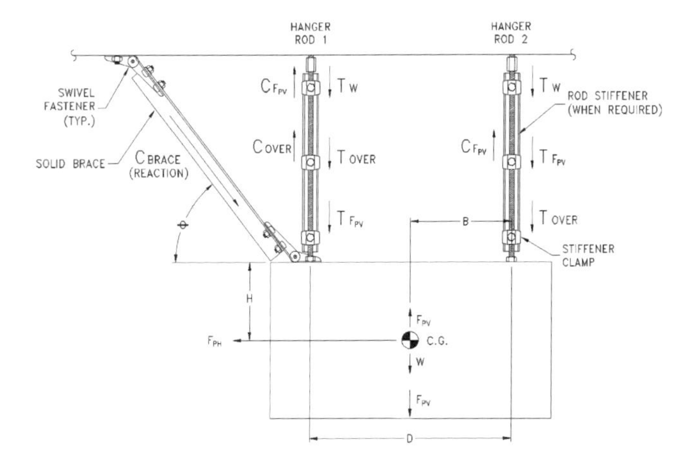

Figure 10-5 Solid-braced equipment connected above the center of gravity with brace in compression.

$$\text{Then } T_{(Rod\ 2)max} = T_w + T_{Fpv} + T_{over}, \quad (10\text{-}7)$$

$$\text{Else } C_{(Rod\ 2)max} = C_{Fpv} - (T_w + T_{over}). \quad (10\text{-}8)$$

Figure 10-6 shows cable and solid-brace systems where the brace connection is below the equipment center of gravity.

Hanger Rod 1

$$\text{If } (T_w + T_{Fpv}) > (C_{Fp} + C_{over}),$$

$$\text{Then } T_{(Rod\ 1)max} = (T_w + T_{Fpv}) - (C_{Fp} + C_{over}), \quad (10\text{-}9)$$

$$\text{Else } C_{(Rod\ 1)max} = (C_{Fp} + C_{over} + C_{Fpv}) - T_w. \quad (10\text{-}10)$$

Hanger Rod 2

$$\text{If } (T_w + T_{over}) > C_{Fpv},$$

$$\text{Then } T_{(Rod\ 2)max} = T_w + T_{over} + T_{Fpv}, \quad (10\text{-}11)$$

$$\text{Else } C_{(Rod\ 2)max} = C_{Fpv} - (T_w + T_{over}). \quad (10\text{-}12)$$

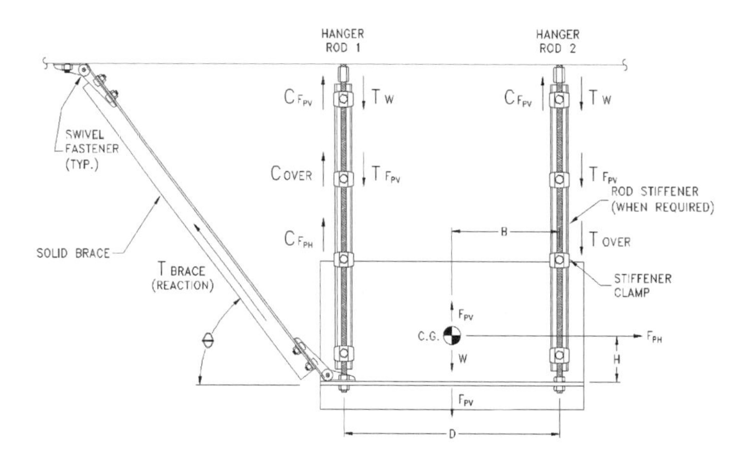

Figure 10-6 Solid-braced equipment connected below the center of gravity with brace in tension.

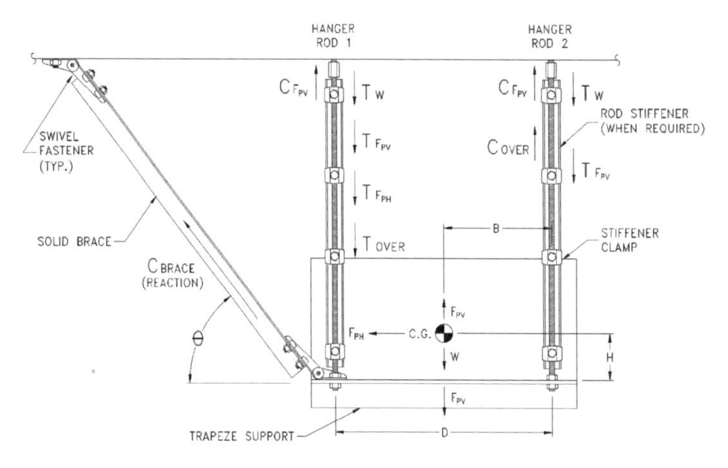

Figure 10-7 Solid-braced equipment connected below the center of gravity with brace in compression.

Figure 10-7 shows solid-brace systems where the brace connection is below the equipment center of gravity.

Hanger Rod 1

$$\text{If } (T_w + T_{Fp} + T_{over}) > C_{Fpv},$$

$$\text{Then } T_{(Rod\ 1)max} = T_w + T_{Fpv} + T_{Fp} + T_{over} \quad (10\text{-}13)$$

$$\text{Else } C_{(Rod\ 1)max} = C_{Fpv} - (T_w + T_{Fp} + T_{over}) \quad (10\text{-}14)$$

Hanger Rod 2

$$\text{If } (T_w + T_{Fpv}) > C_{over},$$

$$\text{Then } T_{(Rod\ 2)max} = (T_w + T_{Fpv}) - C_{over} \quad (10\text{-}15)$$

$$\text{Else } C_{(Rod\ 2)max} = (C_{over} + C_{Fpv}) - T_w \quad (10\text{-}16)$$

where

W = equipment operating weight
L = longest horizontal distance between corner vertical hanger rods
D = shortest horizontal distance between corner vertical hanger rods
H = vertical distance from equipment center of gravity to brace connection
A = distance from vertical hanger rod to equipment center of gravity along length of unit ($A \geq L/2$)
B = distance from vertical hanger rod to equipment center of gravity along width of unit ($B \leq D/2$)
Fp = seismic horizontal force
Fpv = seismic vertical force
N = number of vertical hanger rods
T_w = tensile force due to $W = W(A/L)(B/D)$
$T_{Fp} = C_{Fp}$ = tensile or compressive force due to $Fp = Fp \tan \theta$
$T_{Fpv} = C_{Fpv}$ = tensile or compressive force due to $Fpv = Fpv(A/L)(B/D)$
$T_{over} = C_{over}$ = tensile or compressive force due to equipment overturning = $[(Fp \times H)(2A/L)(2B/D)]/[(D/2)(N/2)]$
$T_{brace} = C_{brace}$ = tensile or compressive force on the brace due to $Fp = Fp/\cos \theta$

For calculated tension loads, check both the hanger rod and its connection to the structure. Threaded hanger rod allowable values are tabulated in Table 10-1. Compressive forces on threaded rods are determined by the unbraced length and upward force on the rod. Refer to Table 7-3 for maximum clamp spacing to attach the rod stiffener to the rod and Tables 10-2 and 10-3 for maximum unbraced rod lengths. For compressive forces, determine if the hanger rod requires a stiffener as discussed in Chapter 7 and shown in Figure 7-12.

VIBRATION-ISOLATED EQUIPMENT

Equipment suspended by spring and/or neoprene vibration isolation hangers should only be braced with cable bracing as discussed in Chapter 7. Refer to Figure 7-17 for correct installation of vibration isolation hangers at sway brace locations.

DESIGN TABLES

Tables 10-1 through 10-6 can be used to determine the vertical hanger rod diameter, maximum unbraced rod length, hanger rod connections to the structure, and cable or solid-brace member size for bracing installed up to a 60° angle from horizontal. Separate tables are included for brace connections to the equipment above the center of gravity and below the center of gravity.

Table 10-1 Threaded Rod Allowable Tension Loads

Threaded Rod Diameter in. (mm)	Allowable Working Load lb (kN)	Allowable Combined Working and Seismic Load lb (kN)
3/8 (10)	610 (2.7)	810 (3.6)
1/2 (13)	1130 (5.0)	1500 (6.6)
5/8 (16)	1810 (8.0)	2410 (10.7)
3/4 (19)	2710 (12.0)	3610 (16.0)
7/8 (22)	3770 (16.7)	5030 (22.3)
1 (25)	4960 (22.1)	6610 (29.4)
1 1/4 (32)	8000 (35.5)	10,660 (47.4)

Table 10-2 Equipment Braced Above Its Center of Gravity

Maximum Equipment Weight, lb (kg)				Seismic Horizontal Force, *Fp* lb (kN)	Hanger Rod Diameter, in. (mm)	Connection of Hanger Rod to Structure		Maximum Unbraced Rod Length in. (mm)	Brace Member Size		Connection to Structure
0.25*g*	0.5*g*	1.0*g*	2.0*g*			Solid	Cable		Solid	Cable	
400 (181)	200 (90)	100 (45)	50 (22)	100 (0.4)	3/8 (10)	A	A	30 (762)	A	A	B
800 (363)	400 (181)	200 (90)	100 (45)	200 (0.9)	1/2 (13)	B	B	39 (990)	A	B	C
1600 (726)	800 (363)	400 (181)	200 (90)	400 (1.8)	1/2 (13)	C	B	27 (685)	B	B	D
2400 (1088)	1200 (544)	600 (272)	300 (136)	600 (2.7)	5/8 (16)	D	C	36 (914)	C	C	E
4000 (1814)	2000 (907)	1000 (453)	500 (226)	1000 (4.4)	3/4 (19)	F	D	41 (1041)	C	C	F
8000 (3628)	4000 (1814)	2000 (907)	1000 (453)	2000 (8.9)	7/8 (22)	G	F	40 (1016)	D	D	H

Table 10-3 Equipment Braced Below Its Center of Gravity

Maximum Equipment Weight, lb (kg)				Seismic Horizontal Force, *Fp* lb (kN)	Hanger Rod Diameter in. (mm)	Connection of Hanger Rod to Structure		Maximum Unbraced Rod Length, in. (mm)	Brace Member Size		Connection to Structure
0.25*g*	0.5*g*	1.0*g*	2.0*g*			Solid	Cable		Solid	Cable	
400 (181)	200 (90)	100 (45)	50 (22)	100 (0.4)	3/8 (10)	B	A	18 (457)	A	A	B
800 (363)	400 (181)	200 (90)	100 (45)	200 (0.9)	1/2 (13)	C	B	24 (610)	A	B	C
1600 (726)	800 (363)	400 (181)	200 (90)	400 (1.8)	1/2 (13)	E	C	27 (686)	B	B	D
2400 (1088)	1200 (544)	600 (272)	300 (136)	600 (2.7)	5/8 (16)	F	E	22 (559)	C	C	E
4000 (1814)	2000 (907)	1000 (453)	500 (226)	1000 (4.4)	3/4 (19)	G	F	26 (660)	C	C	F
8000 (3628)	4000 (1814)	2000 (907)	1000 (453)	2000 (8.9)	7/8 (22)	H	G	25 (635)	D	D	H

Table 10-4 Solid-Brace Members

	Structural Steel Angle—Max. Length of 9 ft, 6 in. (2.9 m), in. (mm)	12 Gage Channel Strut—Max. Length of 9 ft, 6 in. (2.9 m), in. (mm)
A	2 × 2 × 1/8 (51 × 51 × 3)	1 5/8 × 1 5/8 (41 × 41 × 2.7)
B	2 × 2 × 1/4 (51 × 51 × 6)	1 5/8 × 1 5/8 (41 × 41 × 2.7)
C	3 × 3 × 1/4 (76 × 76 × 6)	1 5/8 × 3 1/4 (41 × 83 × 2.7)
D	4 × 4 × 1/4 (102 × 102 × 6)	1 5/8 × 3 1/4 (41 × 83 × 2.7)

Table 10-5 Cable Brace Members (Minimum Breaking-Strength Required)

	Prestretched Steel Cable—Safety Factor of 2, lb (kN)	Standard Steel Cable—Safety Factor of 5, lb (kN)
A	640 (2.9)	1600 (7.1)
B	1600 (7.1)	4000 (17.8)
C	4000 (17.8)	10,000 (44.5)
D	10,000 (44.5)	25,000 (111.2)

Table 10-6 Connections to Structure

	Expansion Anchors into a Concrete Slab Dia. × Embed., in. (mm)	Expansion Anchors into a Concrete Deck Dia. × Embed., in. (mm)	Steel Bolts into a Structural Steel Diameter, in. (mm)	Lag Bolts into a Wood Structure Dia. × Embed., in. (mm)
A	3/8 × 2 1/2 (10 × 64)	3/8 × 3 (10 × 76)	3/8 (10)	3/8 × 3 (10 × 76)
B	1/2 × 3 (13 × 76)	1/2 × 3 (13 × 76)	1/2 (13)	1/2 × 4 (13 × 102)
C	5/8 × 3 1/2 (16 × 89)	3/4 × 5 1/4 (19 × 83)	1/2 (13)	two 1/2 × 4 (two 13 × 102)
D	two 1/2 × 3 (two 13 × 76)	two 1/2 × 3 (two 13 × 76)	5/8 (16)	two 5/8 × 5 (two 16 × 127)
E	two 5/8 × 3 1/2 (two 16 × 89)	two 5/8 × 5 (two 16 × 127)	5/8 (16)	two 5/8 × 5 (two 16 × 127)
F	four 5/8 × 3 1/2 (four 16 × 89)	four 5/8 × 5 (four 16 × 127)	3/4 (19)	four 5/8 × 5 (four 16 × 127)
G	four 3/4 × 4 1/2 (four 19 × 114)	—	7/8 (22)	four 5/8 × 5 (four 16 × 127)
H	—	—	1 (25)	—

The charts assume the following:

$Fpv = 0.2\, S_{DS}\, W_p$

N $\geq$ 4

$H \leq D/2$

$A \leq 3/5\, L$

$B \leq 3/5\, D$

$\theta \leq 60°$

EXCEPTIONS FROM SWAY BRACING

Some equipment may not require sway bracing. The applicable code for the project may list some minimum requirements as noted in Chapter 2. Great care should be taken when bracing exceptions are allowed. Following is a list of installations that the engineer of record should consider before exempting equipment from bracing unless specifically addressed in local codes:

1. In most cases, suspended equipment is connected to piping, ductwork, or electrical conduit, which may or may not be separated by a flexible connector. The installation of the connected system and or flexible connector may not allow the equipment to move freely. For example, if a fan is flexed from ductwork that penetrates a wall only three feet away, the wall will hold the ductwork in place while the fan breaks away from the flexible connector.
2. Equipment may be vital to the operation of the building or part of a life safety or hazardous system.

3. The weight of the equipment in relation to the attached duct, piping, or electrical conduit may be too great for the system connections, resulting in failure of the connections or supports. Equipment separated by flexible connectors is also susceptible to this situation, unless the flexible connection is designed for the full motion of the system and exerts a relatively small force on the system connections.
4. The equipment might swing and impact another system and its support, as shown in Figure 10-8.

EXAMPLE

The following example illustrates how to use the tables outlined in this chapter.

Structure type: concrete slab

Brace type and orientation: cable bracing from above equipment center of gravity

Structural connection: expansion anchors

Seismic input" 0.5*g*

Equipment weight: 750 lb (340 kg)

From Table 10-2, at 0.5*g* and maximum equipment weight of 750 lb (340 kg), the brace requirements are as follows:

Support rod diameter = 1/2 in. (13 mm).

Support rod connection = *C*, one 1/2 in. (13 mm) diameter expansion anchor.

Maximum unbraced rod length = 27 in. (686 mm).

Cable brace member = *B*, prestretched steel cable with a breaking strength of 1600 lb (7.1 kN).

Brace member connection to structure = *D*, two 1/2 in. (13 mm) diameter expansion anchors.

BIBLIOGRAPHY

ASCE. 2005. *ANSI/ASCE Standard 7-05, Minimum design loads for buildings and other structures,* Chapter 6. New York: American Society of Civil Engineers.

MSS. 1983. *ANSI/MSS SP-58-Pipe Hangers and Supports*. Falls Church, VA: Manufacturers Standardization Society of the Valve and Fittings Industry.

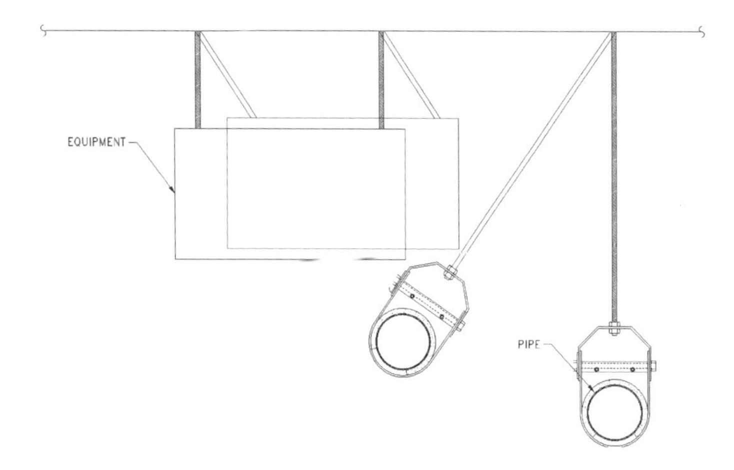

Figure 10-8 Systems impact due to sway.

11 Floor-Mounted Equipment

This chapter focuses on equipment with supports and seismic restraints attached to the floor or roof structure below its center of gravity. Discussion includes rigidly mounted and vibration-isolated systems. Included are seismic design calculations for numerous systems and specific examples. Also included is information on flexible connectors for piping connections. Piping connections to equipment are vulnerable in earthquakes and, although flexible connectors are required by code as noted in Chapter 2, there is very little information in the code regarding the proper design requirements.

RIGIDLY MOUNTED EQUIPMENT

Equipment should be connected directly to the structure using a minimum of four fasteners. Most equipment is provided with mounting holes located near the corners. Fasteners should be sized to fit within the existing mounting holes with no more than 1/8 in. (3.2 mm) total clearance (1/16 in. [1.6 mm] all around) to reduce sliding and eliminate impact loads, as shown in Figure 11-1. Where it is not practical to provide large enough fasteners and the clearance exceeds 1/8 in. (3.2 mm), the remaining gap should be filled with a neoprene grommet, epoxy compound, or welded steel washer, as shown in Figure 11-2.

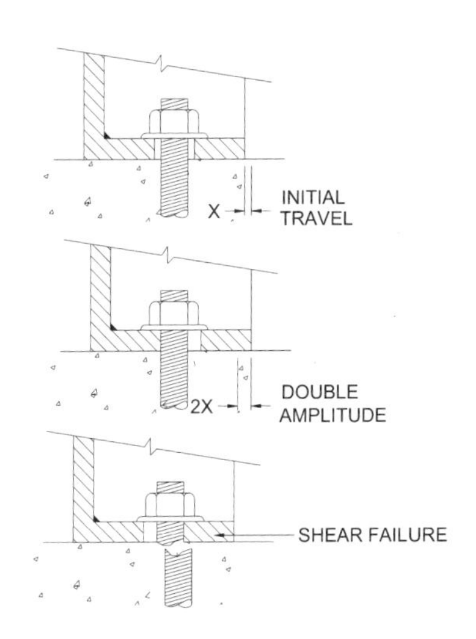

Figure 11-1 Impact shear failure.

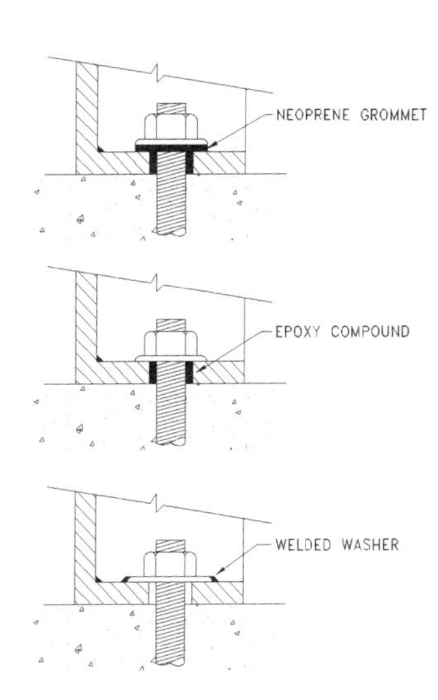

Figure 11-2 Oversized hole.

Equipment provided with inadequate mounting holes can be restrained with anchor brackets attached to the equipment at or near the corners, as shown in Figure 11-3. Where there is no tensile loading because of small seismic forces or no overturning moments as shown in the calculations later in this chapter, single-direction seismic restraints need not be connected to the equipment. Two restraints should be located on each side of the equipment to prevent horizontal motion in any direction, as shown in Figure 11-4. Each restraint should have two anchor bolts to prevent rotation of restraints.

Rotating equipment should be secured to concrete with anchors designed to resist loosening caused by vibration. Chapters 2 and 5 discuss code requirements and limitations on fasteners.

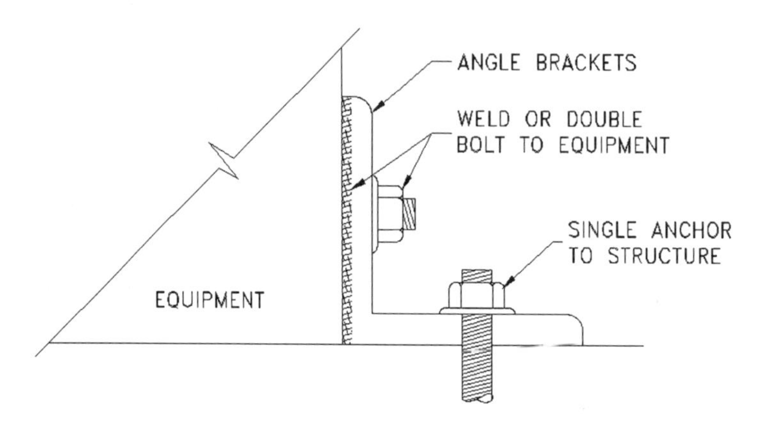

Figure 11-3 Anchor bracket connected to equipment.

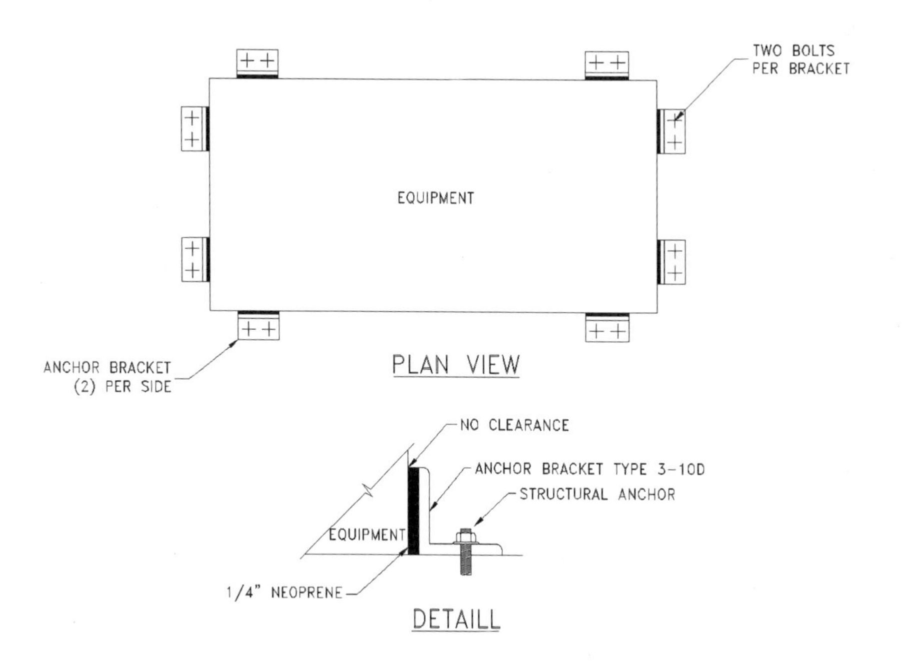

Figure 11-4 Horizontal restraint brackets.

VIBRATION-ISOLATED EQUIPMENT

Vibration isolators and seismic restraints are manufactured in a variety of designs. Before discussing techniques of effective seismic design for vibration-isolated systems, refer to the following definitions and accompanying figures provided in Chapter 3.

Vibration Isolators

Type 3-1:
Elastomeric pads with Type 3-4 bushings. (Figures 3-1 and 3-4)

Type 3-2:
Captive neoprene mounts, where an upward force over 1*g* is restrained by a lower rebound component. (Figure 3-2)

Type 3-5A:
Vibration isolator within a seismic housing where the isolator rests directly on the supporting structure, not on the anchor baseplate. (Figure 3-5A)

Type 3-5B, 3-5C, 3-5D:
Vibration isolator within a seismic housing where the isolator and corresponding load rests on the anchor baseplate. (Figures 3-5B, 3-5C, and 3-5D)

Seismic Snubbers

Type 3-10A:
All-directional, double-acting seismic design. (Figure 3-10A)

Type 3-11:
All-directional, double-acting seismic design often used with dynamic analysis. (Figure 3-11)

Vibration isolators are used to prevent the transmission of vibration from rotating equipment to the structure. They are also used on electrical equipment such as transformers and switchgear. Because isolated equipment must be free to oscillate on the vibration isolators, seismic restraints should be installed with an air gap. The air gap allows the free vibratory oscillation of the equipment until seismic motions of the building cause the restraints to come in temporary contact with the equipment and limit overall differential displacement.

Vibration isolation and seismic restraint are not to be confused with building seismic isolation. Building seismic isolation is used to protect buildings from earthquakes by allowing large displacements. In contrast, seismic restraint of vibration-isolated equipment is designed to contain the equipment and limit displacements.

Most equipment is designed for installation on a floor or housekeeping pad. The equipment may require a supplementary steel frame or concrete filled inertia base to support the equipment on vibration isolators, as shown in Figures 11-5 and 11-6. When seismic restraints are used, the equipment attachment or support frame should be designed for horizontal loads and bending moments created by the seismic restraint. Restrained isolators can be designed with their limit stops built-in.

Where supplemental steel is required to support equipment on vibration isolators, it is important to use complete steel frames, and not independent rails. Independent rails cannot resist the moments created by horizontal seismic loads as shown in photographs 18-36 to 18-39 in Chapter 18, overloading the connection point of the rails to the equipment. Steel frames can resist the moments produced by spring supports and horizontal seismic loads, allowing simple connections to equipment and seismic restraints to function without frame or equipment connection failure.

The main advantage of all-directional, double-acting snubbers is that all snubbers share in the restraint of motion in any direction. However, another important advantage is that overall peak-to-peak displacement is factory preset, as indicated in Figure 11-7, and not

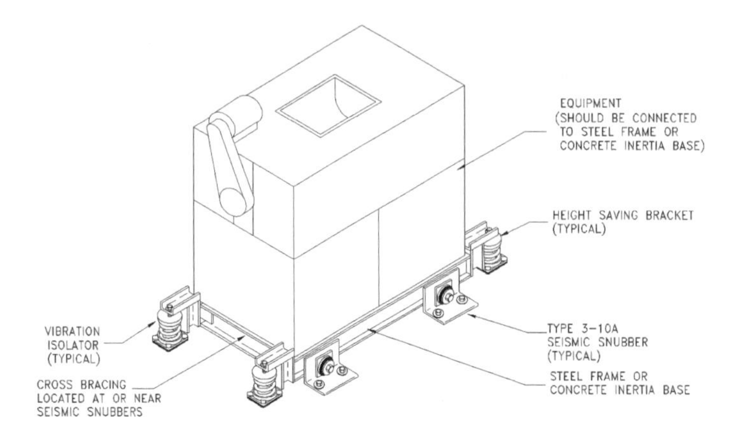

Figure 11-5 Supplemental base—Free-standing springs and independent snubbers.

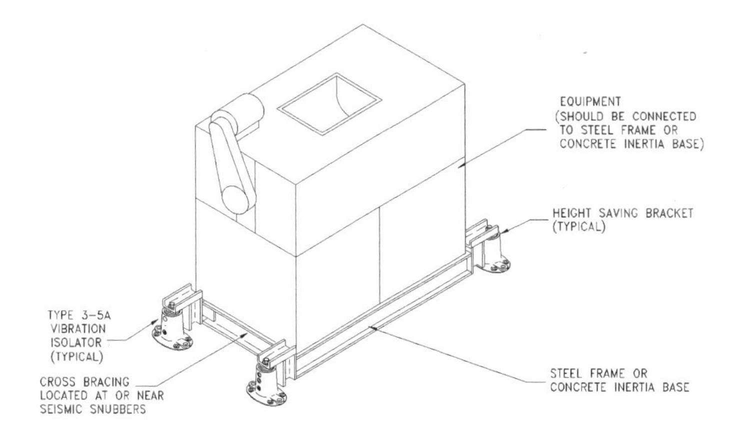

Figure 11-6 Supplemental base—Vibration isolators within seismic restraint housing.

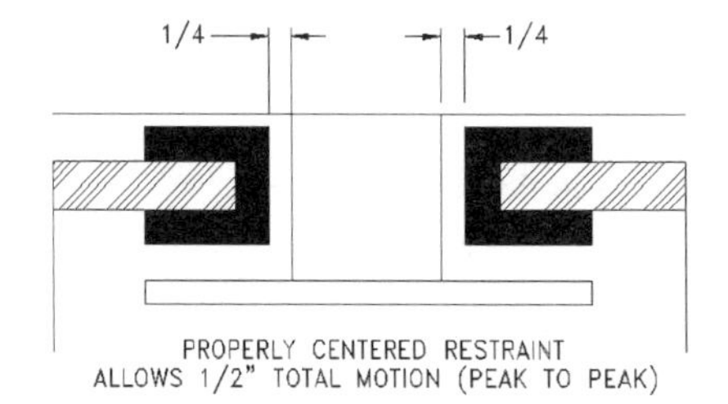

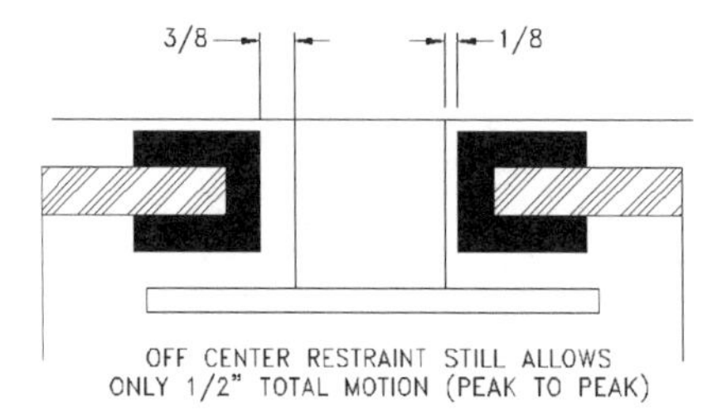

Figure 11-7 Factory preset clearance.

field-set. Single-direction snubbers require field-setting of the air gap, which is difficult to control, as shown in Figure 11-8.

Neoprene bushings prevent steel-to-steel contact that would greatly amplify shock loads. Another advantage is not so obvious yet critically important. The neoprene bushing, as shown in Figure 11-9, is a compressive material that allows for sharing of the load as snubbers are loaded. Hard stops impact one point at a time and do not distribute the load.

Equipment supported on elastomeric pad (Type 3-1) should be restrained by anchor bolts surrounded by neoprene bushings (Type 3-4) to prevent short-circuiting of the vibration isolator. In most cases, equipment mounting holes are not large enough to allow this, so a supplemental support frame designed to accept the seismic loads should be provided between the equipment and the pads. The frame can be built to accommodate anchorage through the pad or connection to separate seismic snubbers (Type 3-10 or 3-11).

Systems that use individual, freestanding isolators can be seismically restrained using separate seismic snubbers (Type 3-10 or 3-11). Separate seismic snubbers require a supplemental support base for connection. Equipment that requires a supplemental support base for vibration isolators, such as pumps and cooling towers, is ideal for this application because the expense for the additional base has already been covered. In addition, because the design

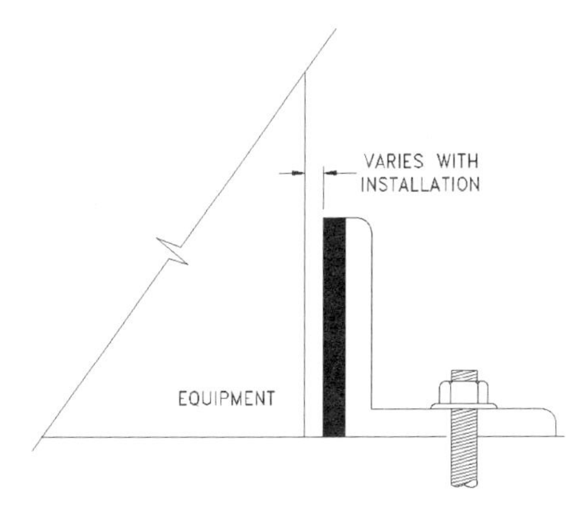

Figure 11-8 Field set clearance.

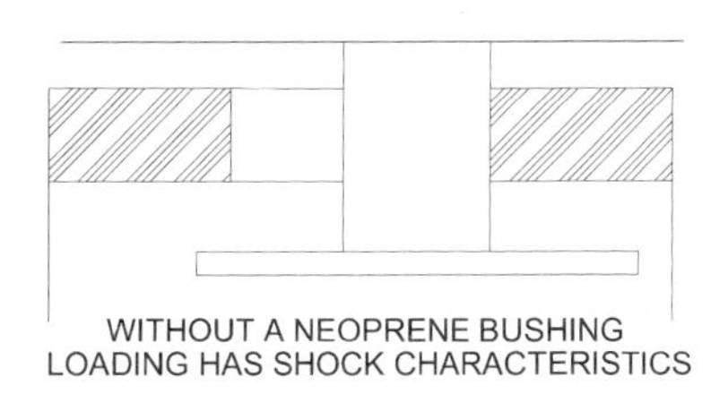

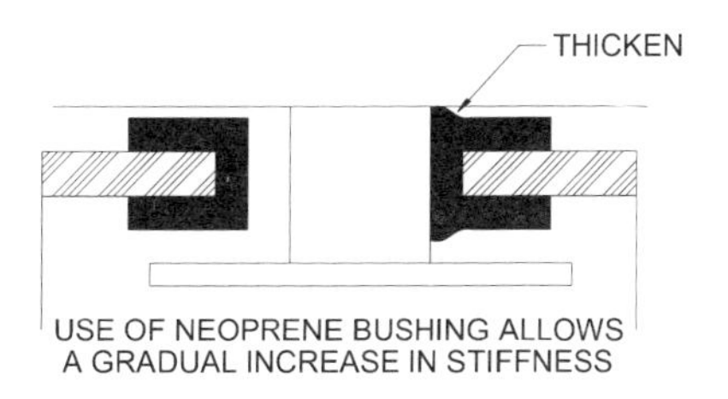

Figure 11-9 Advantage of neoprene bushing.

of separate seismic snubbers is not tied to the individual isolators, the designer can optimize the snubber selection and anchorage.

Restrained isolators (Type 3-2 or 3-5) support and seismically restrain equipment at the same point. Equipment with rigid frames that can be point loaded, such as centrifugal chillers and fans, is ideal for this application because this isolator does not require a supplemental support base. The connection point of the isolator housing to the equipment is critical and should be specifically addressed by the designer, as shown in Figure 11-10. Once again, the equipment manufacturer should verify the capability of the equipment frame to accept the dead loads as well as the seismic vertical and horizontal forces and overturning moments.

As emphasized in Chapter 3, the load capacity of seismic restraints and restrained isolators should be certified by tests or calculations. Structural engineers at the State of California's Office of Statewide Health and Planning Development (OSHPD) have reviewed test reports and calculations for many restraints from numerous isolator manufacturers and have provided a preapproval rating for use in hospitals and public schools in the state. This restraint rating can be used for any project. The initial OSHPD preapproval was the SMACNA *Seismic Restraint Manual Guidelines for Mechanical Systems* in 1978, which has been specified by thousands of engineers throughout the country due to the unbiased review and approval by OSHPD structural engineers. Engineers should require certification of seismic restraints from an independent review board, code enforcement agency, regulatory agency or testing laboratory, such as OSHPD, Underwriters Laboratory (UL), the International Code Council (ICC), Factory Mutual (FM), or other recognized agencies, to ensure public safety in buildings.

Equipment with Internal Vibration Isolation

Equipment provided with internal vibration isolation have shipping bolts that are locked to hold internally isolated components in place and prevent damage during shipment. These bolts must be loosened or removed during installation for the isolation to function. Internally isolated component support frames, restraints, and attachments must be designed for seismic loads defined in the building code for vibration-isolated equipment, and the equipment casing or framework used for attachment of the seismic snubbers must be designed to transfer the snubber loads to the unit anchors and structure.

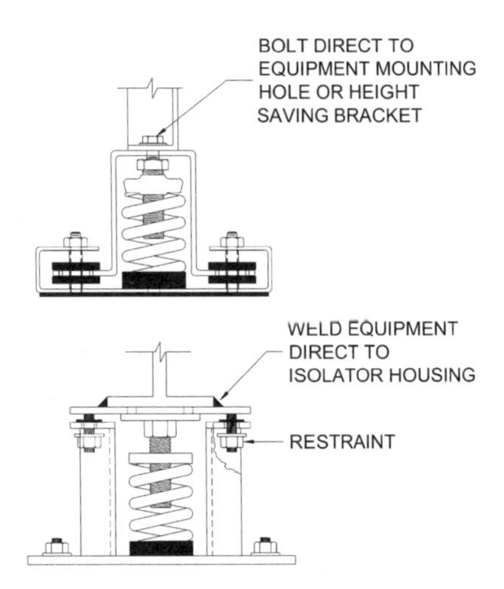

Figure 11-10 Restrained isolator connections to equipment.

CALCULATIONS

Seismic restraint anchorage for floor-mounted equipment is determined by applying the seismic loads at the equipment center of gravity so the horizontal and vertical loads at the restraints are realistic.

In a worst-case analysis for hard-mounted equipment, the horizontal seismic force, F_{ph}, is applied at the equipment center of gravity across the shortest base dimension.

In a worst-case analysis for vibration-isolated equipment, F_{ph} is applied at the equipment center of gravity but directed at a specific angle q that results in the largest tension and compression load on a restrained isolator or snubber. This is called a "worst-angle" overturning analysis. The equipment center of gravity, equipment geometry and location of the restraints are the variables that determine the worst angle q.

The restraint loads determined from the worst-angle overturning analysis are the maximum uplift P_t, maximum compression P_c, and maximum shear P_s. These loads should be used to select the appropriate seismic restraints.

There are three important variations in the calculations that merit further definition. For rigidly mounted equipment, or equipment mounted on Type 3-2 vibration isolators, the calculation includes the weight of the equipment as a restoring force, because the equipment will not put any tension on the anchors until the entire equipment weight at any corner is overcome. For equipment mounted on any other arrangement or restrained by separate seismic snubbers, the calculation does not use the weight of the equipment as a restoring force. The resistance to uplift is dependent on isolator stiffness that varies too much to warrant its use and would be negligible in most cases. Also, because snubber air gaps are field adjustable, variations in actual air gap size are possible. Reduction of restraint forces due to equipment motion across the air gap is not valid.

Because the moment capacity of equipment attachment points is questionable, the *H* dimension of the isolator housing used in the anchorage calculations for Type 3-5 vibration isolators should be the full height of the housing, regardless of limit stop height. This will ensure that the anchorage is designed for the worst-case loading and any moment taken by the equipment attachment will only increase the anchorage safety factor.

The following is a step-by-step procedure for selecting the appropriate floor-mounted equipment. Refer to vibration isolator and seismic snubber definitions listed earlier in this chapter to use this procedure.

Although we have provided an analysis for many different types of vibration isolation and seismic snubber systems, there are some systems that are not specifically addressed, such as seismic snubbers that provide all-directional restraint as a group (Type 3-10B snubbers) and two-directional restraint as a group (Type 3-10C and 3-10D snubbers), etc. In these cases, the manufacturer or designer of the system should provide specific calculations that follow the basic worst-angle principles.

1. Select one of the following overturning calculation formats.
   a. Rigidly mounted equipment.
   b. Equipment restrained with Type 3-2 vibration isolators.
   c. Equipment restrained with Type 3-1 or 3-5 vibration isolators or equipment retrained with Type 3-10 or 3-11 seismic snubbers.
2. Select one of the following anchor bolt connection calculation formats to determine the anchor bolt tension, T_{bolt}, and shear, V_{bolt}.
   a. Rigidly mounted equipment.
   b. Equipment restrained with Type 3-1 vibration isolators.
   c. Equipment restrained with Type 3-2 and 3-5A vibration isolators.
   d. Equipment restrained with Type 3-5B, 3-5C, or 3-5D vibration isolators.
   e. Equipment supported on vibration isolators and restrained with Type 3-10A seismic snubbers.
   f. Equipment supported on vibration isolators and restrained with Type 3-11 seismic snubbers.
3. Select one of the following structural connections to determine if the anchorage is adequate.

a. Concrete expansion anchors.
b. Steel bolts (minimum A307, standard quality bolts).
c. Lag screws.

I. Overturning Calculations

A. Calculations for rigidly mounted equipment (See Figure 11-11)

$$OTM = F_{ph} \times h \tag{11-1}$$

$$RM = (W - F_{pv}) \times d_{min}/2 \tag{11-2}$$

$$T = (OTM - RM)/d_{min} \text{ and } V = F_{ph} \tag{11-3}$$

B. Calculations for equipment restrained with Type 3-2 vibration isolators (See Figure 11-12)

Maximum tension P_t at location 2:

$$P_t = \frac{W - F_{pv}}{N} - \frac{F_{ph}\cos\theta h\left(\frac{b_2}{2}\right)}{I_{yy}} - \frac{F_{ph}\sin\theta h\left(\frac{b_1}{2}\right)}{I_{xx}} \tag{11-4}$$

Maximum compression P_c at location 4:

$$P_c = \frac{W + F_{pv}}{N} + \frac{F_{ph}\cos\theta h\left(\frac{b_2}{2}\right)}{I_{yy}} + \frac{F_{ph}\sin\theta h\left(\frac{b_1}{2}\right)}{I_{xx}} \tag{11-5}$$

Maximum shear P_s at all locations:

$$P_s = \frac{F_{ph}}{N} \tag{11-6}$$

where

$$I_{xx} = \frac{N(N+2)b_1^2}{12(N-2)}, \tag{11-7}$$

$$I_{YY} = \frac{Nb_2^2}{4}, \tag{11-8}$$

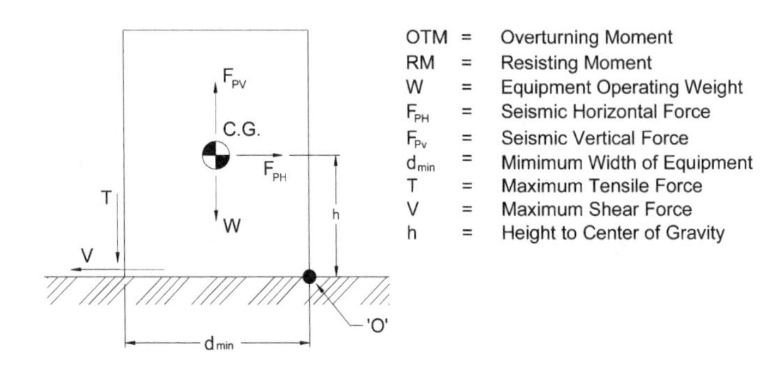

Figure 11-11 Overturning calculations—rigidly mounted equipment.

$$\theta = \tan^{-1}\left(\frac{I_{yy}b_1}{I_{xx}b_2}\right), \tag{11-9}$$

The equations can be simplified as follows if $N = 4$.

$$P_t = \frac{W - F_{pv}}{4} - \frac{F_{ph}\cos\theta h}{2b_2} - \frac{F_{ph}\sin\theta h}{2b_1} \tag{11-10}$$

$$P_c = \frac{W + F_{pv}}{4} + \frac{F_{ph}\cos\theta h}{2b_2} + \frac{F_{ph}\sin\theta h}{2b_1} \tag{11-11}$$

$$P_s = \frac{F_{ph}}{4} \tag{11-12}$$

where

$$\theta = \tan^{-1}\left(\frac{b_2}{b_1}\right). \tag{11-13}$$

C. Calculations for equipment restrained with Type 3-1 or 3-5 vibration isolators or equipment restrained with Type 3-10 or 3-11 seismic snubbers (See Figure 11-13)

Maximum tension P_t at location 2:

$$P_t = \frac{-F_{pv}}{N} - \frac{F_{ph}\cos\theta h\left(\frac{b_2}{2}\right)}{I_{yy}} - \frac{F_{ph}\sin\theta h\left(\frac{b_1}{2}\right)}{I_{xx}} \tag{11-14}$$

Maximum compression P_c at location 4:

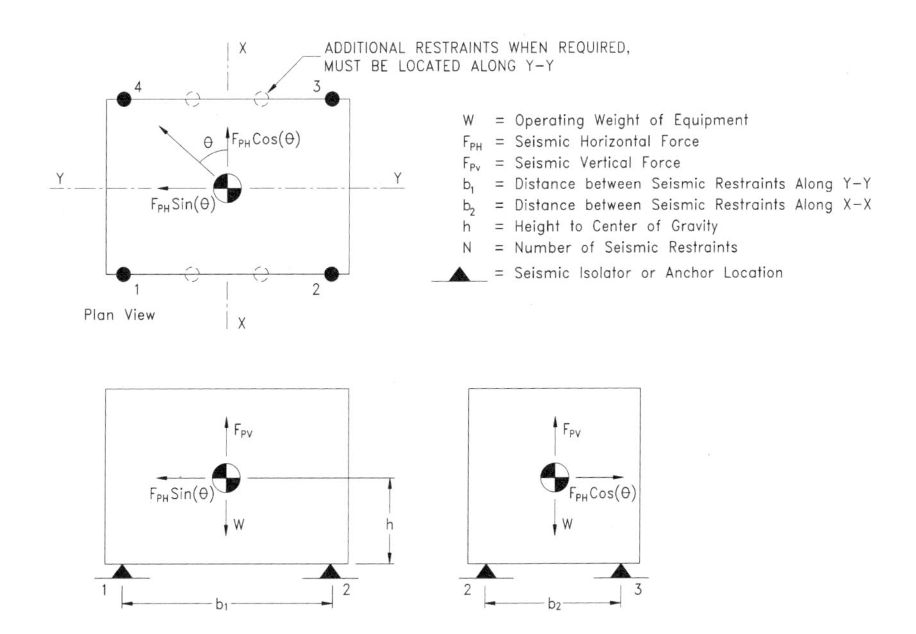

Figure 11-12 Overturning calculation—Type 3-2 vibration isolators.

$$P_c = \frac{F_{pv}}{N} + \frac{F_{ph}\cos\theta h\left(\frac{b_2}{2}\right)}{I_{yy}} + \frac{F_{ph}\sin\theta h\left(\frac{b_1}{2}\right)}{I_{xx}} \tag{11-15}$$

Maximum shear P_s at all locations:

$$P_s = \frac{F_{ph}}{N} \tag{11-16}$$

where

$$I_{xx} = \frac{N(N+2)b_1^2}{12(N-2)}, \tag{11-17}$$

$$I_{yy} = \frac{Nb_2^2}{4}, \tag{11-18}$$

$$\theta = \tan^{-1}\left(\frac{I_{yy}b_1}{I_{xx}b_2}\right). \tag{11-19}$$

The equations can be simplified as follows if $N = 4$.

$$P_t = \frac{-F_{pv}}{4} - \frac{F_{ph}\cos\theta h}{2b_2} - \frac{F_{ph}\sin\theta h}{2b_1} \tag{11-20}$$

$$P_c = \frac{F_{pv}}{4} + \frac{F_{ph}\cos\theta h}{2b_2} + \frac{F_{ph}\sin\theta h}{2b_1} \tag{11-21}$$

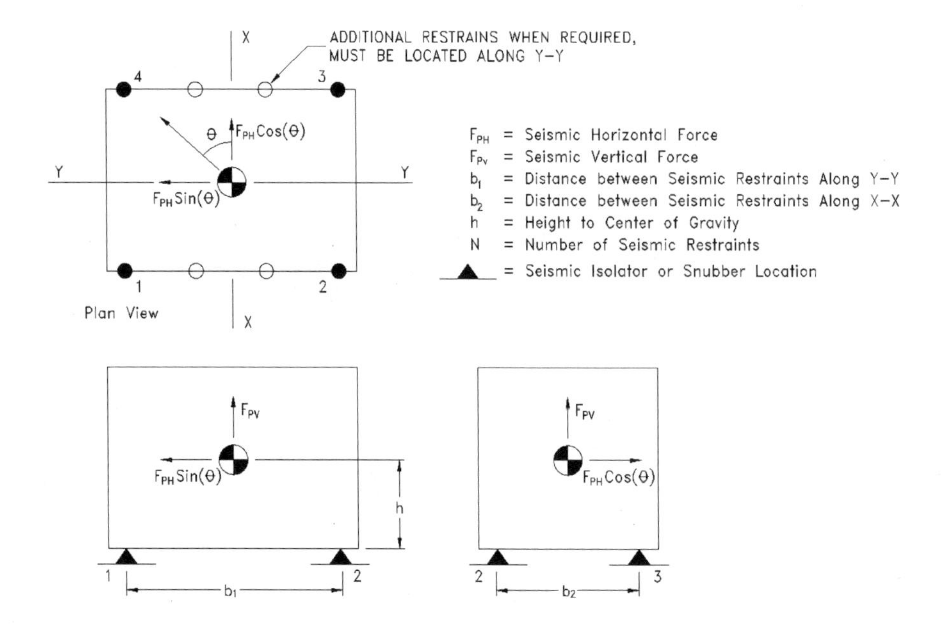

Figure 11-13 Overturning calculation—Type 3-1or 3-5 vibration isolators or Type 3-10 or 3-11 seismic snubbers.

$$P_s = \frac{F_{ph}}{4} \quad (11\text{-}22)$$

where

$$\theta = \tan^{-1}\left(\frac{b_2}{b_1}\right) \quad (11\text{-}23)$$

II. Anchor Bolt Calculations

A. Rigidly mounted equipment (See Figure 11-14)

Determine the anchor bolt tension T_{bolt} and shear V_{bolt}.

Analysis of anchor bolt reactions:

$$T_{bolt} = \frac{T_b}{\left(\frac{n}{2}\right)} \quad (11\text{-}24)$$

and

$$V_{bolt} = \frac{V_b}{n}. \quad (11\text{-}25)$$

B. Equipment restrained with Type 3-1 vibration isolators (See Figure 11-15)

Determine the anchor bolt tension T_{bolt} and shear V_{bolt}.

Analysis of anchor bolt reactions:

$$T_{bolt} = \frac{P_t}{n} \quad (11\text{-}26)$$

and

$$V_{bolt} = \frac{P_s}{n}. \quad (11\text{-}27)$$

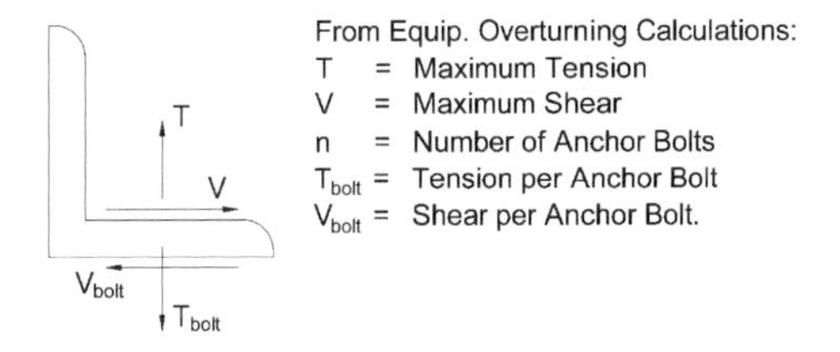

Figure 11-14 Anchor bolt calculations—rigidly mounted equipment.

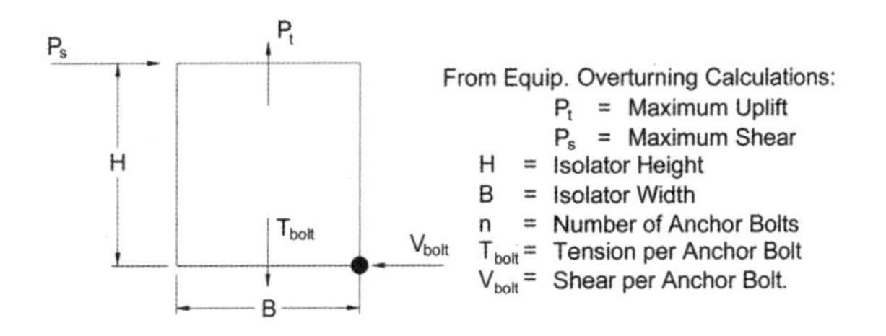

Figure 11-15 Anchor bolt calculation—Type 3-1 vibration isolators.

C. Equipment restrained with Type 3-2 or 3-5A vibration isolators (See Figure 11-16)

Determine the anchor bolt tension T_{bolt} and shear V_{bolt} for a two-bolt anchor arrangement.

Analysis of anchor bolt reactions with seismic loads applied on the narrow end of the isolator:

$$T_{bolt} = \frac{P_S H}{n\left(\frac{B}{2}\right)} + \frac{P_S}{n} \tag{11-28}$$

and

$$V_{bolt} = \frac{P_S}{n}. \tag{11-29}$$

D. Equipment restrained with Type 3-5B, 3-5C, or 3-5D vibration isolators (See Figure 11-17)

Determine the anchor bolt tension T_{bolt} and shear V_{bolt} for a four-bolt anchor arrangement.

Analysis of anchor bolt reactions with seismic loads applied on the narrow end of the isolator:

$$T_{bolt} = \frac{P_S H}{\left(\frac{n}{2}\right)\left[a + b + \frac{a^2}{(a+b)}\right]} + \frac{\left(P_t - \frac{W}{N}\right)}{n} \tag{11-30}$$

and

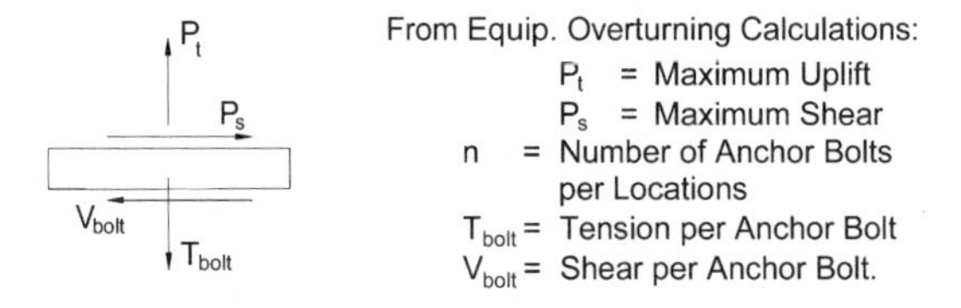

Figure 11-16 Anchor bolt calculation—Type 3-2 or 3-5A vibration isolators.

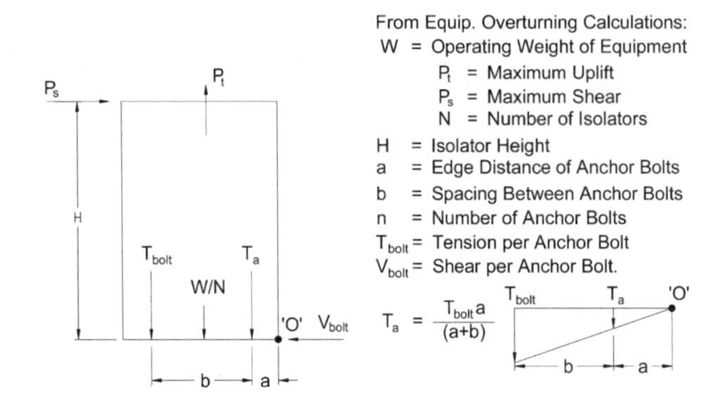

Figure 11-17 Anchor bolt calculation—Type 3-5B, 3-5C, or 3-5D vibration isolators.

$$V_{bolt} = \frac{P_S}{n} \quad (11\text{-}31)$$

E. Equipment supported on vibration isolators and restrained with Type 3-10A seismic snubbers (See Figure 11-18)

Determine the anchor bolt tension T_{bolt} and shear V_{bolt} for a two-bolt anchor arrangement.

Analysis of anchor bolt reactions with seismic loads applied on the narrow end of the snubber:

$$T_{bolt} = \frac{P_S H + P_t B}{n\left(\frac{B}{2}\right)} \quad (11\text{-}32)$$

and

$$V_{bolt} = \frac{P_S}{n}. \quad (11\text{-}33)$$

F. Equipment supported on vibration isolators and restrained with Type 3-11 seismic snubbers (See Figure 11-19)

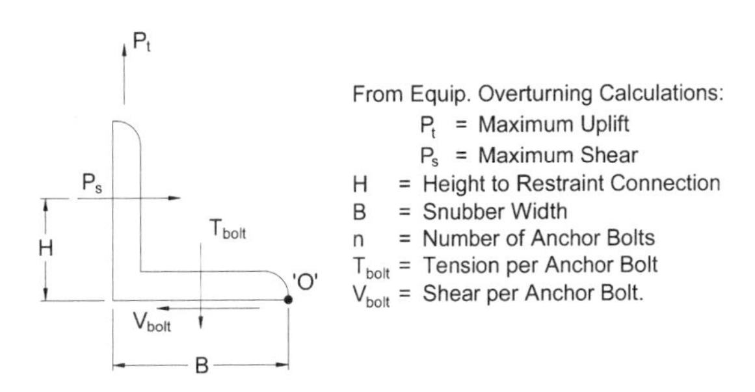

Figure 11-18 Anchor bolt calculation—Type 3-10A seismic snubber.

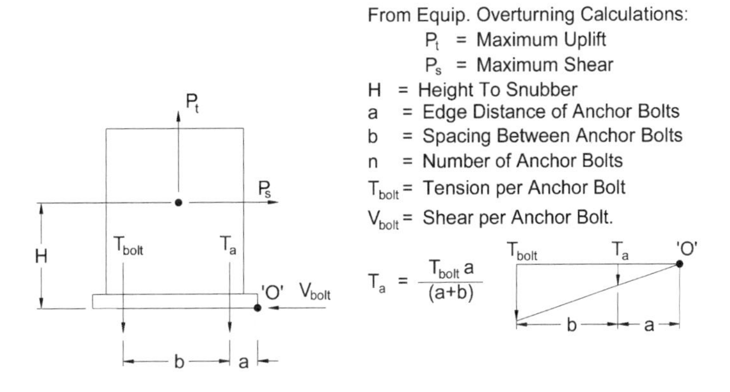

Figure 11-19 Anchor bolt calculation—Type 3-11 seismic snubber.

Determine the anchor bolt tension T_{bolt} and shear V_{bolt} for a four-bolt anchor arrangement.

Analysis of anchor bolt reactions with seismic loads applied on the narrow end of the snubber:

$$T_{bolt} = \frac{P_s H}{\left(\frac{n}{2}\right)\left[a + b + \frac{a^2}{(a+b)}\right]} + \frac{P_t}{n} \qquad (11\text{-}34)$$

and

$$V_{bolt} = \frac{P_s}{n} \qquad (11\text{-}35)$$

III. Structural Connections

A. Concrete expansion anchors

Combine anchor bolt reactions with allowable values using the proper interaction formula. Refer to the anchor bolt manufacturer's test report for allowable shear and tension values and proper interaction formula.

Interaction formula 1

$$\left(\frac{T_{bolt}}{T_{allow}}\right) + \left(\frac{V_{bolt}}{V_{allow}}\right) \le 1.0 \qquad (11\text{-}36)$$

Interaction formula 2

$$\left(\frac{T_{bolt}}{T_{allow}}\right)^{5/3} + \left(\frac{V_{bolt}}{V_{allow}}\right)^{5/3} \le 1.0 \qquad (11\text{-}37)$$

B. Steel bolts (minimum A307, standard quality bolts)

Combine tensile and shear stresses on the bolts from the calculated anchor bolt reactions using the interaction formula and bolt areas discussed in Chapter 5.

Tensile stress $f_t = T_{bolt}/A_t$ and (11-38)

Shear stress $f_v = V_{bolt}/A_k$ (11-39)

where A_t and A_k are the tensile and minimum root areas of the bolt.

Allowable sheer stress $F_v = 10{,}000 \text{ psi} \times 133^{*} = 13{,}333 \text{ psi}$

*1/3 allowable stress increase for seismic loads.

Allowable tensile stress F_t: If $26{,}000 - 1.8f_v \le 20{,}000$ psi

then, $F_t = (26{,}000 - 1.8f_v) \times 1.33$ psi

else, $F_t = 20{,}000 \times 1.33 = 26{,}667$ psi.

If $f_t \le F_t$ and $f_v \le F_v$, then anchor bolts are satisfactory.

C. Lag Screws

Combine anchor bolt reactions with allowable withdrawal and lateral lag screw values using the interaction formula discussed in Chapter 5.

$$\text{Maximum load on lag screw } P_\alpha = \sqrt{T_{bolt}^2 + V_{bolt}^2} \qquad (11\text{-}40)$$

$$\text{Allowable load on lag screw } P_{\alpha}allow = \frac{WZ}{W\cos^2\alpha + Z\sin^2\alpha} \tag{11-41}$$

where

W = lag screw allowable withdrawal load,

Z = lag screw allowable lateral load,

$\alpha = \tan^{-1}\left(\frac{T_{bolt}}{V_{bolt}}\right)$

If $P_{\alpha} \leq P_{\alpha}allow$, then lag screw is satisfactory.

IV. Eccentric Overturning Calculations

Some equipment may load the seismic restraint system eccentrically because of the plan offset distance between the center of gravity of the equipment and the geometric center of the seismic restraints or anchors. If either offset distance exceeds 20% of the overall distance between restraints, an eccentric worst-angle overturning calculation should be performed in lieu of the calculations discussed earlier. The anchor bolt and structural connection analysis do not change.

A. Calculations for equipment restrained with Type 3-2 vibration isolators (See Figure 11-20)

Maximum tension P_t at location 2:

$$P_t = \frac{W - F_{pv}}{N} - \left(\frac{W - F_{pv}}{b_1 b_2}\right)\left[\left(\frac{b_1}{2}\right)e_x + \left(\frac{b_2}{2}\right)e_y - e_x e_y\right] - \frac{F_{ph}\cos\theta_{t,c}h\left(\frac{b_2}{2}\right)}{I_{yy}} - \frac{F_{ph}\sin\theta_{t,c}h\left(\frac{b_1}{2}\right)}{I_{xx}} \tag{11-42}$$

Maximum compression P_c at location 4:

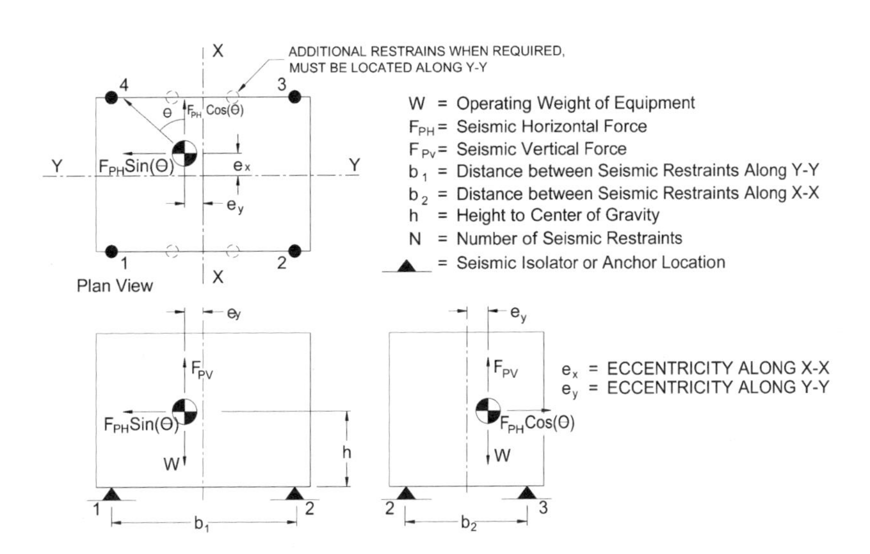

Figure 11-20 Eccentric overturning calculation—Type 3-2 vibration isolators.

$$P_c = \frac{W+F_{pv}}{N} + \left(\frac{W+F_{pv}}{b_1 b_2}\right)\left[\left(\frac{b_1}{2}\right)e_x + \left(\frac{b_2}{2}\right)e_y - e_x e_y\right] + \frac{F_{ph}\cos\theta_{t,c} h\left(\frac{b_2}{2}\right)}{I_{yy}} + \frac{F_{ph}\sin\theta_{t,c} h\left(\frac{b_1}{2}\right)}{I_{xx}} \tag{11-43}$$

Maximum shear P_s at location 2:

$$P_{sx} = \frac{F_{ph}\cos\theta_s}{N} + \frac{F_{ph}\cos\theta_s e_y b_2 + F_{ph}\sin\theta_s e_x b_1}{2(I_{xx}+I_{yy})} \tag{11-44}$$

$$P_{sy} = \frac{F_{ph}\sin\theta_s}{N} + \frac{F_{ph}\cos\theta_s e_y b_1 + F_{ph}\sin\theta_s e_x b_2}{2(I_{xx}+I_{yy})} \tag{11-45}$$

$$P_s = \sqrt{P_{sx}^2 + P_{sy}^2} \tag{11-46}$$

where

$$I_{xx} = \frac{N(N+2)b_1^2}{12(N-2)}, \tag{11-47}$$

$$I_{yy} = \frac{Nb_2^2}{4}, \tag{11-48}$$

$$\theta_{t,c} = \tan^{-1}\left(\frac{I_{yy}b_1}{I_{xx}b_2}\right), \tag{11-49}$$

$$\theta_s = \tan^{-1}\left(\frac{e_x}{e_y}\right). \tag{11-50}$$

B. Calculations for equipment restrained with Type 3-1 or 3-5 vibration isolators or equipment restrained by Type 3-10 or 3-11 seismic snubbers (See Figure 11-21)

Maximum tension P_t at location 2:

$$P_t = \frac{-F_{pv}}{N} - \left(\frac{-F_{pv}}{b_1 b_2}\right)\left[\left(\frac{b_1}{2}\right)e_x + \left(\frac{b_2}{2}\right)e_y - e_x e_y\right] - \frac{F_{ph}\cos\theta_{t,c} h\left(\frac{b_2}{2}\right)}{I_{yy}} - \frac{F_{ph}\sin\theta_{t,c} h\left(\frac{b_1}{2}\right)}{I_{xx}} \tag{11-51}$$

Maximum compression P_c at location 4:

$$P_{tc} = \frac{F_{pv}}{N} + \left(\frac{-F_{pv}}{b_1 b_2}\right)\left[\left(\frac{b_1}{2}\right)e_x + \left(\frac{b_2}{2}\right)e_y - e_x e_y\right] + \frac{F_{ph}\cos\theta_{t,c} h\left(\frac{b_2}{2}\right)}{I_{yy}} + \frac{F_{ph}\sin\theta_{t,c} h\left(\frac{b_1}{2}\right)}{I_{xx}} \tag{11-52}$$

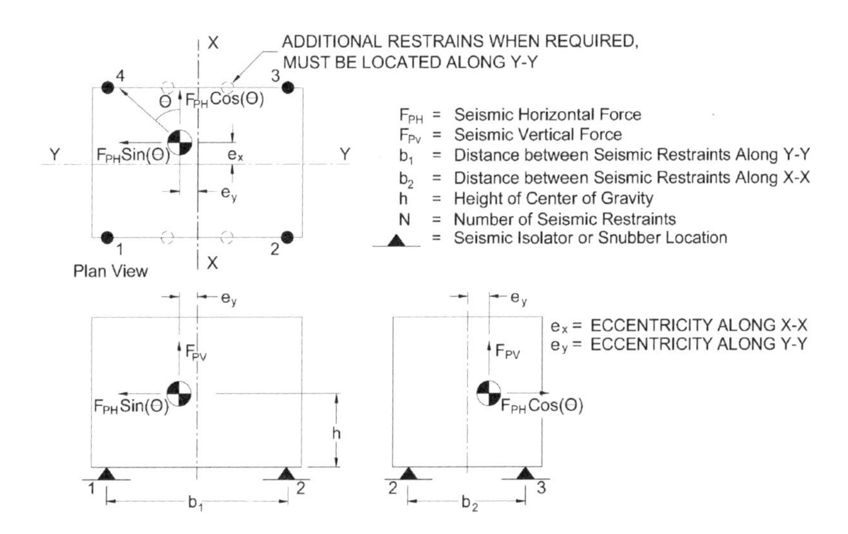

Figure 11-21 Eccentric overturning calculation—Type 3-1 or 3-5 vibration isolators.

Maximum shear P_s at location 2:

$$P_{sx} = \frac{F_{ph}\cos\theta_s}{N} + \frac{F_{ph}\cos\theta_s e_y b_2 + F_{ph}\sin\theta_s e_x b_1}{2(I_{xx} + I_{yy})} \tag{11-53}$$

$$P_{sy} = \frac{F_{ph}\sin\theta_s}{N} + \frac{F_{ph}\cos\theta_s e_y b_1 + F_{ph}\sin\theta_s e_x b_2}{2(I_{xx} + I_{yy})} \tag{11-54}$$

$$P_s = \sqrt{P_{sx}^2 + P_{sy}^2} \tag{11-55}$$

where

$$I_{xx} = \frac{N(N+2)b_1^2}{12(N-2)}, \tag{11-56}$$

$$I_{yy} = \frac{Nb_2^2}{4}, \tag{11-57}$$

$$\theta_{t,c} = \tan^{-1}\left(\frac{I_{yy}b_1}{I_{xx}b_2}\right), \tag{11-58}$$

and

$$\theta_s = \tan^{-1}\left(\frac{e_x}{e_y}\right). \tag{11-59}$$

EXAMPLE CALCULATIONS

The following are examples of how to perform seismic overturning and anchorage calculations for a chiller and an air-handling unit supported on vibration isolation. Each piece of equipment is designed for two different seismic acceleration inputs. The resulting tension, compression, and shear loads should be checked against the capacity of the seismic restraint or housing for proper sizing. The anchorage to the structure is the final design step.

Both examples assume the difference between the center of gravity of the equipment and the geometric center of the seismic restraints is negligible.

Example 1 (See Figure 11-22)

Equipment type: chiller
Isolator/restraint: vibration isolator Type 3-5C with four-bolt connection
Structure type: steel beam

Operating weight W = 12,000 lb (5443 kg)
Maximum distance between vibration isolators along y-y axis b_1 = 120 in. (3048 mm)
Maximum distance between vibration isolators along x-x axis b_2 = 48 in. (1219 mm)
Height to center of gravity h = 40 in. (1016 mm)
Number of vibration isolators N = 4

A. Seismic horizontal force $F_{ph} = 0.075 \times W = 900$ lb (4 kN)
Seismic vertical force $F_{pv} = 0$

From overturning calculation format I.C:
Maximum tension P_t at location 2:

$$P_t = \frac{0}{4} - \frac{900\cos(21.8°)40}{2(48)} - \frac{900\sin(21.8°)40}{2(120)} = 404 \text{ lb } (-1.8 \text{ kN})$$

Maximum compression P_c at location 4:

$$P_c = \frac{0}{4} - \frac{900\cos(21.8°)40}{2(48)} + \frac{900\sin(21.8°)40}{2(120)} = 404 \text{ lb } (1.8 \text{ kN})$$

Maximum shear P_s at all locations:

$$P_s = \frac{900}{4} = 225 \text{ lb } (1 \text{ kN})$$

where

$$\theta = \tan^{-1}\left(\frac{120}{48}\right) = 21.8°.$$

Check these loads against the OSHPD preapproval or manufacturer's tested or calculated capacities.

From anchor bolt connection calculation format IID and vibration isolator dimensions of H = 7 in. (178 mm), a = 0.75 in. (19 mm), b = 3.5 in. (89 mm), and n = 4:

$$T_{bolt} = \frac{(225)(7)}{\left(\frac{4}{2}\right)\left[(0.75) + (3.5) + \frac{(0.75)^2}{(0.75 + 3.5)}\right]} + \frac{\left(404 - \frac{12000}{4}\right)}{4} = -469(\text{ lb} \rightarrow 0)$$

and

$$V_{bolt} = \frac{225}{4} = 57 \text{ lb } (0.25 \text{ kN})$$

From structural calculation format IIIB:

Using four 1/2 in. (13 mm) diameter A307 quality bolts with $A_t = 0.142$ in.^{2} (92 mm^2) and $A_k = 0.126$ in.^{2} (81 mm^2):

$$f_v = \frac{57 \text{ lb}}{0.126 \text{ in.}^2} = 452 \text{ psi (0 MPa)} \leq F_v = 13333 \text{ psi (91.9 MPa)}$$

$$f_t = \frac{0 \text{ lb}}{0.142 \text{ in.}^2} = 0 \text{ psi (0 MPa)}$$

$f_v \leq F_v$ and $f_t \leq F_t$; therefore, anchorage is satisfactory.

B. Seismic horizontal force $F_{ph} = 0.90 \times W = 10,800$ lb (4899 kg)
Seismic vertical force $F_{pv} = 0.30 \times W = 3600$ lb (1633 kg)

From overturning calculation format IC:
Maximum tension P_t at location 2:

$$P_t = \frac{-3600}{4} - \frac{10800\cos(21.8°)40}{2(48)} - \frac{10800\sin(21.8°)40}{2(120)}$$
$$= -5747 \text{ lb } (-25.6 \text{ kN})$$

Maximum compression P_c at location 4:

$$P_c = \frac{3600}{4} + \frac{10800\cos(21.8°)40}{2(48)} + \frac{10800\sin(21.8°)40}{2(120)}$$
$$= 5747 \text{ lb (25.6 kN)}$$

Maximum shear P_s at all locations:

$$P_s = \frac{10800}{4} + 2700 \text{ lb (12 kN)}$$

where

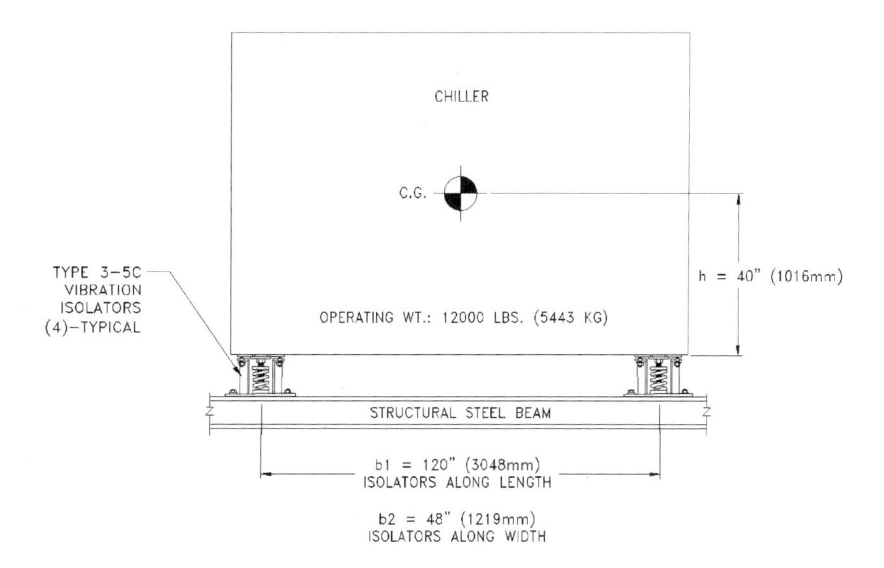

Figure 11-22 Example 1—seismic restraint layout.

$$\theta = \tan^{-1}\left(\frac{120}{48}\right) = 21.8°.$$

Check these loads against the OSHPD preapproval or manufacturer's tested or calculated capacities.

From anchor bolt connection calculation format IID and vibration isolator dimensions of $H = 7$ in. (178 mm), $a = 0.75$ in. (19 mm), $b = 3.5$ in. (89 mm), and $n = 4$:

$$T_{bolt} = \frac{(2700)(7)}{\left(\frac{4}{2}\right)\left[(0.75) + (3.5) + \frac{(0.75)^2}{(0.75 + 3.5)}\right]} + \frac{\left(5747 - \frac{12000}{4}\right)}{4} = 2843 \text{ lb (12.7 kN)}$$

and

$$V_{bolt} = \frac{2700}{4} = 675 \text{ lb (3 kN)}$$

From structural calculation format IIIB:

Using four 5/8 in. (16 mm) diameter A307 quality bolts with $A_t = 0.226$ in.^{2} (146 mm^2) and $A_k = 0.202$ in.^{2} (130 mm^2):

$$f_v = \frac{675 \text{ lb}}{0.202 \text{ in.}^2} = 3342 \text{ psi (23 MPa)} \leq F_v = 13333 \text{ psi (91.9 MPa)}$$

$$f_t = \frac{2843 \text{ lb}}{0.202 \text{ in.}^2} = 12580 \text{ psi (86.7 MPa)} \leq F_t = [26000 - 1.8(3342)] \times 1.33$$

$$F_t = 19{,}985 \text{ psi} \times 1.33 = 26{,}579 \text{ (183.2 MPa)}$$

$f_v \leq F_v$ and $f_t \leq F_t$; therefore, anchorage is satisfactory.

Example 2 (See Figure 11-23)

Equipment type: vertical air-handling unit

Isolator/restraint: vibration isolator and separate seismic snubber type 3-10A with two-bolt connection

Structure type: concrete slab

Operating weight $W = 3500$ lb (1588 kg)

Maximum distance between seismic restraints along y-y axis $b_1 = 72$ in (1829 mm)

Maximum distance between seismic restraints along x-x axis $b_2 = 60$ in. (1524 mm)

Height to center of gravity $h - 36$ in. (914 mm)

Number of seismic restraints $N = 6$

A. Seismic horizontal force $F_{ph} = 0.075 \times W = 263$ lb (1.2 kN)
   Seismic vertical force $F_{pv} = 0$

From overturning calculation format IC:

Maximum tension P_t at location 2:

$$P_t = \frac{0}{6} - \frac{263\cos(51.3°)(36)\left(\frac{60}{2}\right)}{5400} - \frac{263\sin(51.3°)(36)\left(\frac{72}{2}\right)}{5184}$$
$$= -84 \text{ lb } (0.4 \text{ kN})$$

Maximum compression P_c at location 4:

$$P_c = \frac{0}{6} + \frac{263\cos(51.3°)(36)\left(\frac{60}{2}\right)}{5400} + \frac{263\sin(51.3°)(36)\left(\frac{72}{2}\right)}{5184}$$
$$= 84 \text{ lb } (0.4 \text{ kN})$$

Maximum shear P_s at all locations:

$$P_s = \frac{263}{6} = 44 \text{ lb } (0.2 \text{ kN})$$

where

$$I_{xx} = \frac{6(6+2)72^2}{12(6-2)} = 5184 \text{ in.}^2 (\ 3.3 \times 10^6 \text{ mm}^2),$$

$$I_{xx} = \frac{6(60)^2}{4} = 5400 \text{ in.}^2 (\ 3.5 \times 10^6 \text{ mm}^2),$$

$$\theta = \tan^{-1}\left(\frac{(5400)(72)}{(5184)(60)}\right) = 51.3°.$$

Check these loads against the OSHPD preapproval or manufacturer's tested or calculated capacities.

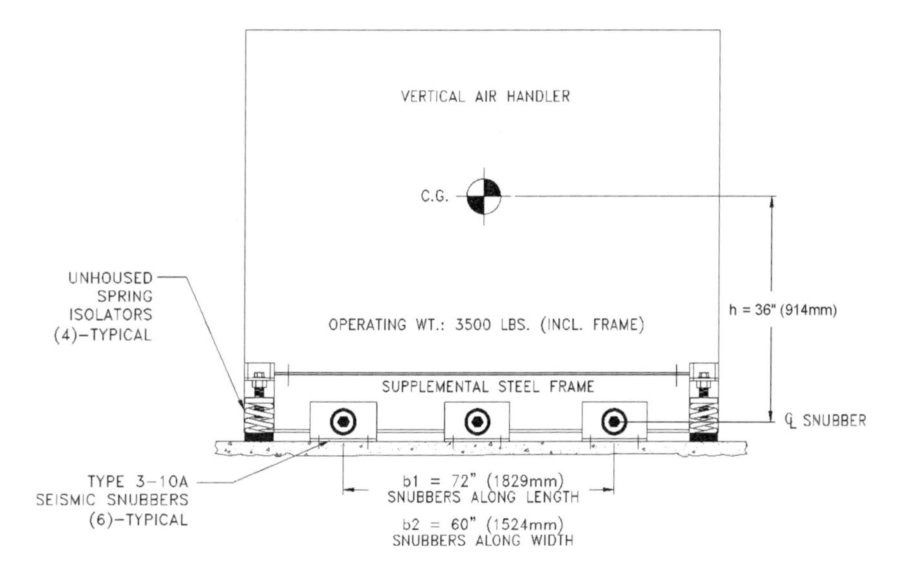

Figure 11-23 Example 2—seismic restraint layout.

From anchor bolt connection calculation format IIE and seismic restraint dimensions of $H = 2.5$ in. (64 mm), $B = 4$ in. (102 mm), and $n = 2$:

$$T_{bolt} = \frac{(44)(2.5) + (84)(4)}{2\left(\frac{4}{2}\right)} = 112 \text{ lb } (0.5 \text{ kN})$$

and

$$V_{bolt} = \frac{44}{2} = 22 \text{ lb } (0.1 \text{ kN}).$$

From structural calculation format IIIA:
Using two 3/8 in. (10 mm) diameter expansion anchors,

$$\left(\frac{112}{400}\right)^{5/3} + \left(\frac{22}{700}\right)^{5/3} = 0.12 \leq 1.0$$

Note: T_{allow} and V_{allow} from Table 5-3 in Chapter 5.

Therefore, restraint anchorage is satisfactory.

B. Seismic horizontal force $F_{ph} = 0.90 \times W = 3150$ lb (14 kN)
Seismic vertical force $F_{pv} = 0.30 \times W = 1050$ lb (4.7 kN)

From overturning calculation format IC:
Maximum tension P_t at location 2:

$$P_t = \frac{-1050}{6} - \frac{3150\cos(51.3°)(36)\left(\frac{60}{2}\right)}{5400} - \frac{3150\sin(51.3°)(36)\left(\frac{72}{2}\right)}{5184}$$
$$= -1183 \text{ lb } (-5.3 \text{ kN})$$

Maximum compression P_c at location 4:

$$P_c = \frac{1050}{6} + \frac{3150\cos(51.3°)(36)\left(\frac{60}{2}\right)}{5400} + \frac{3150\sin(51.3°)(36)\left(\frac{72}{2}\right)}{5184}$$
$$= 1183 \text{ lb } (5.3 \text{ kN})$$

Maximum shear P_s at all locations:

$$P_s = \frac{3150}{6} = 525 \text{ lb } (2.3 \text{ kN})$$

where

$$I_{xx} = \frac{6(6+2)72^2}{12(6-2)} = 5184 \text{ in.}^2 (3.3 \times 10^6 \text{ mm}^2),$$

$$I_{xx} = \frac{6(60)^2}{4} = 5400 \text{ in.}^2 (3.5 \times 10^6 \text{ mm}^2),$$

$$\theta = \tan^{-1}\left(\frac{(5400)(72)}{(5184)(60)}\right) = 51.3°.$$

Check these loads against the OSHPD preapproval or manufacturer's tested or calculated capacities.

From anchor bolt connection calculation format IIE and seismic restraint dimensions of $H = 2.5$ in. (64 mm), $B = 4$ in. (102 mm), and $n = 2$:

$$T_{bolt} = \frac{(525)(2.5) + (1183)(4)}{2\left(\frac{4}{2}\right)} = 1511 \text{ lb } (6.7 \text{ kN})$$

and

$$V_{bolt} = \frac{525}{2} = 263 \text{ lb } (1.2 \text{ kN}).$$

From structural calculation format IIIA:
Using two 5/8 in. (16 mm) diameter expansion anchors,

$$\left(\frac{1511}{1600}\right)^{5/3} + \left(\frac{263}{1800}\right)^{5/3} = 0.95 \le 1.0$$

Note: For T_{allow} and V_{allow} from Table 5-3 in Chapter 5, special inspection is required.

Therefore, restraint anchorage is satisfactory.

FLEXIBLE PIPING CONNECTORS FOR EQUIPMENT

There are many different types of flexible connectors for piping connections to equipment. Spherical rubber connectors are often used for vibration and noise isolation purposes, but can also be used for seismic displacement. Braided-steel copper hoses can be used, too, if they are designed to accept the appropriate displacement and are oriented properly. Sixty-degree V or parallel leg loops are another and often better choice because they move in all planes. As noted in Chapter 2, there are at least 10 different requirements in the IBC that require evaluation of differential piping motion caused by seismic motion of the building. Of all the requirements, those for piping connections to equipment may be the most critical. Piping very seldom breaks in an earthquake, but small equipment nozzles and cast-iron equipment connections have a history of failure, and have contributed to a large number of equipment failures. Proper design of flexible connectors will prevent problems and is a requirement that has been repeatedly accentuated in the new building codes.

Although they are required by code, there is very little guidance regarding types of flexible connectors. The design engineer can look at the overall system flexibility, or movement requirements can be addressed directly with flexible connectors to simplify the solution and limit dependence on system flexibility. There is a new extended commentary on ASCE 7-10 that includes some important design suggestions to help the design engineer integrate flexible connections to meet code requirements.

Spherical rubber connectors may be used for vibration isolation and seismic protection at equipment connections where temperature, pressure, fluid type, system pressure and movements are acceptable. They are not generally intended or recommended for installation in ceiling spaces. Some spherical rubber connectors are designed to extend to a pressure balance length and can restrain the piping pressure thrust without adding control rods. The connector's angular movement capability and low angular stiffness can be utilized to limit piping loads and moments on equipment connections. On vibration-isolated systems, connector extension that would otherwise short-circuit seismic snubbers can be avoided if the connectors are preextended to their pressure balance length during installation.

Spherical connectors are available in many different elastomers and reinforcement materials. The correct elastomer for the application should be used. Sulfur-cured neoprene may be acceptable for some applications, but peroxide-cured EPDM may be a better choice. Spherical connectors are currently manufactured with nylon, polyester or aramid fiber reinforcement, each with different strengths, temperature capabilities, and operating lifespans. Some connectors have pressure balance lengths and can be used without control rods for maximum motion capability. Others need control rods or cables to prevent overextension, limiting their usefulness. The design engineer should investigate the options and select the proper connector to provide a safe installation.

Braided-stainless-steel or bronze hoses can be used on equipment located outside of mechanical rooms, in ceiling spaces or when noise transmission is not important. The hoses should be longer than current industry pump connector standards and close pitched to provide a low stiffness working against the inertia and stiffness of the connected piping. For pump installations, long hoses installed at the pump in the vertical pipe risers are good for protecting the equipment connections from piping loads and moments. Braided hoses should only be used for transverse motion. Subjecting braided hoses to torsion or axial movement is not recommended.

Braided hose 60° V loops can be used to protect equipment connections from differential piping motion in all directions. Long lengths and close-pitched hose are essential. The resulting extremely low all-directional stiffness and zero-pressure thrust can eliminate the need for hard anchors and is ideal for expansion compensation as well as seismic design. Figure 11-24 includes details of a spherical rubber connector, spool-type connector, ball-joint connector, braided flexible stainless-steel hose and a 90° and 60° flexible V loop. Manufactures with tested or calculated force and movement data are important for the approval process.

Figure 11-25 illustrates the differences in a solid pipe drop to equipment, a pipe drop with a spherical rubber connector, a pipe drop with a braided hose, and a pipe drop with a 60° flexible V loop. Because the piping header is braced to the deck overhead and the equipment is anchored to the floor, the building sway and interstory drift causes differential displacement to occur between the piping header and the equipment and should be considered in calculations where flexible connectors are used. The solid pipe drop will bend to accommodate the displacement but will develop the highest loads on the pipe, pipe braces, and equipment connections. The solid pipe drop also transfers a moment to the equipment when displaced that can lead to equipment connection failure. Spherical rubber connectors may accept the differential displacement and reduce the moment on the equipment connection and reduce the bending loads from the pipe. A long braided hose, selected to accommodate the full displacement, can also reduce the force on the equipment connection, with flexible support of the piping to allow the braided hose to move laterally without axial restraint. A 60° or 180° loop installed at the equipment connection may be the best selection of all. They allow large displacement in any direction and virtually eliminate loads on the equipment connections.

Spool-type hand-built rubber expansion joints are another good solution for piping over 24 in. Recommended locations are mechanical rooms or industrial spaces. Because spool-type joints are not pressure balanced, axial-pressure thrust must be restrained by piping anchors or by control rods, which limit their axial and transverse motion. However, they offer flexibility and sometimes are the only solution.

Using spool-type rubber expansion joints for axial-motion compensation requires piping anchors designed to resist pressure thrust loads. Allowing piping pressure thrust loads on equipment connections is not recommended. Control rods used on these expansion joints to resist pressure thrust will eliminate axial-movement capability and may limit lateral-movement capability.

The angular offset and rotational capability of ball joints can be used in multiple combinations where fluid type, temperature, pressure, or motion exceeds the capabilities of more commonly available flexible connectors. But ball joint breakaway forces can be high and can exceed equipment allowable nozzle loads. Ball joints may not be appropriate for equipment connections.

Broken connections on small, in-line reheat coils shut down a major hospital in the Northridge earthquake. In response, OSHPD added specific rules for in-line duct devices and

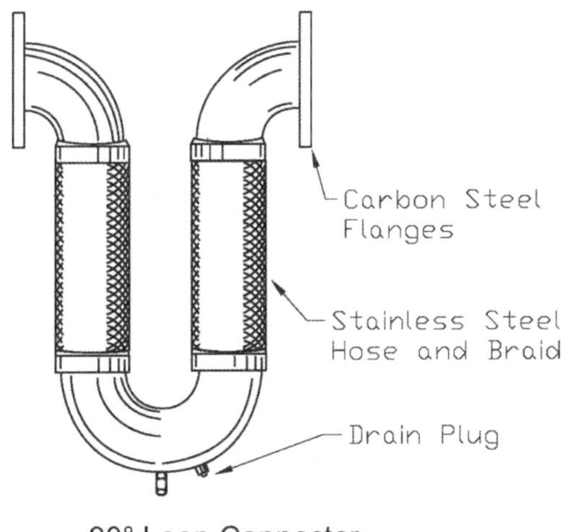

90° Loop Connector

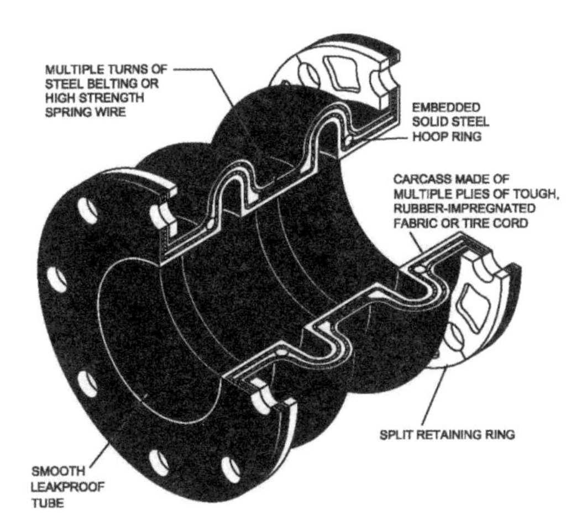

Spool Type Connector

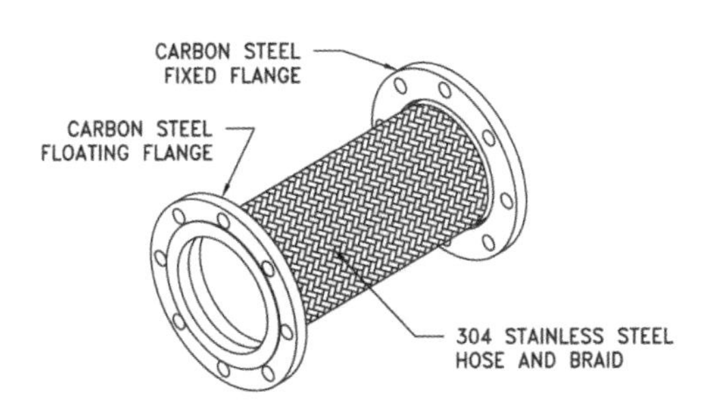

Braided Flexible Hose

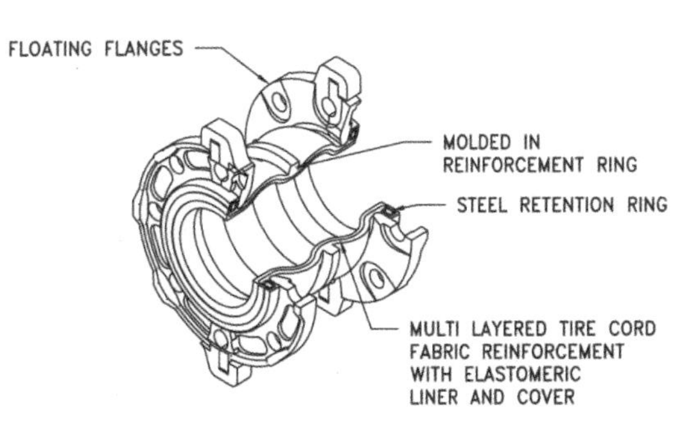

Spherical Rubber Connector

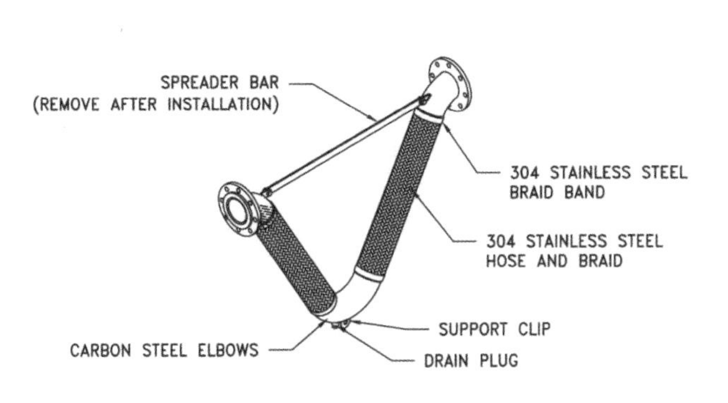

60° Flexible V Loop

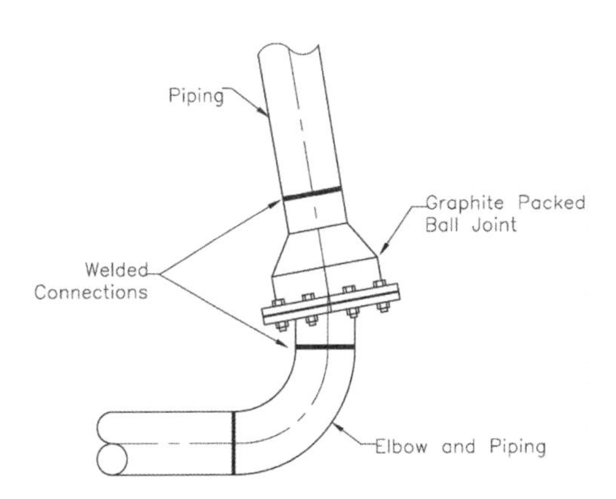

Ball Joint Connector

Figure 11-24 Flexible connectors for equipment.

attached piping in the preapproved seismic bracing manuals. These rules were eventually added to ASCE 7 and IBC. See Chapter 2 for more information.

There are often hundreds of small-diameter pipe connections to coils in variable-air-volume (VAV) terminal units and fan-coil units installed in suspended ductwork in a building. These small piping connections are often overlooked. There are discussions about the code language, components, and equipment, but the code intent is clear. Failure of the smallest coil tubing can cause extensive water damage. Water leaking fromsmall broken lines accumulates as it flows and can cause a nuisance or a breakdown of the equipment.

There are several things to consider when selecting a flexible connector for small, in-line coil piping connections. There may be large differential displacements between the duct coils and connected piping. This is especially true if the piping, the duct, or both are unbraced. The coils have very small copper tubing cantilevered off the side that is easily broken. Unfortunately, there is no information from most coil manufacturers on acceptable nozzle loads. Industry standard braided-bronze hose is far too short and stiff to protect the coil connection. Copper braided, flexible-hose 60° V loops with published test data should be strongly considered to eliminate this breakage.

Figure 11-26 shows the addition of flexible-hose V loops to the piping connections of unbraced, in-line coils.

Figure 11-27 shows the addition of flexible-hose V loops to the piping connections of braced, in-line coils. The bracing is attached to the coil section using strut trapeze members at top and bottom that are connected by threaded rods. This assembly can be designed to provide the independent support required by code. Then the sway bracing is attached to the support rod connections. See Chapter 7 for more information on duct bracing.

The design engineer must know three things to specify a proper flexible connector for a particular installation:

1. The equipment manufacturer must provide the acceptable nozzle loads (allowable piping connection loads) on the equipment. When asked, virtually all equipment manufacturers will answer "zero." Although some flexible-connector manufacturers can provide flexible connectors with very low stiffness, it is impractical to reduce nozzle loads to "zero." Equipment manufacturers can test nozzles easily and design engineers should demand published acceptable nozzle loads on all equipment that they specify.
2. The flexible-hose manufacturer must provide certified stiffness data. Stiffness increases as pressure increases despite what some manufacturers say in their sales literature. In a highly flexible hose, an increase in pressure will dramatically increase stiffness. Testing is essential, as is proper flexible-connector design.
3. There must be some estimate made of maximum potential differential motion of the piping at the equipment connection. This can be based on the amount of interstory drift, which varies with structure type, or the pendulum action of unbraced systems. Often, some input is required from the project structural engineer.

The design engineer should be careful when reviewing substitutions and alternate products. Many flexible connector products on the market are manufactured based on old designs that have been repeatedly modified and cheapened.

BIBLIOGRAPHY

ASCE. 2010. *ASCE 7-10 Minimum Design Loads for Buildings and Other Structures*. Reston,VA: American Society of Civil Engineers.

Mason Industries. 2009. *Seismic Restraint of Suspended Piping, Ductwork and Electrical Systems*. Hauppauge, NY.

ICC. 2009. *2009 International Building Code® (IBC®)*. Washington, DC: International Code Council.

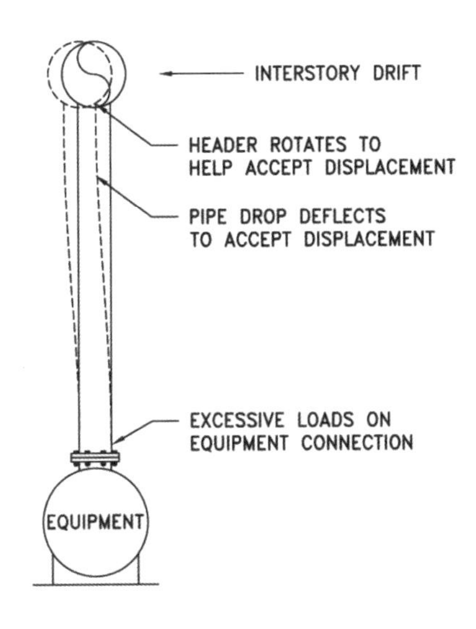

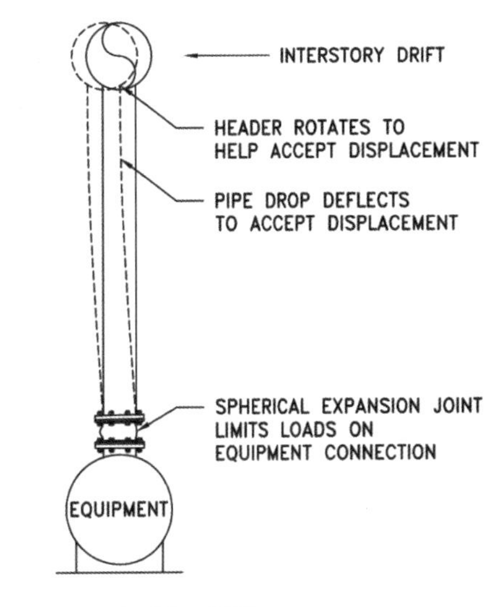

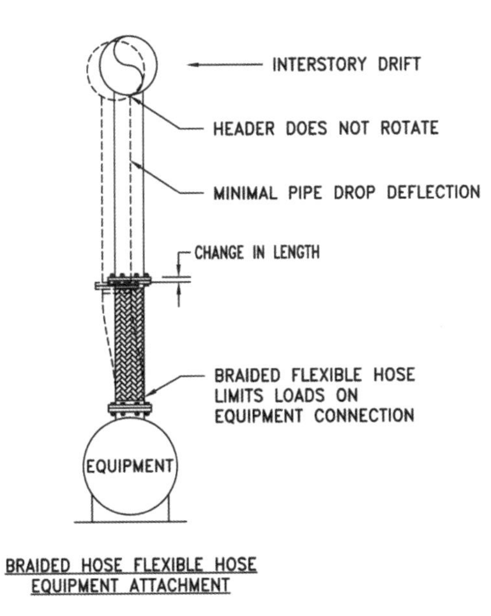

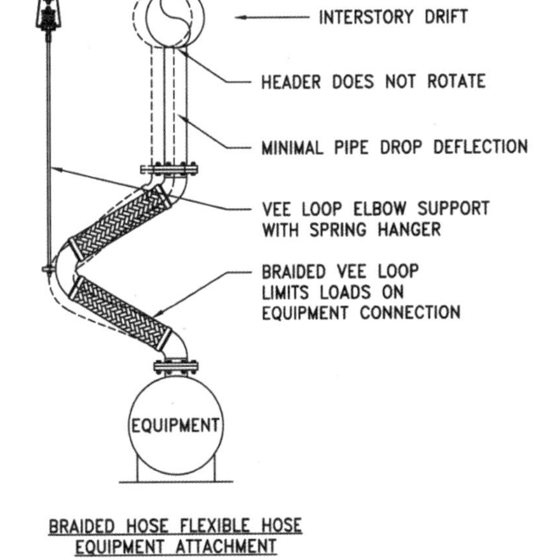

Figure 11-25 Flexible connectors on an unbraced, in-line coil.

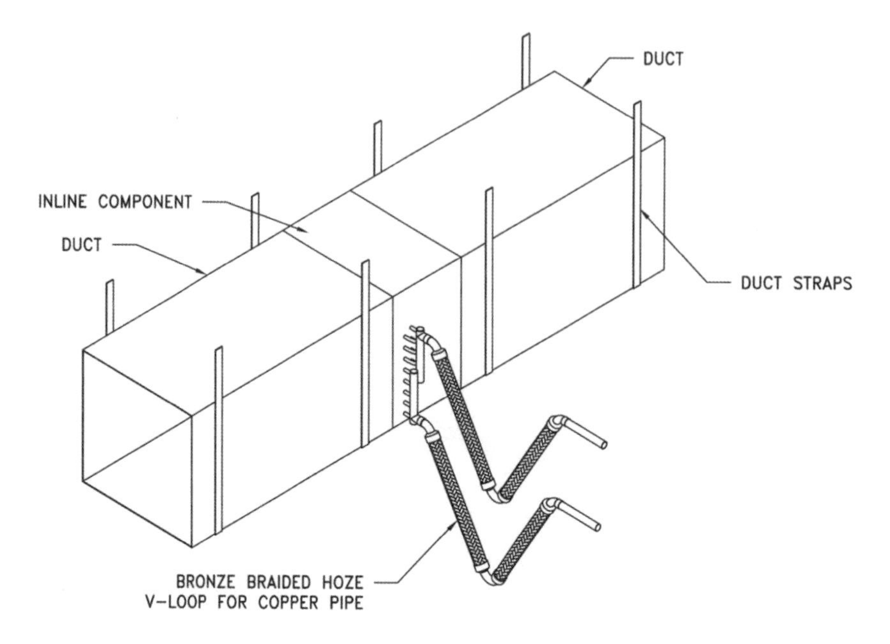

Figure 11-26 Flexible connectors on an unbraced inline coils

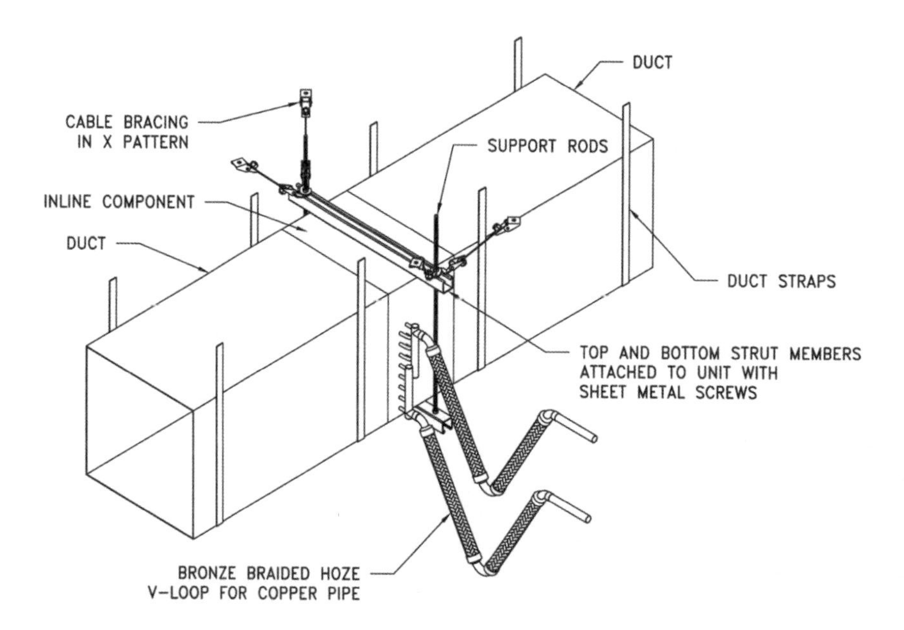

Figure 11-27 Flexible connectors on a braced, in-line coil.

12 Wind or Seismic Loading?

Equipment mounted on the roof must be analyzed for both seismic and wind loads. Each piece of equipment has different proportions and must be analyzed individually. On some projects, wind loads govern for a cooling tower, but seismic loads govern for rooftop air-handling units. The seismic restraint engineer determines which loading is greater and prepares restraint calculations to satisfy those requirements.

Seismic loads begin with the determination of F_p as discussed in Chapters 2 and 3. These loads are directly related to the weight of the equipment and somewhat dependent on the shape and size.

Wind loads are directly related to size and shape. One of the most widely used publications for wind load designing is the American Society of Civil Engineers (ASCE) standard ANSI/ASCE 7-05, *Minimum Design Loads for Buildings and Other Structures*. This design manual has in-depth procedures for designing all forms of wind loads, from buildings to billboard signs.

In this chapter, we only present the information required to do a wind load analysis on a nonstructural building component, such as a piece of rooftop equipment, and not the wind loads for designing the structure itself. The methods and calculations shown here are for a conservative approach. The basic equation for equipment wind loads is as follows:

$$F = q_z G A_f C_f \tag{12-1}$$

where

F = wind load force, lb (kN)

q_z = velocity pressure at height z from Equation 12-2

G = gust factor from Table 12-2

A_f = projected area normal to the wind, ft^2 (m^2)

C_f = force coefficient from Table 12-4

$$q_z = 0.00256\, K_z\, I\, V^2\ (\text{lb/ft}^2) \text{ or}$$
$$q_z = 0.613\, K_z\, I\, V^2\ (\text{N/m}^2) \tag{12-2}$$

where

K_z = velocity pressure exposure coefficient from Table 12-3

V = velocity, mph (m/s) from Figure 12-2

I = importance factor from Table 12-1

Note: If the height of the building $h \leq 60$ ft (18.3m) and $A_f < 0.1Bh$, then F must be increased by a factor of 1.9. The factor may be decreased linearly from 1.9 to 1.0 as the value of A_f is increased from $0.1Bh$ to Bh.

Table 12-1 Importance Factors*

Nature of Occupancy	Category	Importance Factor for Nonhurricane Regions and Hurricane-Prone Regions with V = 85 to 100 mph (38 to 45 m/s) and Alaska	Importance Factor for Hurricane-prone regions with V > 100 mph (45 m/s)
Buildings with low hazard to humans	I	0.87	0.77
Buildings not covered in categories I, III, and IV	II	1.00	1.00
Buildings with substantial hazard to human life	III	1.15	1.15
Buildings designated as essential facilities	IV	1.15	1.15

*Reprinted from ASCE 7-05, Tables 1-1 and 6-1, with permission.

Table 12-2 Wind Exposures*

Exposure B	Urban and suburban areas, wooded areas, or other terrain with numerous, closely spaced obstructions having the size of single-family dwellings or larger. Use of this exposure shall be limited to those areas for which terrain representative of Exposure B prevails in the upwind direction for a distance of at least 2600 ft (792 m) or 20 times the height of the building or other structure, whichever is greater. Exception: for buildings where h ? 30 ft (9.1 m), the upwind distance may be reduced to 1500 ft (457 m). Gust factor $G = 0.85$, $\alpha = 7.0$, and $z_g = 1200$ ft (366 m).
Exposure C	Exposure C shall apply for all cases where exposures B or D do not apply. Gust factor $G = 0.85$, $\alpha = 9.5$, and $z_g = 900$ ft (274 m).
Exposure D	Flat, unobstructed areas and water surfaces outside hurricane prone regions. This category includes smooth mud flats, salt flats, and unbroken ice. Exposure D prevails in the upwind direction for a distance greater than 5000 ft (1254 m) or 20 times the height of the building or structure, whichever is greater. Exposure D shall extend downwind for a distance of 600 ft (200 m) or 20 times the height of the building in Exposures B and C. Gust factor $G = 0.85$, $\alpha = 11.5$, and $z_g = 700$ ft (213 m).

*Reprinted from ASCE 7-05, with permission.

A more simplified method to determine a wind load is to use the following equation:

$$F = S\, C_f A_f \tag{12-3}$$

where

F = wind load force, lb (kN)

S = precalculated velocity pressure that incorporates the gust factor G along with q_z from Equation 12-2

C_f = force coefficient from Table 12-4

A_f = projected area normal to the wind, ft^2 (m^2)

These values have been calculated for velocities of 85, 90, 110, 130, and 150 mph (38, 40, 49, 58, and 67 m/s) and can be found in Table 12-5.

Note: If the height of the building $h \leq 60$ ft (18.3m) and $A_f < 0.1Bh$, then F must be increased by a factor of 1.9. The factor may be decreased linearly from 1.9 to 1.0 as the value of A_f is increased from $0.1Bh$ to Bh.

Table 12-3 Velocity Pressure Exposure Coefficient K_z

Height z above ground level, ft (m)	Exposure (see Note 1 thru 5)			
	B		C	D
	Case 1	Case 2	Cases 1 & 2	Cases 1 & 2
15 (4.6)	0.70	0.57	0.85	1.03
20 (6.1)	0.70	0.62	0.90	1.08
25 (7.6)	0.70	0.66	0.94	1.12
30 (9.1)	0.70	0.70	0.98	1.16
40 (12.2)	0.76	0.76	1.04	1.22
50 (15.2)	0.81	0.81	1.09	1.27
60 (18)	0.85	0.85	1.13	1.31
70 (21.3)	0.89	0.89	1.17	1.34
80 (24.4)	0.93	0.93	1.21	1.38
90 (27.4)	0.96	0.96	1.24	1.40
100 (30.5)	0.99	0.99	1.26	1.43
120 (36.6)	1.04	1.04	1.31	1.48
140 (42.7)	1.09	1.09	1.36	1.52
160 (48.8)	1.13	1.13	1.39	1.55
180 (54.9)	1.17	1.17	1.43	1.58
200 (61)	1.20	1.20	1.46	1.61
250 (76.2)	1.28	1.28	1.53	1.68
300 (91.4)	1.35	1.35	1.59	1.73
350 (106.7)	1.41	1.41	1.64	1.78
400 (121.9)	1.47	1.47	1.69	1.82
450 (137.4)	1.52	1.52	1.73	1.86
500 (152.4)	1.56	1.56	1.77	1.89

Notes:

1. Linear interpolation for intermediate values of height z is acceptable.
2. Case 1: a. All components and cladding
   b. Main wind force resisting system in low-rise buildings designed using figure 6-10 of ASCE 7-05

   Case 2: a. All main wind force resisting systems in buildings except those in low- rise buildings designed using figure 6-10 of ASCE 7-05
   b. All main wind force resisting systems in other structures
3. The velocity pressure exposure coefficient K_z shall be calculated from the following equations:
   $K_z = 2.01\ (z/z_g)^{2/\alpha}$ for 15 ft $\leq z \leq z_g$
   or
   $K_z = 2.01\ (15/z_g)^{2/\alpha}$ for $z < 15$ ft
   Note: height z shall not be taken as less than 30 ft (9.1 m) for Case 1 in Exposure B.
4. Exposure categories, α, and z_g are defined in Table 12-2.
5. Reprinted from ASCE 7-05, Table 6-3, with permission.

Table 12-4 Force Coefficient Cf for Chimneys, Tanks, Rooftop Equipment, and Similar Structures

Cross-Section	Type of Surface	H/D Values 1	7	25
Square (wind normal to face)	All	1.3	1.4	2.0
Square (wind along diagonal)	All	1.0	1.1	1.5
Hexagonal or Octagonal	All	1.0	1.2	1.4
Round ($D\sqrt{q_z} > 2.5$)	Moderately smooth	0.5	0.6	0.7
($D\sqrt{q_z} > 5.3$, D in m, q_z in N/m^2)	Rough ($D'/D \cong 0.02$)	0.7	0.8	0.9
	Very rough ($D'/D \cong 0.08$))	0.8	1.0	1.2
Round ($D\sqrt{q_z} \le 2.5$) ($D\sqrt{q_z} \le 5.3$, D in m, q_z in N/m^2)	All	0.7	0.8	1.2

Notes:

1. The design wind force shall be calculated based on the area of the equipment projected on a plane normal to the wind direction. The force shall be assumed to act parallel to the wind direction.
2. Linear interpolation is permitted for H/D values other than those shown.
3. Legend:
   D = diameter of circular cross section or least horizontal dimension for square, hexagonal, or octagonal cross-sections, ft (m)
   D' = depth of protruding elements, such as spoilers and ribs, ft (m)
   H = height of equipment or structure, ft (m)
   q_z = velocity pressure evaluated at height z above ground, lb/ft^2 (N/m^2)
4. Reprinted from ASCE 7-05, figure 6-21, with permission.

Table 12-5 allows you to select a height z versus an exposure category and wind speed. Values in Table 12-5 are for importance factor I = 1.15. For I = 1.0, multiply the value by 0.87, and for I = 0.87, multiply the value by 0.76. Interpolation between values is acceptable.

EXAMPLE Calculate the wind loading for an air-handling unit that is 27 ft (8.23 m) long, 7.5 ft (2.29 m) tall, and 4 ft (1.22 m) wide. The building is an essential facility with an Exposure C and a basic wind speed of 110 mph (49.2 m/s). The building height, Z, above ground level is 100 ft (30.5 m). See Figure 12-1.

The wind load can be calculated by using Equations 12-1 or 12-3.

$$F = q_z G C_f A_f$$

or

$$F = S C_f A_f$$

Using the precalculated velocity pressure S from Table 12-5, we find that at 110 mph (49.2 m/s), Exposure C, and 100 ft (30.5 m) elevation, S = 38.1 lb/ft^2 (1826 N/m^2).

The force coefficient, C_f, from Table 12-4 is 1.3. This is for a square or rectangular piece of equipment with the wind normal to the face and a height/width ratio close to 1.0.

$$A_f = 27 \text{ ft} \times 7.5 \text{ ft} = 202.5 \text{ ft}^2 \ (18.8 \text{ m}^2)$$

Inputting these values into Equation 12-3 gives

$$F = 38.1 \text{ lb/ft}^2 \times 1.3 \times 202.5 \text{ ft}^2,$$

$$F = 10{,}030 \text{ lb } (44.6 \text{ kN}).$$

See the example in Chapter 13 for the design of seismic restraints and connections to the structure for this unit.

Table 12-5 S Values, lb/ft² (N/m²)

Exposure Category	Wind Speed, mph (m/s)	Height z above ground level, ft (m) 0-15 (0-4.6)	30 (9.1)	50 (15.2)	100 (30.5)	200 (60.9)	300 (91.4)	400 (121.9)	500 (152.4)
B Case 1	85 (38)	12.7 (608)	12.7 (608)	14.6 (699)	17.9 (857)	21.7 (1039)	24.4 (1168)	26.6 (1274)	28.2 (1350)
	90 (40.2)	14.2 (680)	14.2 (680)	16.4 (785)	20.1 (962)	24.3 (1163)	27.3 (1307)	29.8 (1427)	31.6 (1513)
	110 (49.2)	21.2 (1015)	21.2 (1015)	24.5 (1173)	30 (1436)	36.3 (1738)	40.9 (1958)	44.5 (2131)	47.2 (2260)
	130 (58.1)	29.6 (1417)	29.6 (1417)	34.3 (1642)	41.9 (2006)	50.7 (2428)	57.1 (2734)	62.2 (2978)	66 (3160)
	150 (67.1)	39.4 (1886)	39.4 (1886)	45.6 (2183)	55.7 (2667)	67.6 (3237)	76 (3639)	82.8 (3964)	87.8 (4204)
B Case 2	85 (38)	10.3 (493)	12.7 (608)	14.6 (699)	17.9 (857)	21.7 (1039)	24.4 (1168)	26.6 (1274)	28.2 (1350)
	90 (40.2)	11.5 (551)	14.2 (680)	16.4 (785)	20.1 (962)	24.3 (1163)	27.3 (1307)	29.8 (1427)	31.6 (1513)
	110 (49.2)	17.3 (828)	21.2 (1015)	24.5 (1173)	30 (1436)	36.3 (1738)	40.9 (1958)	44.5 (2131)	47.2 (2260)
	130 (58.1)	24.1 (1154)	29.6 (1417)	34.3 (1642)	41.9 (2006)	50.7 (2428)	57.1 (2734)	62.2 (2978)	66 (3160)
	150 (67.1)	32.1 (1537)	39.4 (1886)	45.6 (2183)	55.7 (2667)	67.6 (3237)	76 (3639)	82.8 (3964)	87.8 (4204)
C Cases 1 & 2	85 (38)	15.6 (749)	17.6 (843)	19.5 (936)	22.5 (1077)	26.4 (1264)	28.3 (1358)	30.3 (1451)	32.2 (1545)
	90 (40.2)	17.6 (843)	19.5 (936)	22.5 (1077)	25.4 (1217)	29.3 (1404)	32.2 (1545)	34.2 (1639)	36.1 (1732)
	110 (49.2)	25.4 (1217)	29.3 (1404)	33.2 (1592)	38.1 (1826)	44 (2107)	47.9 (2294)	50.8 (2435)	53.7 (2575)
	130 (58.1)	36.1 (1732)	41 (1966)	45.9 (2200)	53.7 (2575)	61.5 (2950)	67.4 (3231)	71.4 (3418)	75.2 (3605)
	150 (67.1)	47.9 (2294)	54.7 (2622)	61.5 (2950)	71.4 (3418)	82.1 (3933)	89.8 (4308)	94.8 (4551)	99.7 (4776)
D Cases 1 & 2	85 (38)	18.5 (890)	21.5 (1030)	23.5 (1124)	25.4 (1217)	29.3 (1405)	31.3 (1498)	33.2 (1592)	34.2 (1639)
	90 (40.2)	20.5 (983)	23.5 (1124)	25.4 (1217)	29.3 (1405)	32.2 (1545)	35.2 (1685)	37.1 (1779)	38.1 (1826)
	110 (49.2)	31.3 (1498)	35.2 (1685)	38.1 (1826)	43 (2060)	48.9 (2341)	52.7 (2529)	54.7 (2621)	56.7 (2715)
	130 (58.1)	44 (2107)	48.9 (2341)	53.7 (2576)	60.6 (2903)	68.4 (3277)	73.3 (3512)	77.2 (3699)	80.1 (3840)
	150 (67.1)	57.6 (2762)	65.5 (3137)	71.3 (3417)	80.1 (3840)	90.9 (4355)	97.7 (4682)	102.6 (4917)	106.5 (5104)

1. Linear interpolation for intermediate values of z is acceptable.
2. For values of height z greater than 500 ft (152.4 m), use Equation 12-2 with the values derived from the equations in Note 2 of Table 12-3.
3. Exposure categories are defined in Table 12-2.
4. Precalculated S values are for an importance factor of 1.15. For $I = 1.0$, multiply these values by 0.87. For $I = 0.87$, multiply these values by 0.76.
5. Table adapted from ASCE 7-95, Table C6-3, with permission.

After determining the wind forces, one can design and calculate the transfer of those loads to the structure. Examples include a roof curb for an air handler, structural steel for equipment such as cooling towers and air-cooled condensers, or roof rails for small centrifugal exhaust fans. Please see the following chapters:

- Chapter 2 for building code determination
- Chapter 5 for connections to structure
- Chapter 13 for rooftop air-handling equipment
- Chapter 14 for cooling towers
- Chapter 15 for exhaust and mushroom fans

BIBLIOGRAPHY

ASCE. 2005. *ANSI/ASCE Standard 7-05, Minimum design loads for buildings and other structures,* Chapter 6. New York: American Society of Civil Engineers.

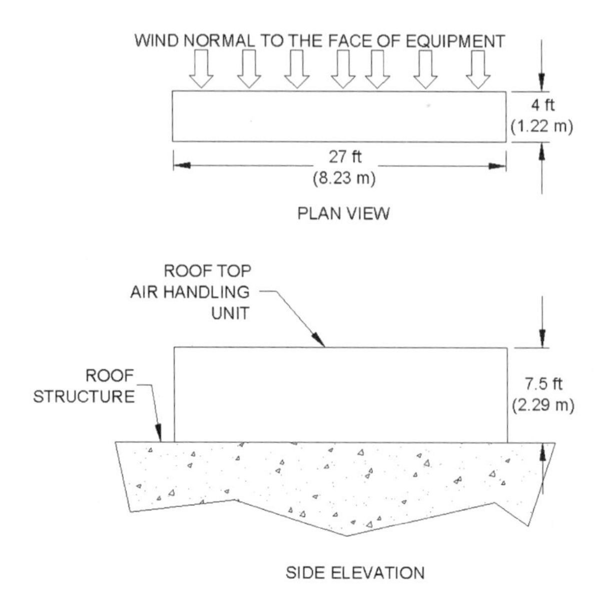

Figure 12-1 Example—wind load calculation.

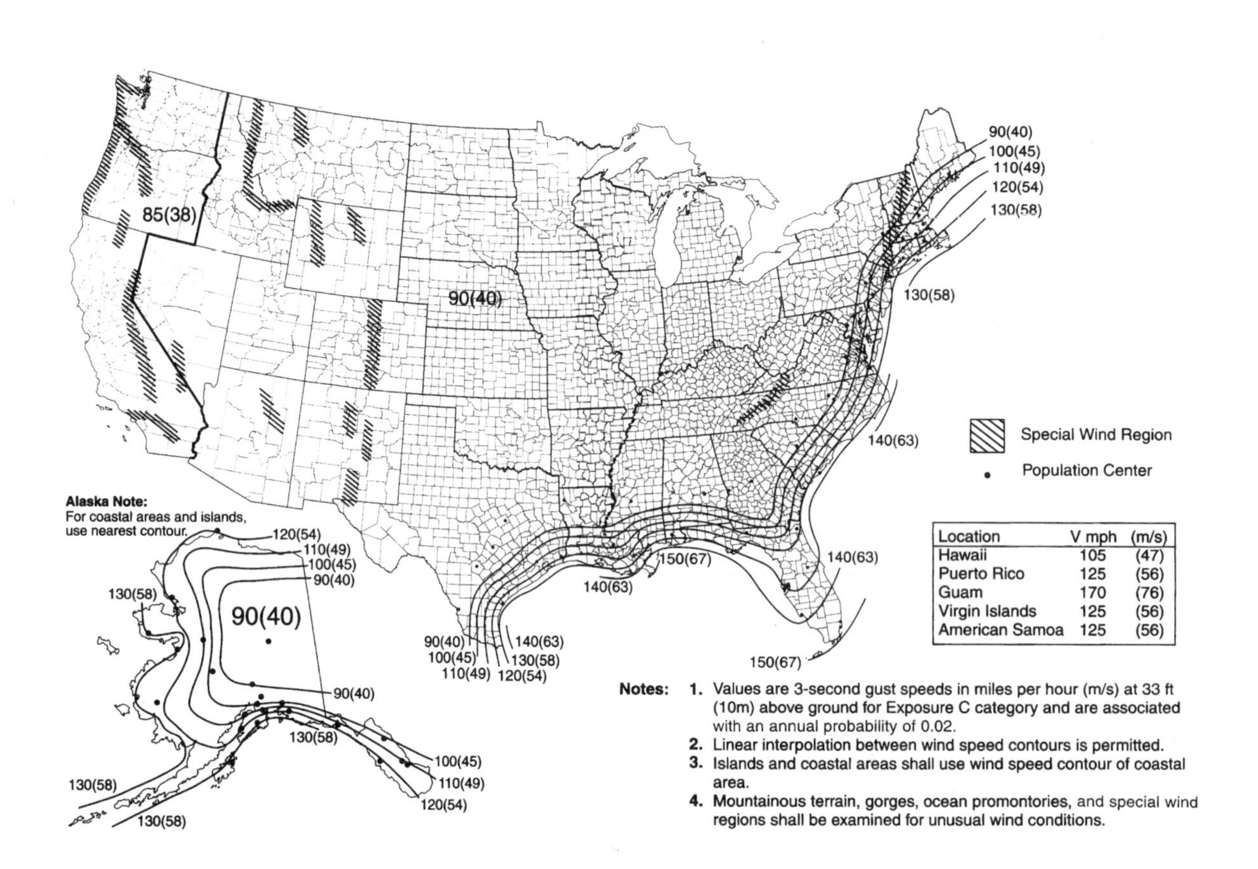

Figure 12-2 Basic wind velocity. (Reprinted from ASCE 7-05, with permission)

13 Rooftop Air-Handling Units

There are three styles of curbs commonly used to support rooftop air-handling units (RTUs). They include a standard sheet metal curb, a curb-mounted vibration isolation system that sits on top of a standard sheet metal curb, and a complete vibration isolation curb. Site-fabricated steel and wood curbs are not covered in this chapter and should be designed by a qualified design professional.

In addition to the inherent gravity loads, curbs are subject to both seismic and wind forces. (Refer to Chapter 12 to determine which force is the governing force.) Curbs must be designed to transfer all applicable loads to the roof structure with sufficient connection of the unit to the curb and the curb to the structure.

In Chapter 2, the 2006 International Building Code® (IBC®) was presented. There were two points in that section that impact roof curbs. First, a term of the height within the structure divided by the height of the structure has been added in the calculation of F_p. Because curbs are usually on the roof, this term will always increase the F_p value. Secondly, all manufacturers will be required to supply a certificate of compliance stating that their products can safely resist the F_p seismic forces.

Figures 13-1 to 13-14 show several types of curbs and typical connections. Note that by using either through bolts or lag bolts in the horizontal position (as in Figures 13-8 to 13-13), there is no tensile force on the bolt. Some rooftop air handlers have base rail channels with no overhang and require additional steel to provide one (see Figure 13-8). Other units have flat bottoms, and they, too, require additional steel (see Figure 13-10). See Figure 13-11 for units restrained with cable through the lifting lugs of the unit.

STANDARD SHEET METAL CURBS

Standard sheet metal curbs are generally offered by the air handler manufacturer or can be purchased separately from other vendors. They are typically provided in heights ranging from 12 in. (305 mm) to 24 in. (610 mm). Figure 13-1 shows a typical cross-sectional view of a sheet metal curb, which is generally constructed out of 12 to 18 gage steel. These curbs distribute the weight of the air-handling unit uniformly over the entire perimeter of the curb. This uniform loading limits the strength of the curb as its unbraced length increases. The unbraced length is the distance measured between crossbraces or between the end section and a crossbrace. Unless the curb is provided with crossbracing at midlength, the unbraced length of the curb is its overall length. In general, the unbraced length of a curb should not exceed 8 ft (2.4 m). Crossbraces are steel sections that transfer loads from the top of the curb to the bottom at the roof connection points and run perpendicular to the sides of the curb. The connection of the end sections and the crossbraces to the side members is the key to the restraint capabilities of a curb. Most curb ends and crossbraces are connected to the side members with sheet metal angles and screws. This arrangement must be capable of withstanding the horizontal loads from earthquakes and normal everyday lateral forces. Figure 13-2 shows poorly and correctly detailed corner connections.

CURB-MOUNTED VIBRATION ISOLATION SYSTEMS

Curb-mounted vibration isolation systems are manufactured from aluminum or steel and sit directly on top of a standard sheet metal curb. They are designed to overhang the sheet metal curb to provide a weatherseal and must be connected to the sheet metal curb for positive attachment.

Most aluminum vibration isolation curbs are not adjustable and must be designed properly to keep the unit level. Since they are not adjustable, they usually do not have a built-in vertical restraint system. Without a vertical restraint system, these curbs cannot be subject to either uplift or overturning forces. Most have built-in bumpers in the corners for wind and small seismic disturbances, but because they are essentially open steel springs between two pieces of aluminum, they should not be used in seismic zones with an F_p greater than 0.2*g*. Even if additional restraints are added to the curb, the designer must be aware that the aluminum alloy usually has a tensile strength of between 40 and 60% and a yield strength of 80% of ASTM A 36 steel. Figure 13-3 shows a typical cross-sectional detail of an aluminum vibration isolation system.

For seismic zones with F_p greater than 0.2*g*, a steel curb-mounted vibration isolation system is recommended. This system should be designed with complete spring adjustment capability and seismic restraint in all vertical and horizontal directions. The seismic restraints should be located at or near the sides, ends, and/or crossbraces of the standard sheet metal curb below. All connections of unit to vibration isolation system, vibration isolation system to sheet metal curb, and sheet metal curb to roof structure must be concentrated near the seismic restraints. A practical limit in size of the steel-curb-mounted vibration isolation systems in high seismic zones and/or in essential facilities is no more than 12 ft (3.7 m) long with units weighing no more than 3000 lb (1360 kg). See Figure 13-4A for a steel curb cap.

COMPLETE VIBRATION ISOLATION CURB

Complete vibration isolation curbs incorporate gravity load support, seismic restraint, and vibration isolation all in one. They are made of either formed or structural steel sections. The key to their seismic capabilities is the built-in vertical restraint systems and cross bracing at spring supports.

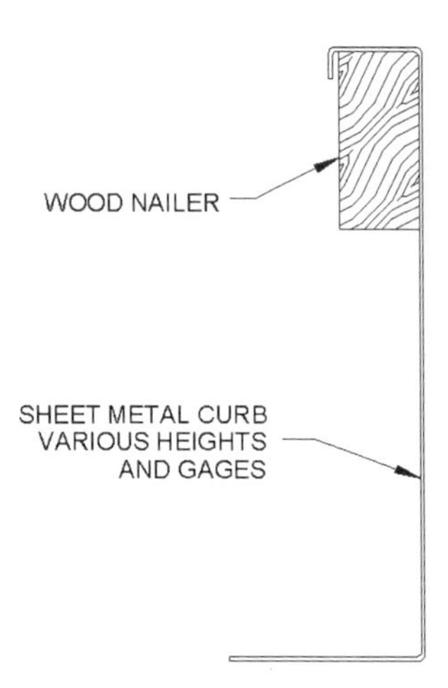

Figure 13-1 Standard sheet metal curb.

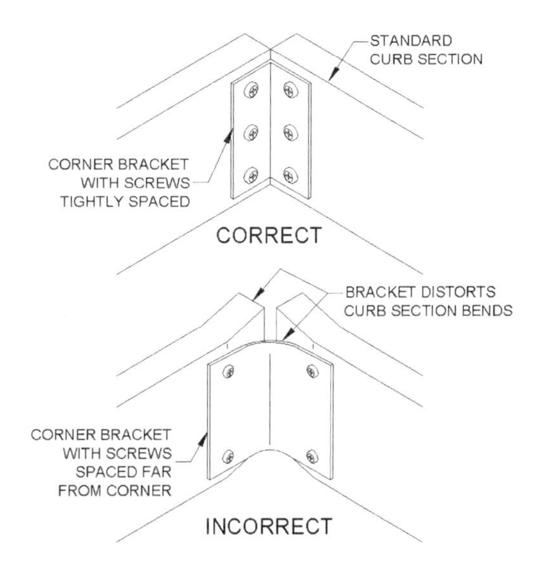

Figure 13-2 Standard sheet metal curb corner connections.

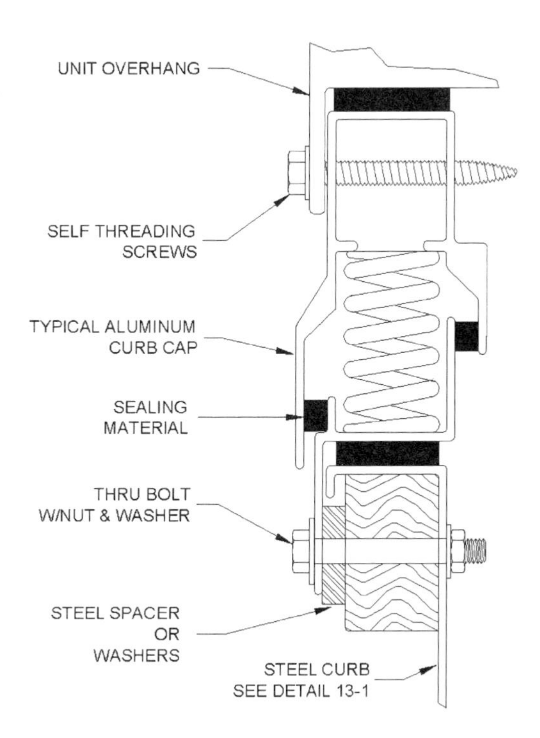

Figure 13-3 Aluminum vibration isolation curb.

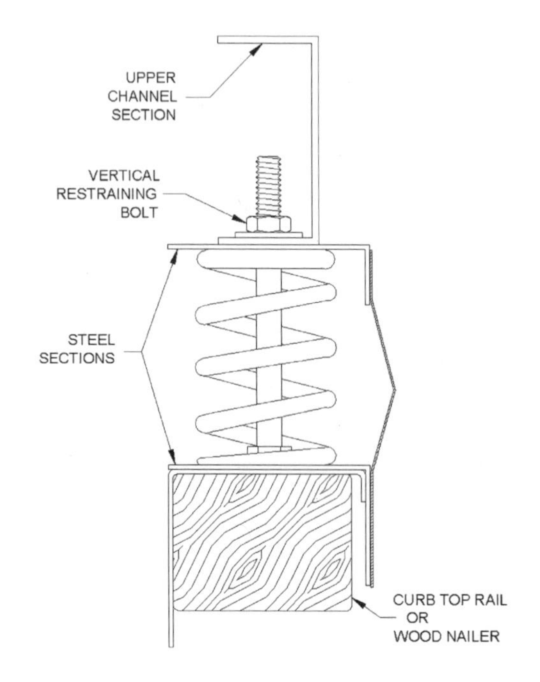

Figure 13-4A Steel curb cap.

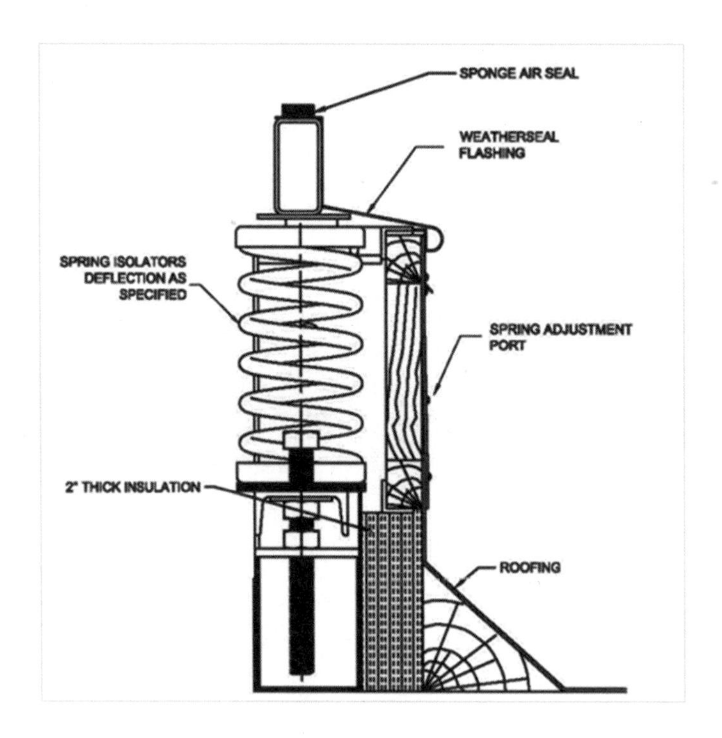

Figure 13-4B Complete vibration isolation curb.

Vibration isolation curbs are designed with vertical support and restraint systems that both provide a platform to transfer the unit gravity loads and act as a restraint to resist horizontal, uplift, and/or overturning seismic or wind forces. There are two ways these curbs are seismically rated. One is to design for a maximum lineal horizontal load per unit length versus maximum distance between restraints. The other is to take the rated load for the restraint system and base the seismic rating on the number of restraints required. This allows the curbs to be used in any seismic area, because additional supports or restraints can be added as required to handle the exact seismic loading. Figure 13-4B shows a typical cross-sectional view of a complete vibration isolation curb.

Other advantages of the complete vibration isolation curb include limiting seismic responsibility to one party and lowering overall height of the curb. The design of the curb and obtaining its connections from a single manufacturer prevent gray area discussions over who is responsible for which portions of the system. In addition, curb-mounted vibration isolation systems can reach combined heights of 24 in. (610 mm). Complete systems can limit the overall height to 14 1/2 in. (368 mm). Additional reductions in height can be designed but are subject to a minimum distance from roof-to-spring port openings in keeping with roofing or project specifications.

EXAMPLE

Design the anchorage for a 5000 lb (22.24 kN) air-handling unit where $S_{DS} = 0.35$ The unit is 27 ft (8.23 m) long, 7.5 ft (2.29 m) tall, and 4 ft (1.22 m) wide. The curb is a 12 in. tall, 14 gage standard sheet metal curb. The rooftop unit overhang is approximately 2 in. (51 mm), and the curb is anchored to a concrete deck. See Figure 12-1.

Calculate the seismic force F_p.

$$F_p = \frac{0.4ap\ SDS\ Wp\left(1 + 2\frac{z}{h}\right)}{Rp/Ip} \qquad (13\text{-}1)$$

$$F_p = 0.4\frac{(2.5)(0.35)(5000)\left(1 + 2\frac{1}{1}\right)}{\left(\frac{2}{1}\right)} = 2625 \text{ lb } (11.67 \text{ kN})$$

F_p does not need to be greater than

$$F_p = 1.6\ S_{DS}\ I_p\ W_p \qquad (13\text{-}2)$$

$$F_p = 1.6(0.35)(1.0)(5000) = 2800 \text{ lb } (12.45 \text{ kN})$$

and F_p shall not be taken as less than

$$F_p = 0.3\ S_{DS}\ I_p\ W_p \qquad (13\text{-}3)$$

$$Fp = 0.3(0.35)(1.0)(5000) = 525 \text{ lb } (2.33 \text{ kN})$$

The unit must also be designed for a concurrent vertical force F_{pv} of the following:

$$F_{pv} = \pm\ 0.2\ S_{DS}\ W_p \qquad (13\text{-}4)$$

$$F_{pv} = \pm\ 0.2\ (0.35)(5000) = 350 \text{ lb } (1.55 \text{ kN})$$

Calculate the overturning moment (OTM).

$$(13\text{-}5) \quad \text{OTM} = F_p \times H_{cg} \qquad (13\text{-}5)$$

$$\text{OTM} = 2625 \text{ lb } (11.67 \text{ kN}) \times 90 \text{ in. } (2.29 \text{ m})/2 = 118125 \text{ in.-lb } (13.35 \text{ kN·m})$$

Calculate the resisting moment (RM).

$$\text{(13-6)} \quad RM_{min} = (W_p - F_{pv}) \times d_{min} / 2 \qquad (13\text{-}6)$$

$$RM_{min} = (5000 \text{ lb} - 350 \text{ lb}) \times 48 \text{ in.}/2 = 111{,}600 \text{ in.-lb } (12.61 \text{ kN·m})$$

Calculate the tensile force T_B at the top of the curb using RM_{min}.

$$T_B = (OTM - RM_{min})/d_{min} \qquad (13\text{-}7)$$

$$T_B = (118{,}125 \text{ in.-lb} - 111{,}600 \text{ in.-lb})/48 \text{ in.} = 136 \text{ lb } (0.6 \text{ kN}) \text{ (for one side of the curb)}$$

Calculate the tensile force per bolt connecting the unit to the curb.

$$T_{B1} = T_B/\text{number of bolts on one side of curb} \qquad (13\text{-}8)$$

$$T_{B1} = 136 \text{ lb}/8 \text{ bolts} = 17 \text{ lb } (0.076 \text{ kN}) \text{ per bolt}$$

Calculate the shear force V_B at the top of the curb.

$$V_B = 1/2 \times F_p/\text{number of bolts on one side of the curb} \qquad (13\text{-}9)$$

(Assume 16 bolts on curb, 8 on each side.)

$$V_B = 0.5 \times 2625 \text{ lb}/8 \text{ bolts} = 164 \text{ lb } (0.73 \text{ kN})$$

Calculating the Attachment of the Unit to the Curb

Using the equations in Chapter 5, the maximum load P_B on the lag screw can be found using the resultant force.

$$P_B = \sqrt{T_{B1}^2 + V_B^2}$$

$$P_B = \sqrt{17^2 + 164^2} = 165 \text{ lb } (0.73 \text{ kN})$$

Because the overhang allows for the horizontal bolting of the overhang into the wood nailer of the curb, the lag bolts need only be checked for shear.

Checking Table 5-6 from Chapter 5, a 3/8 diameter lag bolt with a 3 in. penetration has an allowable shear load of 415 lb (1.85 kN) for 3 in. (76 mm) embedment. Values shall be reduced by half for 1 1/2 in. (38 mm) embedment.

$$P_{allow} = 207 \text{ lb } (0.92 \text{ kN})$$

Since $P_{allow} > P_B$, the anchorage is adequate.

Seismic Force Check on the Standard Sheet Metal Curb

Since the seismic force $F_p = 2625$ lb (11.67 kN) and there is 54 ft (16.5 m) of curb length available to resist the horizontal load, the allowable load per lineal length of curb is:

Allowable lineal load = 2625 lb (11.67 kN)/54 ft (16.5 m) = 48.6 lb/ft (0.71 kN/m).

The curb must be designed and certified to resist the above forces. The minimum number of crossbraces required, based on a maximum unbraced length of 8 ft (2.4 m), would be three. This will keep the unbraced length of curb to 6 3/4 ft (2 m).

Anchoring the Curb to the Structural Deck

The curb will be anchored to a lightweight-concrete-filled deck. Because the curb will be anchored on the upper side of a concrete deck, and the flutes cannot be seen, we will use 1/2 in. diameter through-bolts with fishplates. Assume 10 bolts—five per side, one in each corner, and one at the end of each crossbrace.

Calculate the shear force in the bolts F_V.

$$F_V = F_p / \text{number of bolts} = 2625 \text{ lb } (11.67 \text{ kN})/10 = 262.5 \text{ lb } (1.17 \text{ kN})$$

Calculate the shear stress in the bolts f_V:

$$f_V = F_V/\text{root area of bolt}$$

$$f_V = 262.5\ \text{lb}/0.126\ \text{in.}^2 = 2083\ \text{psi}\ (14.36\ \text{MPa})$$

$$f_{Vallow} = 10{,}000\ \text{psi}\ (68.9\ \text{MPa})$$

Because $f_{Vallow} > f_V$; the shear stress on the bolts is adequate.

Calculate the tensile force in the bolts T_{B2}.

$$T_{B2} = T_B/\text{number bolts on one side of the curb}$$

$$T_{B2} = 136\ \text{lb}\ (0.6\ \text{kN})/5\ \text{bolts} = 27.2\ \text{lb}\ (0.12\ \text{kN})$$

Calculate the tensile stress f_{TB2}.

$$f_{TB2} = T_{B2}/\text{tensile stress area of the bolt}$$

$$f_{TB2} = 27.2\ \text{lb}\ (0.12\ \text{kN})/0.142\ \text{in.}^2 = 192\ \text{psi}\ (1.3\ \text{MPa})$$

Calculate the allowable tensile stress f_{Tallow}.

Because this bolt is in both tension and shear, we cannot use standard allowable stresses. f_{Tallow} must conform to the following equation:

$$f_{Tallow} = 26 - 1.8 f_V \leq 20.$$

Increasing 33% for short-term seismic use, the equation becomes

$$f_{Tallow} = 34.6 - 1.8 f_V \leq 26.6.$$

Substituting, we get

$$f_{Tallow} = 34.6 - 1.8\ (2.083\ \text{ksi}) = 30.8\ \text{ksi},$$

which is not less than 26.6 ksi, so use 26.6 ksi as the allowable tensile stress on the bolts.

Because fTB2 > f_{Tallow}, the bolts are adequate in tension.

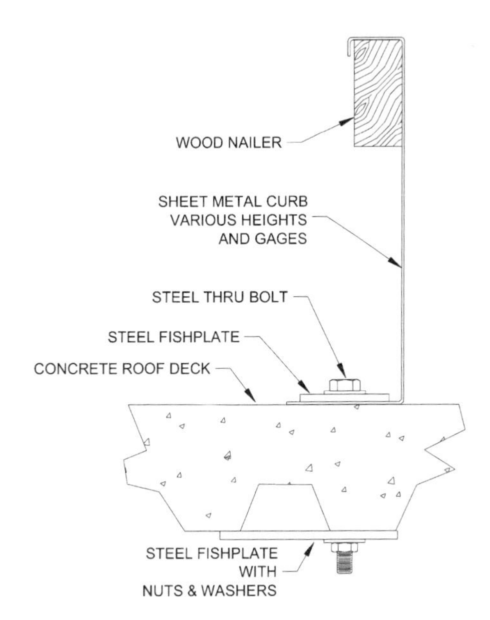

Figure 13-5 Concrete attachment with nuts, through bolts, and fishplates.

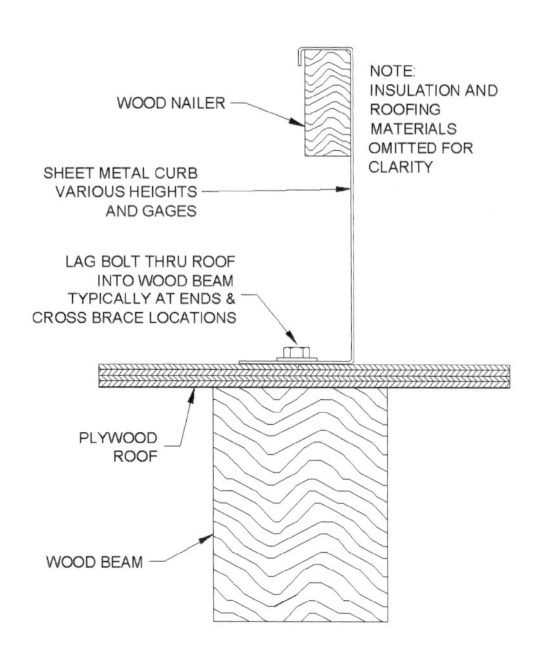

Figure 13-6 Standard curb attached to wood frame system.

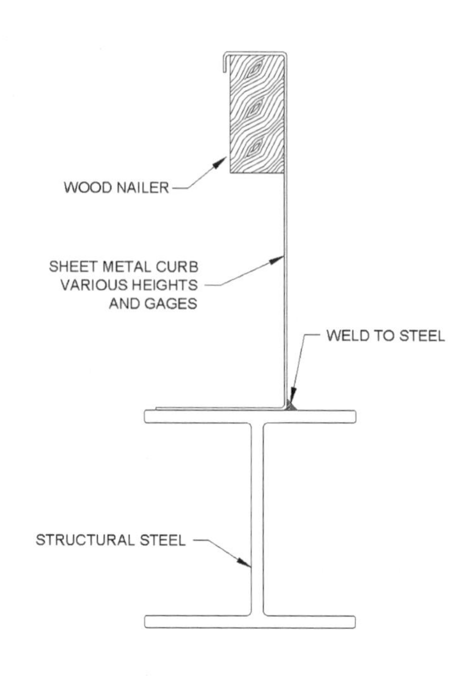

Figure 13-7 Curb welded to building steel.

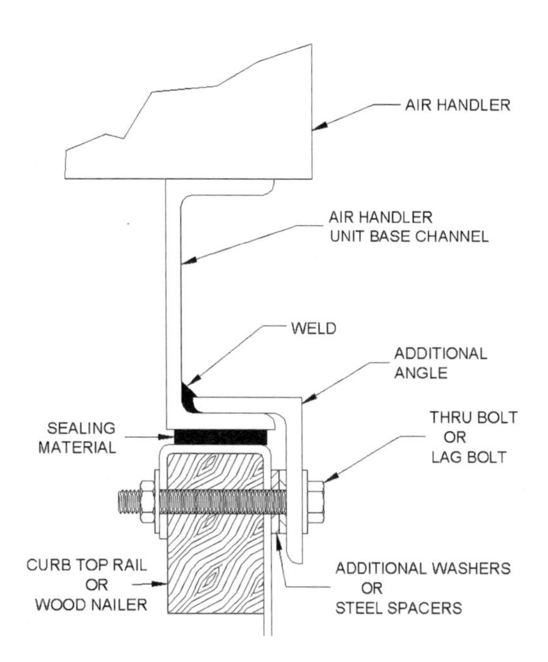

Figure 13-8 Additional angles for RTUs with channel frames.

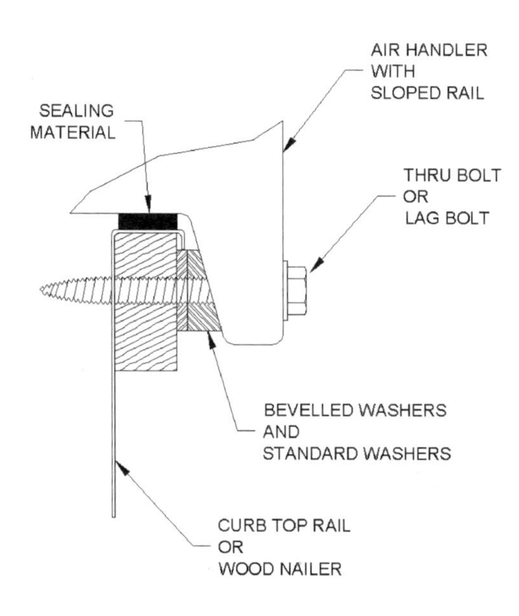

Figure 13-9 Sloped unit overhang attachment to curb.

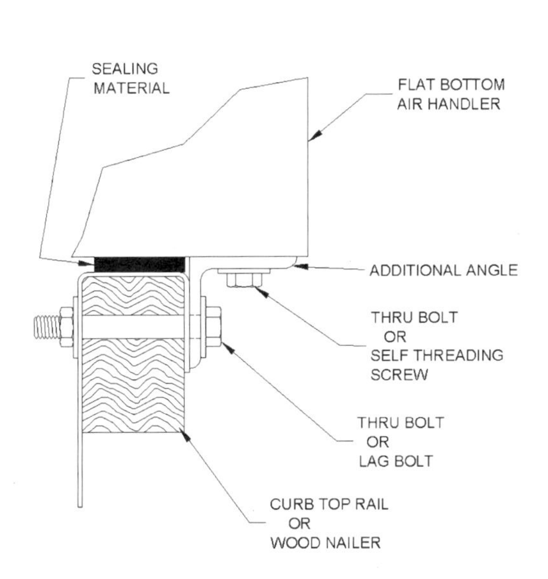

Figure 13-10 Additional angle for flat bottom units.

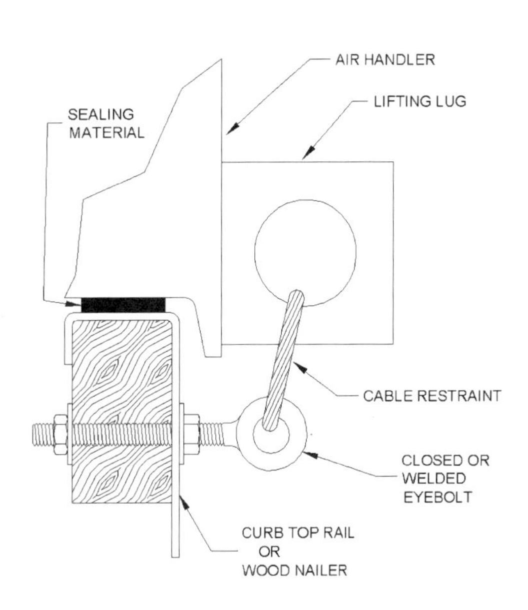

Figure 13-11 Cable tie down.

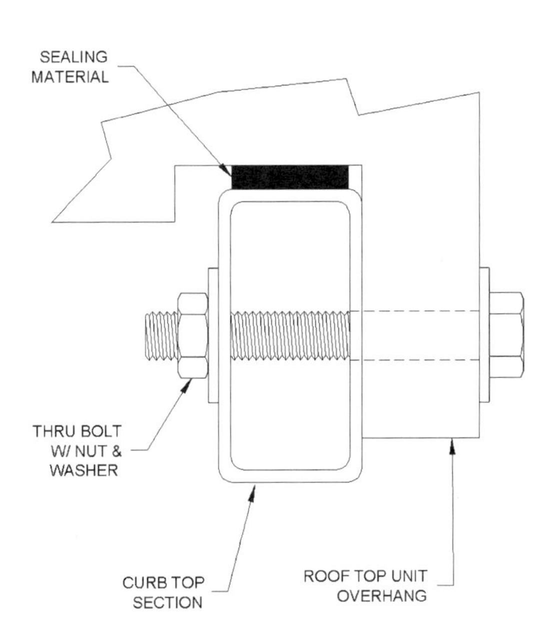

Figure 13-12 Unit overhang through-bolted to curb top section.

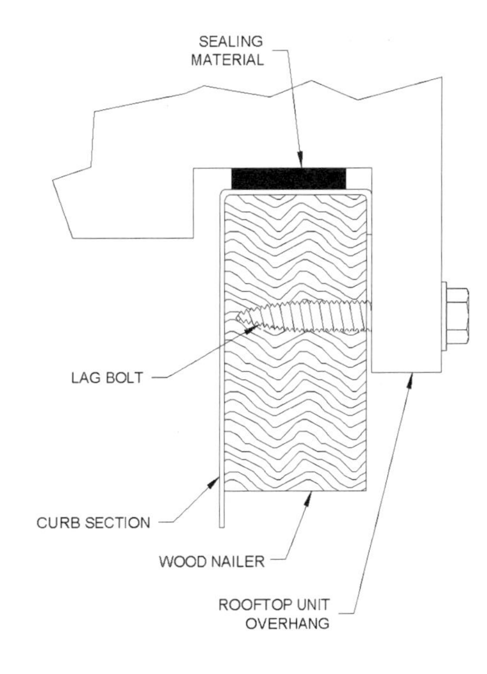

Figure 13-13 Unit overhang lag bolted to curb.

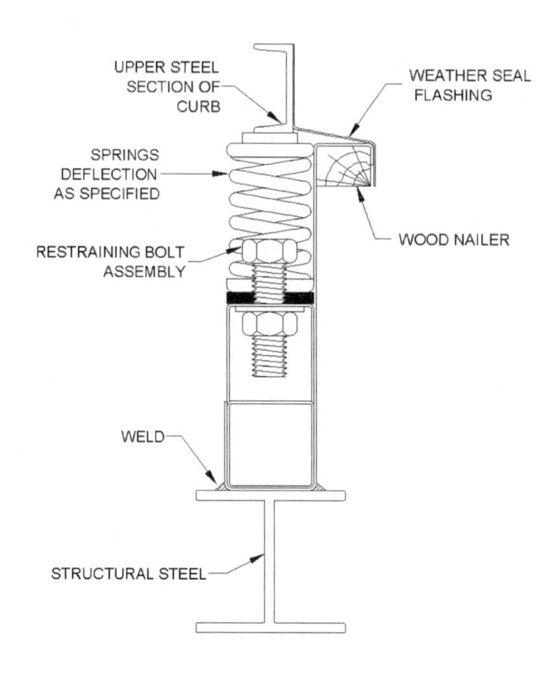

Figure 13-14 Curb-mounted steel vibration isolation system.

14 Cooling Towers and Condensers

The two largest pieces of equipment on a roof that are not curb mounted are usually cooling towers and condensers. Both units require openings to allow air to circulate and cool the liquid. Because these units have large fans to move air, they are usually vibration isolated. (Refer to the vibration isolation schedule in Chapter 48 of the 2011 *ASHRAE Handbook—HVAC Applications*.) In addition to seismic forces, both are subject to wind forces because of their large surface area and height. (Refer to Chapter 12 to determine which force governs.)

COOLING TOWERS

Cooling towers include steel, wood, and fiberglass construction. Some wood units are built on site. These wooden towers need to be designed and analyzed as a building structure and not as a piece of equipment. Most cooling towers are designed to be uniformly supported over the length of the unit. This means a separate steel substructure might be required between the tower piers or isolated supports. It should be designed to transfer the loads from the tower to the support with additional cross bracing as required. Check the certified tower submittal prints or with the tower manufacturer for recommended support steel requirements and allowable deflections.

Most cooling towers are supported by continuations of columns that penetrate the roof or by concrete piers that may run the entire length of the unit. The connections of the supporting steel to the vertical steel or concrete piers should be part of the seismic design for proper transfer of the load.

Vibration isolation rails consist of either angles or channels with springs between the rails. They have some form of vertical restraint but are usually not designed for horizontal forces. These rails may require horizontal cross bracing to keep them from falling over and should be subject to the same seismic certification as all other equipment.

Piping from chillers or pumps penetrates the roof and connects to the cooling towers. In some cases, it is attached or hung from the cooling tower support steel. The weight of the water-filled pipe should be included in the seismic design.

Figures 14-1 and 14-2 show cooling towers attached to steel substructures, isolated and nonisolated. Figures 14-3 and 14-4 show cooling towers attached to concrete piers or rails, isolated and nonisolated. Design and allowable loads for fasteners can be found in Chapter 5. An example for the design of the seismic restraints for a cooling tower can be found in Example 1. See Figure 14-5 for a fiberglass tower attached to a concrete slab.

AIR-COOLED CONDENSERS

Air-cooled condensers are different from cooling towers in that they have legs that allow air circulation. These legs are designed for direct attachment to the roof structure, which must be designed for seismic load transfer to the building structure. Air-cooled condensers are sometimes mounted on rails made from pressure-treated wood. See Figures 14-6 and 14-7 for supporting on piers, isolated and nonisolated. See Figures 14-8 and 14-9 for mounting on structural steel, isolated and nonisolated. See Figure 14-10 for mounting on pressure-treated wood.

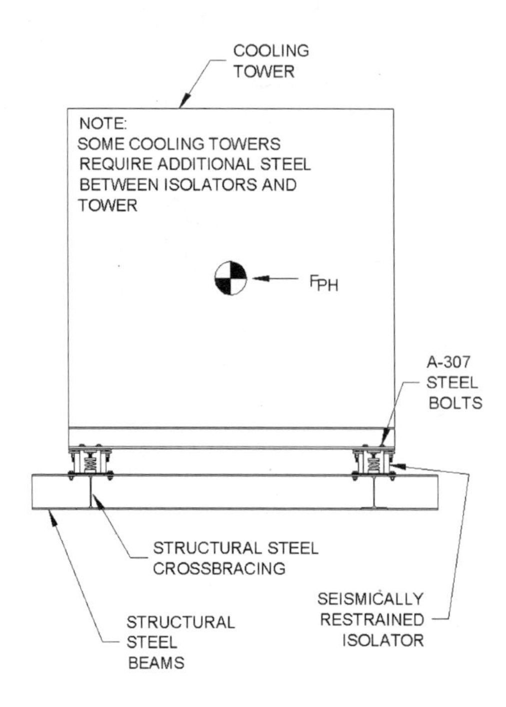

Figure 14-1 Cooling tower supported by restrained vibration isolators resting on steel.

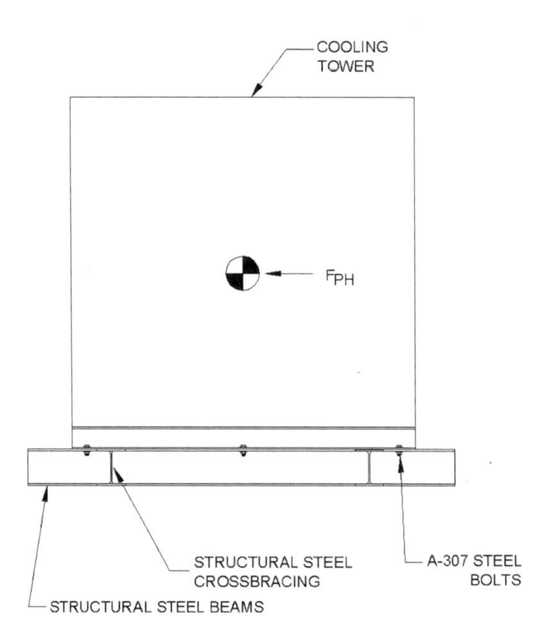

Figure 14-2 Cooling tower rigidly mounted on structural steel.

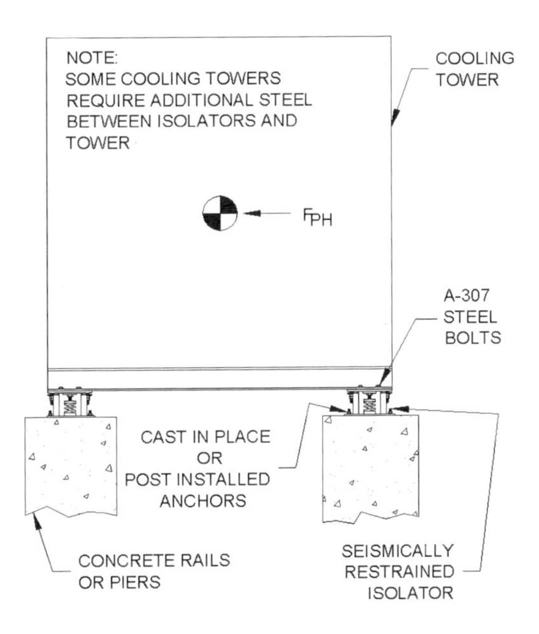

Figure 14-3 Isolated cooling tower attached to concrete piers or continuous concrete abutments.

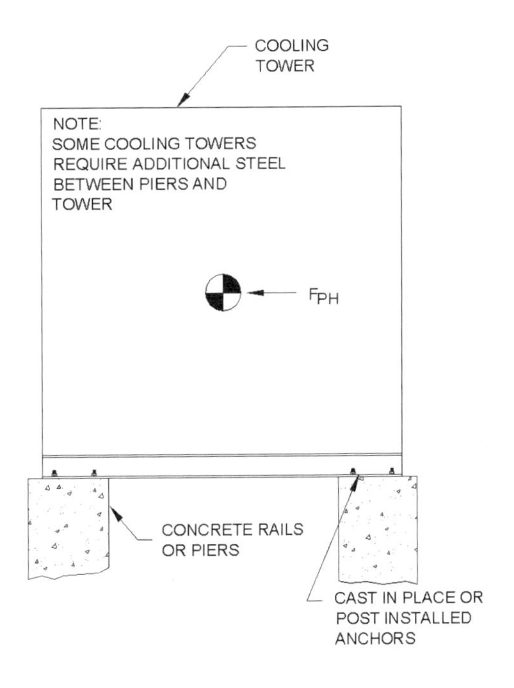

Figure 14-4 Cooling tower rigidly attached to concrete piers or continuous concrete abutments.

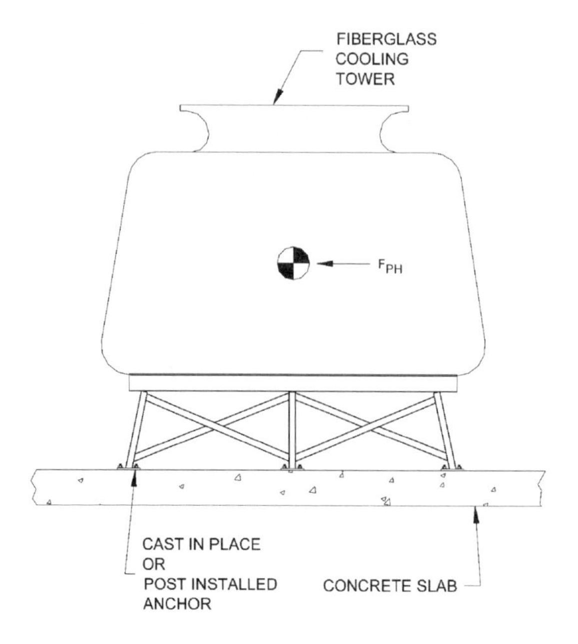

Figure 14-5 Fiberglass cooling tower rigidly mounted on concrete.

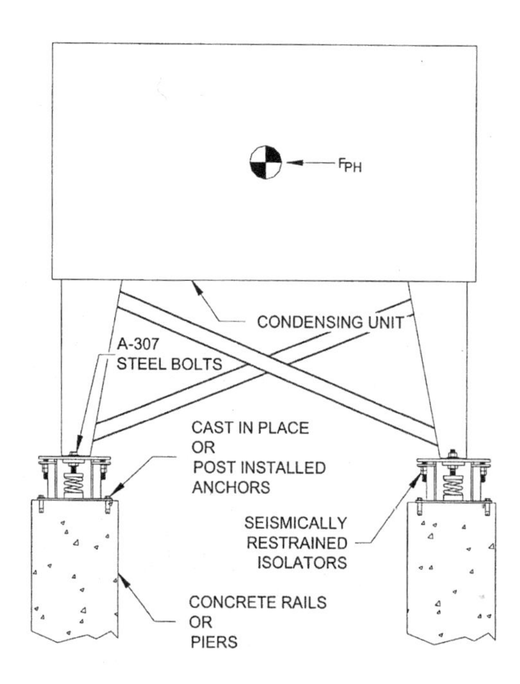

Figure 14-6 Isolated condenser attached to concrete piers or continuous concrete abutments.

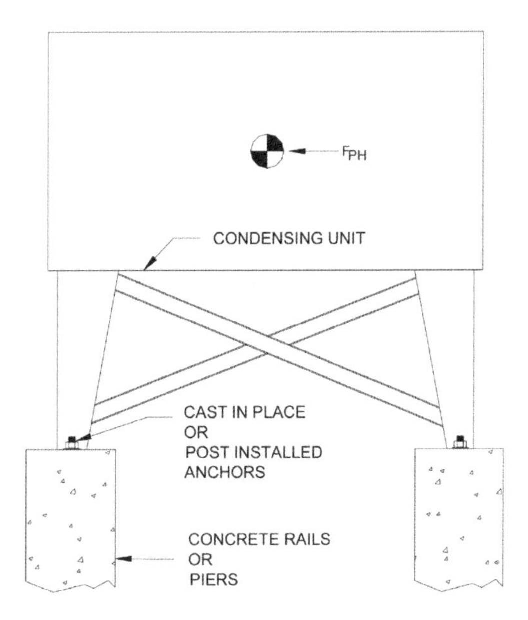

Figure 14-7 Condenser rigidly mounted on concrete piers or continuous concrete abutments.

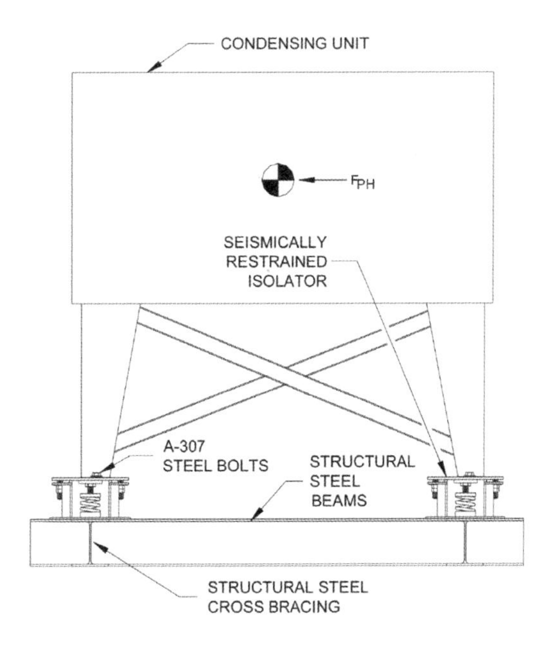

Figure 14-8 Isolated condenser mounted on structural steel base.

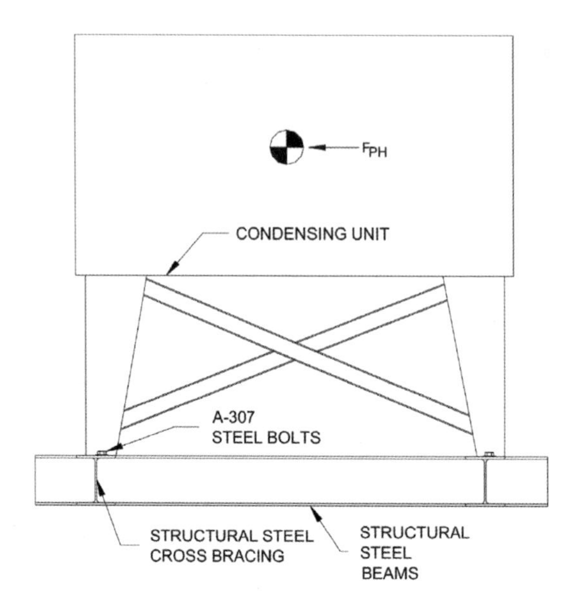

Figure 14-9 Condenser rigidly mounted on structural steel base.

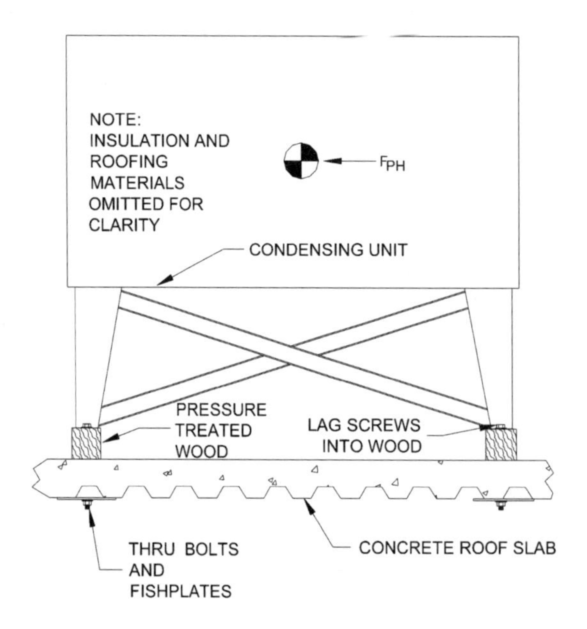

Figure 14-10 Condenser rigidly mounted on treated wood.

EXAMPLE 1

A cooling tower is 24 ft (7.3 m) long, 12 ft (3.65 m) wide, and 16 ft (4.8 m) tall. The operating weight of the tower is 34,500 lb (15 680 kg). The building has an S_{DS} of 1.2. The tower is not isolated and attached with 7/8 in. (22 mm) diameter bolts to a W 12 × 53 I beam. See Figure 14-2.

Calculate the seismic forces F_p.

$$F_p = \frac{0.4 a_p S_{DS} W_p \left(1 + 2\frac{z}{h}\right)}{\frac{R_p}{I_p}}$$

$$F_p = \frac{0.4(1)(1.2)(34,500)\left(1 + 2\frac{1}{1}\right)}{\frac{2}{1}} = 24,840 \text{ lb } (110.5 \text{ kN})$$

F_p does not need to be greater than:

$$F_p = 1.6\, S_{DS}\, I_p W_p$$

$$F_p = 1.6\,(1.2)(1.0)(34500) = 66,240 \text{ lb } (294.65 \text{ kN})$$

and F_p shall not be taken as less than:

$$F_p = 0.3\, S_{DS}\, I_p\, W_p$$

$$F_p = 0.3(1.2)(1.0)(34500) = 12,420 \text{ lb } (55.25 \text{ kN})$$

The unit must also be designed for a concurrent vertical force F_{pv} of the following:

$$F_{pv} = \pm\, 0.2\, S_{DS}\, W_p$$

$$F_{pv} = \pm\, 0.2(1.2)(34500) = 8280 \text{ lb } (36.83 \text{ kN})$$

Calculate the overturning moment (OTM).

$$\text{OTM} = F_p \times H_{c.g.}$$

OTM = 24,840 lb (110.5 kN) × 96 in. (2.44 m) = 2,384,640 in.-lb (269.62 kN·m)

Calculate the resisting moment (RM)

$$\text{RM} = (W_p - F_{pv}) \times d_{min}/2$$

RM = (34,500 lb (153.5 kN) – 8280 lb (36.83 kN)) × 144 in. (3.65 m)/2
= 1,887,840 in.-lb (212.92 kN·m)

Calculate the tensile force T_b.

$$T_b = (\text{OTM} - \text{RM})/d_{min}$$

T_b = (2,384,640 in.-lb [269.62 kN·m] – 1,887,840 in.-lb [212.92 kN·m])
/144 in. (3.65 m)

$$T_b = 3450 \text{ lb } (15.35 \text{ kN})$$

along each long side of the cooling tower

Calculate the shear force V_b.

$$V_b = F_p$$

$$V_b = 24{,}840 \text{ lb } (110.5 \text{ kN})$$

Calculate the shear force per bolt V_{b1}.

$$V_{b1} = V_b/\text{number of bolts}$$

$$V_{b1} = 24{,}840 \text{ lb } (110.5 \text{ kN})/8 \text{ bolts} = 3105 \text{ lb } (13.81 \text{ kN})$$

Calculate the shear stress per bolt f_v.

$$f_v = V_{b1}/\text{root area of the 7/8 in. diameter bolt}$$

$$f_v = 3105 \text{ lb } (13.81 \text{ kN})/0.429 \text{ in.}^2 \ (276.8 \text{ mm}^2) = 7238 \text{ psi } (49.9 \text{ MPa})$$

$$f_{vallowable} = 10{,}000 \text{ psi } (68.9 \text{ Mpa})$$

Because $f_{vallowable} > f_v$, the bolts are adequate for shear stress.

Calculate the tensile force per bolt T_{b1}.

$$T_{b1} = T_b/\text{number of bolts on one side of the tower}$$

$$T_{b1} = 3450 \text{ lb } (15.35 \text{ kN})/4 \text{ bolts} = 863 \text{ lb } (3.8 \text{ kN})$$

Calculate the tensile stress per bolt f_{Tb1}.

$$f_{TB1} = T_{b1}/\text{tensile stress area of the bolt}$$

$$f_{TB1} = 863 \text{ lb } (3.8 \text{ kN})/0.462 \text{ in.}^2 \ (298 \text{ mm}^2) = 1868 \text{ psi } (12.87 \text{ MPa})$$

Calculate the allowable tensile stress ($f_{Tallowable}$).

Because these bolts are in both tension and shear, we cannot use the standard allowable stresses and unity calculations. $f_{Tallowable}$ must conform to the following equation:

$$f_{Tallowable} = 26 - 1.8 f_v \leq 20$$

Increasing 33% for short-term seismic use, the equation becomes

$$f_{Tallowable} = 34.6 - 1.8 f_v \leq 26.6$$

$$f_{Tallowable} = 34.6 - 1.8 \ (7238 \text{ psi } [49.9\text{MPa}]) = 21{,}571 \text{ psi } (148.7 \text{ MPa})$$

Because $f_{Tb1} = 1868$ psi (12.87 MPa) and is less than 21,571 psi (148.7 MPa), the bolts are adequate in tension.

15 Rooftop Fans

Fans are often located on rooftops. Different types of fans may be mounted on different types of supports, but all installations share one common concern. Because attachment of the fans to the structure requires penetration of the weatherproof membrane, careful planning, coordination, and proper weatherproofing are essential.

Rooftop fans are much smaller than cooling towers, condensers, or rooftop package units and have very little impact on the design of the roof. Fan supports and attachments are not always completely detailed on plans. An inexperienced installer may simply mount a fan on wood sleepers that are set in roofing mastic, directly on the roof membrane. While this limits the potential for water leaks by eliminating membrane penetrations, roofing mastic is not an acceptable seismic attachment.

Because the structure is often hidden under a layer of insulation and roofing, attachment to the roof structure is very difficult unless dimensions and detailing are coordinated with the equipment supplier and vibration isolation vendor.

All penetrations of the roof membrane must be sealed. The installer must apply a layer of nonhardening elastomeric sealer under the equipment leg or isolator base plate before it is set in place. The anchor bolt requires sealing as well.

WOOD SLEEPERS

Sleepers must be attached to the roof structure. Roofing cement or mastic is not an acceptable attachment. Sleepers can be screwed or bolted (Figure 15-1). All penetrations of the roof membrane must be sealed with a nonhardening sealer to prevent water leaks. Allowable loads on lag screws may be found in Table 5-6 in Chapter 5.

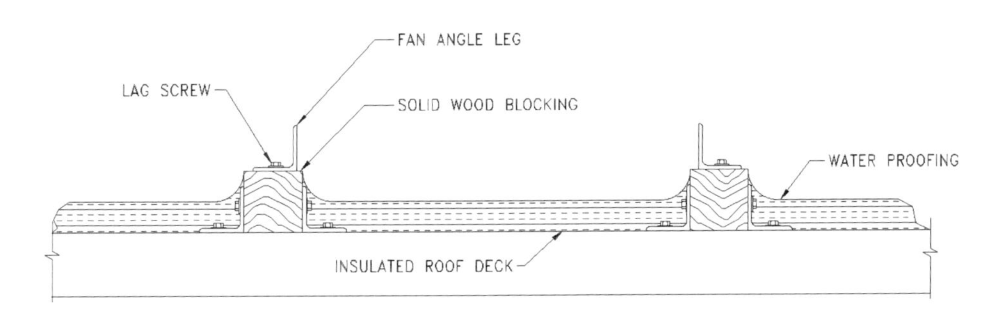

Figure 15-1 Fan mounted on wood sleepers.

WOOD PLATFORMS

Because lag screws must be attached to solid wood members and have no allowable load rating when attached only to a plywood cover, main wood members or solid blocking must be laid out carefully during construction of the platform. If the fan is to be attached directly to the platform, the wood members must be located to match the equipment anchorage holes. If the fan is to be mounted on vibration isolation, the wood members must match the anchorage locations of the restrained isolators or restraints. In either case, the penetrations must be weatherproofed with a nonhardening sealer to prevent water leaks. Allowable loads on lag screws may be found in Table 5-6 in Chapter 5.

CURBS

Curb-mounted fans must be attached to the curb or curb isolation system. Most curb-mounted equipment has an exterior curb lip or flashing that will offer a certain amount of lateral resistance; however, direct attachment is required to prevent overturning caused by wind or seismic loads.

Attachment of the fan to the curb is usually done with sheet metal screws. The screws must be long enough to penetrate the wood nailer and sheet metal curb as shown in Figures 15-2 and 15-3. The curb must be attached to the structure as shown in Figures 15-4, 15-5, and 15-6. Expansion anchors can be used for attaching the curb to a concrete slab or deck. Lag screws can be used for attaching the curb to a wood platform or structural member. Curbs can be bolted to steel structural members or welded. In all cases, attachment should be made at the corners of the curb.

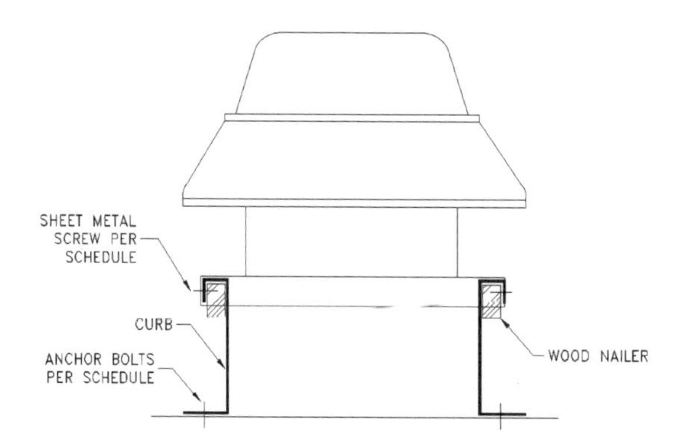

Figure 15-2 Seismic anchorage detail for roof exhausters.

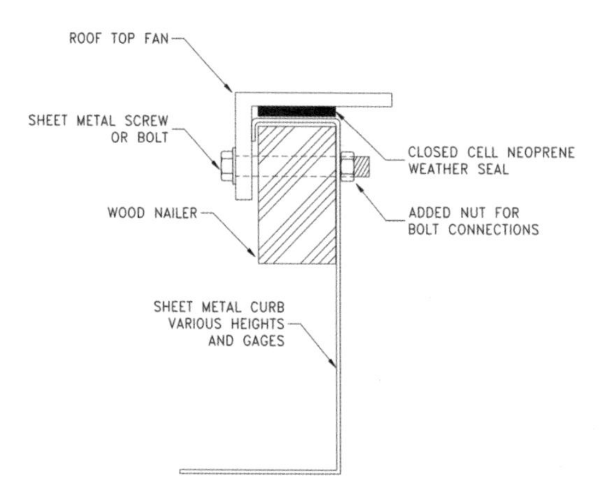

Figure 15-3 Fan attachment to curb.

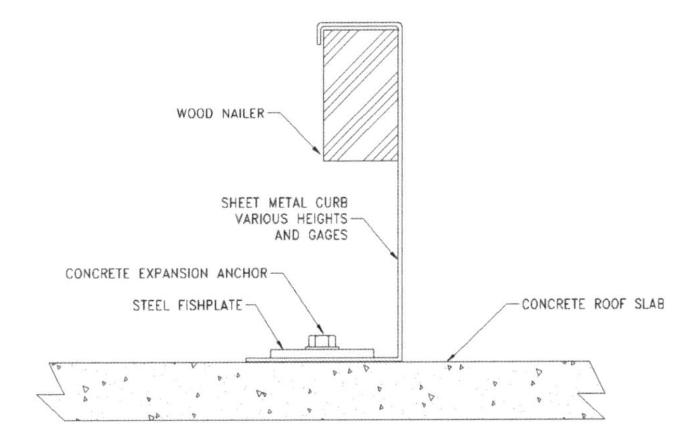

Figure 15-4 Attachment of curb to concrete slab.

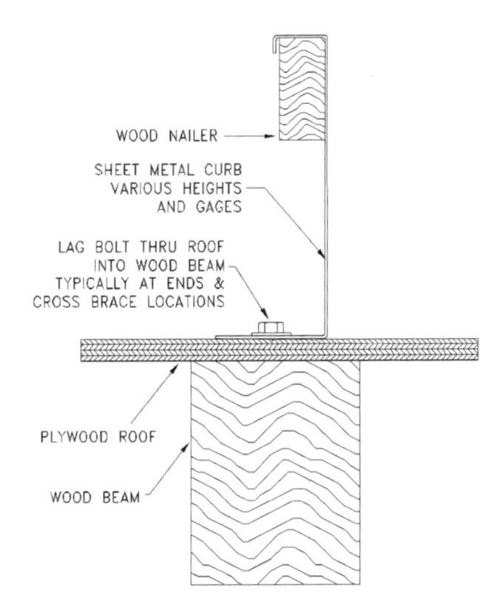

Figure 15-5 Attachment of curb to a wood structure.

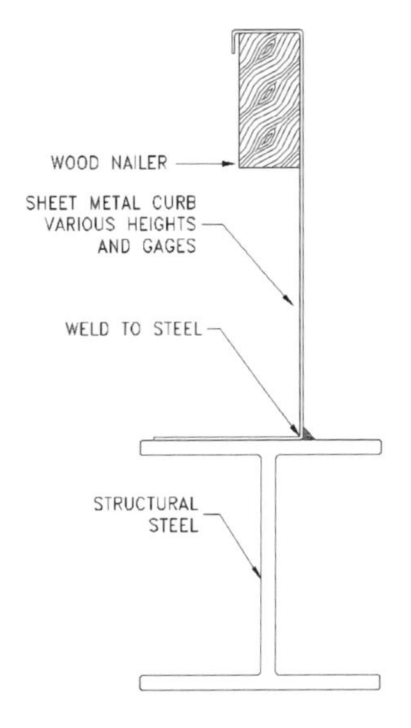

Figure 15-6 Attachment of a fan curb to a steel structure.

CONCRETE PADS AND PIERS

Post-installed anchors can be used to attach fans to concrete pads and piers. The dimensions of the pads or piers should be coordinated with the equipment and isolation supplier. Pads or piers must be anchored to the roof deck or structural steel. Since the expansion anchors penetrate the weatherproof coating, a nonhardening sealant should be used to prevent leaks. Concrete pads should be designed as noted in Chapter 6. Allowable loads on concrete post-installed anchors can be found in Chapter 5, Tables 5-2, 5-3A, 5-3B, 5-4A and 5-4B, and 5-5.

STEEL POSTS AND PLATFORMS

Rooftop fans may be mounted on steel frames that are supported on steel posts that are welded to the structure as shown in Figure 15-7. Fans can usually be welded or bolted directly to steel members. The use of raised posts or platforms minimizes the number of roof penetrations.

ANCHORAGE DESIGN LOADS

Rooftop fans are subject to both wind loads and seismic loads. Refer to Chapter 12 to determine which is the worst-case loading.

EXAMPLE

Design the anchorage using lag screws for a roof-mounted centrifugal fan, where $Ss = 0.50$. Refer to Figure 15-8. The fan is lag screwed to wood beams that are lag screwed to angles welded to the building steel. The seismic loading is larger than the wind loading, so it will be used for the example. The building has an Occupancy Category of II, the Ss is 1.00, it is a site class D, and the building is 40 ft (12.2 m) tall.

Weight of the fan is 1000 lb (4.45 kN), $I_p = 1.0$, $z = 40$, h=40, $a_p = 2.5$, $R_p = 6.0$, H_{cg} = 60 in. (1524 mm)

Calculate the seismic force F_p.

$$F_p = \frac{0.4a_p S_{DS}(1 + 2\frac{z}{h})W_p}{R_p \S I_p}$$

where Fp shall not be greater than $F_p = 1.6\ S_{DS} I_p W_p$

and shall not be less than $F_p = 0.3\ S_{DS} I_p W_p$

and $S_{DS} = 2/3\ S_{MS}$ where $S_{MS} = 2/3\ F_a \times S_s$

$F_a = 1.4$ from Table 2-1 in Chapter 2.

$S_{MS} = 2/3(1.4 \times 1) = 0.93$

$S_{DS} = 2/3(0.93) = 0.62$

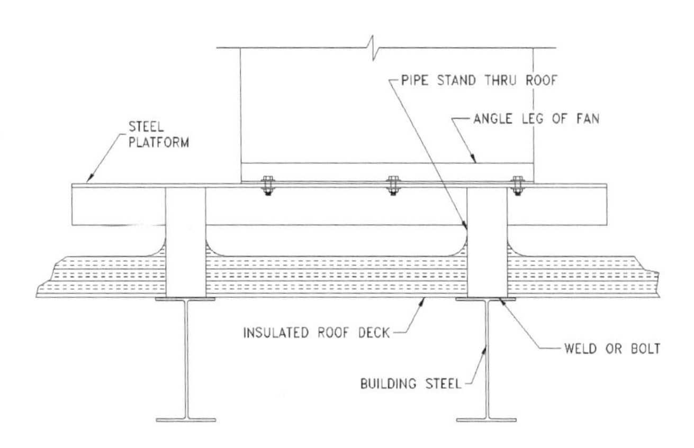

Figure 15-7 Attachment of a steel pedestal.

$$F_p = 0.4(2.5)(0.62)(1+2)1000 = 310 \text{ lb } (1.37 \text{ kN})$$

F_p shall not be greater than F_p = 1.6 (0.62) (1) (1000) = 992 lb (4.4 kN)

or less than F_p = 0.3 (0.62) (1) (1000) = 186 lb (0.83 kN)

The component must also be designed or a concurrent force in the vertical direction equal to $F_{pv} = 0.2\ S_{DS}\ W_P$

F_{pv} = 0.2 (0.62) (1000) = 124 lb (0.55 kN)

Calculate the overturning force (OTM).

$$\text{OTM} = F_p \times H_{cg}$$

$$\text{OTM} = 310 \text{ lb } (1.37 \text{ kN}) \times 60 \text{ in. } (1524 \text{ mm}) = 18{,}600 \text{ in.-lb } (2108 \text{ kN·mm})$$

Calculate the resisting moment (R_M).

$$R_M = (W_P - F_{PV}) \times d_{min}/2$$

$$R_M = (1000 - 124) \times 30/2 = 13{,}140 \text{ in.-lb } (1486 \text{ kN·mm})$$

Calculate the tensile force T_B using R_M.

$$T_B = (\text{OTM} - R_M)/d_{min}$$

$$T_B = (18{,}600 - 13{,}140)/30 = 182 \text{ lb } (0.81 \text{ kN})$$

Calculate the shear force V_B:

$$V_B = F_p/\text{no. of bolts}$$

$$V_B = 500 \text{ lb}/4 \text{ bolts } (2.23 \text{ kN}/4) = 125 \text{ lb } (0.5575 \text{ kN})$$

Checking Table 5-6 in Chapter 5, we find values for a 3/8 in. diameter lag screw with a 3 in. embedment. Therefore the allowable loads are:

Allowable withdrawal load W = 509 lb (2.26 kN)

Allowable lateral load Z = 189 lb (0.84 kN)

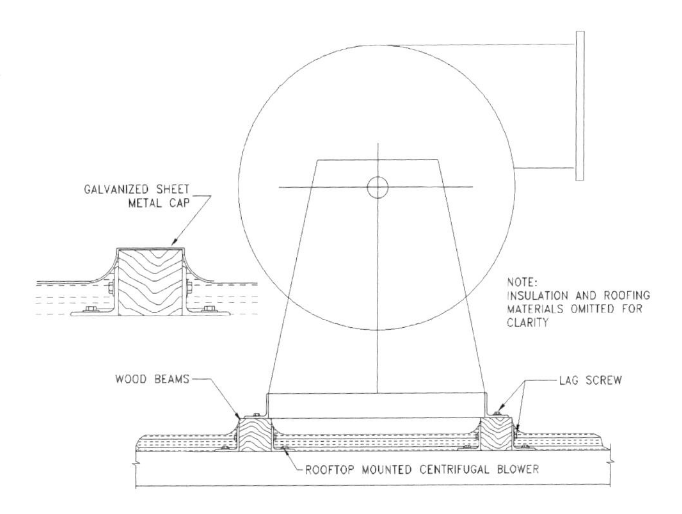

Figure 15-8 Rooftop-mounted centrifugal blower on wood sleeper.

Calculating the Attachment of the Fan Angle to Wood Beam

Using the equations from Chapter 5, the maximum load P_{B1} on the lag screw can be found using the resultant force of the individual forces.

$$P_{B1} = \sqrt{T_B^2 + V_B^2}$$

$$P_{B1} = 182^2 + 125^2$$

$$P_{B1} = 220 \text{ lb } (0.98 \text{ kN})$$

The maximum allowable load $P_{allowable1}$ is determined from the following two equations:

$$\alpha = \tan^{-1}(T_B/V_B)$$

$$P_{allowable1} = W \times Z/W \cos^2\alpha + Z \sin^2\alpha$$

Substituting values, we get:

$$\alpha = \tan^{-1}(182/125) = 56°$$

$$P_{allowable1} = (509 \times 189)/\ 509 \cos^2 56° + 189 \sin^2 56° = 332 \text{ lb } (1.47 \text{ kN})$$

Because $P_{allowable1} > P_{B1}$, the connection of the fan angle to the wood is adequate.

Calculating the Attachment of the Wood Beam to the Building Steel

The worst loading is when the angles are situated as shown in Figure 15-8. The 3/8 × 3 in. (10 ×76 mm) embedded lag screws are now in the horizontal position. The T_B load from the previous section is now the shear load along with F_p A secondary F_p now becomes the tensile or withdrawal load on the lag screws. Therefore,

$$T_{B2} = 125 \text{ lb } (0.5575 \text{ kN})$$

$$V_{B2} = 125 \text{ lb } (0.5575 \text{ kN})$$

$$V_{B3} = 182 \text{ lb } (0.81 \text{ kN})$$

$$P_{B2} = \sqrt{T_{B2}^2 + V_{B2}^2 + V_{B3}^2} = 254 \text{ lb } (1.13 \text{ kN})$$

Calculate $P_{allowable2}$:

$$\alpha = \tan^{-1}(T_{B2}/V_{B3}) = 34°$$

$$P_{allowable2}\ (509 \times 189)/509 \cos^2 34° + 189 \sin^2 34° = 235.2 \text{ lb } (1.04 \text{ kN})$$

Because $P_{allowable2} < P_{B2}$, the connection is *not* adequate. Longer embedment, larger diameter, or additional lag screws are required.

16 Bomb Blast Design

INTRODUCTION

Dynamic loadings on structures caused by bomb blasts have been associated with wartime conditions only. While the threat of a full-scale war is diminishing, there is an increase in the danger of structural damage through acts of terrorism. Although more limited, the results are often devastating. Damage to the structural integrity of a building and the corresponding loss of functionality can be described as ranging from a complete structural collapse and, consequently, a total loss of all function, through partial to no structural damage with continued function. This chapter concentrates on the latter, where through the protection of equipment and auxiliary systems against shock from bomb blast, building services can be continued.

The type of shock isolation system should be established in an early stage of the structural design, in close cooperation with the structural, mechanical, and electrical disciplines.

SOURCES OF DYNAMIC IMPACT

Figure 16-1 shows how bomb blast forces are transmitted to a structure. If the blast is above ground, the forces are primarily transmitted by air and, if below ground, primarily as a shock wave.

Weapons exploding near or below ground level normally show effects that are greater in magnitude and duration than similar bursts in midair. An explosion near the surface generates a ground shock effect by a direct transfer of the weapon's energy into the ground and by an interaction of the expanding air blast pressure wave along the ground surface. This last phenomenon, called an "air-induced ground shock," occurs when the air blast pressure compresses the ground surface, resulting in a stress wave that propagates downward and outward along the surface. Together with the "direct-induced ground shock" and depending on the site's geology, vertical and usually stronger horizontal shock waves impact the structure's foundation and transmit the forces to the rest of the structure.

This dynamic impact is similar to an earthquake. However, bomb blast is characterized by high structural accelerations or high *g*-forces of short duration (milliseconds), whereas an earthquake is characterized by the usually larger horizontal motions of relatively long duration. The long duration increases the possibility of damaging resonance effects.

SHOCK OR DESIGN RESPONSE SPECTRA

For structures with a protective purpose, the design process consists of both a static and dynamic analysis. The static portion concerns the loadings during normal day-to-day operation. The dynamic portion demonstrates the protective capability of the structure against air pressure and ground shock loadings. This analysis provides time-history data of displacements, velocities, and accelerations. The results of a dynamic analysis, showing

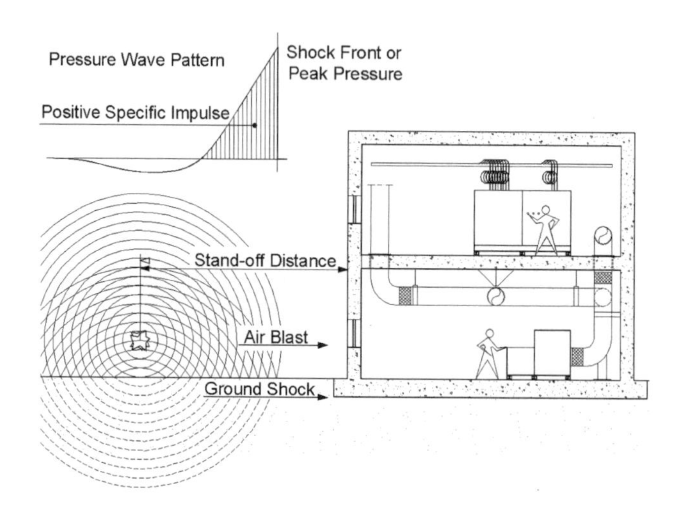

Figure 16-1 Diagram of bomb blast forces transmitted to a structure.

the maximum results, are plotted on a triparte log diagram as a function of frequency. This is done for all three axes.

To establish the required level of protection, the structural response spectra need to be transformed into design or shock response spectra (DRS or SRS). This is done by expressing the response of a single degree of freedom spring-mass system to the established structural motion in terms of the maximum relative displacement between structure and supported equipment and system. The corresponding pseudovelocity and absolute acceleration, or g-force, are then defined as a factor ($2\pi \times f_n$), respectively, $(2\pi \times f_n)^2$ times the frequency-dependent maximum relative displacement.

Design response spectra are normally simplified to an envelope of three straight lines running through the maximum occurring displacement, velocity, and acceleration (D_o, V_o, A_o). In Figure 16-2, a fictitious shock spectrum is shown for a bomb blast application. For comparison purposes, a 0.2 *g* earthquake design response spectrum has been included as well.

For high-frequency systems, the spring will act relatively stiffly. When the structure moves, the spring forces the supported mass to move with it, and the mass receives the same shock acceleration. Rigidly supported systems fall into this category.

However, due to an amplification of motion, design accelerations can increase by a factor of 1.5 to 4 times the shock acceleration, especially around the intermediate frequencies. These are the frequencies where the shock velocity shows an almost constant value.

In general, the amplification for the displacement is less than that for the velocity, which is, in turn, less than that for the acceleration.

For low-frequency systems, the response approaches an asymptote of constant relative displacement. This can be seen as the combination of a very large mass and a very flexible spring. When the structure moves relatively fast, the mass does not have the time to move with it. Thus, the maximum spring deflection will be the same as the maximum structural displacement. The mass remains motionless and experiences only small accelerations.

SHOCK ISOLATION DEVICES

Shock isolation devices come in various shapes, materials, and characteristics (Figure 16-3).

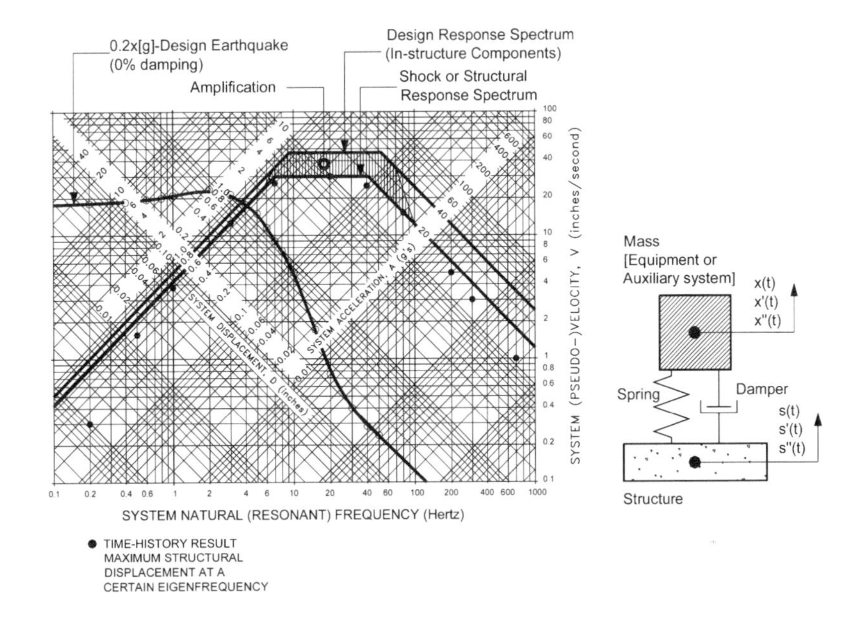

Figure 16-2 Shock spectrum for a bomb blast application.

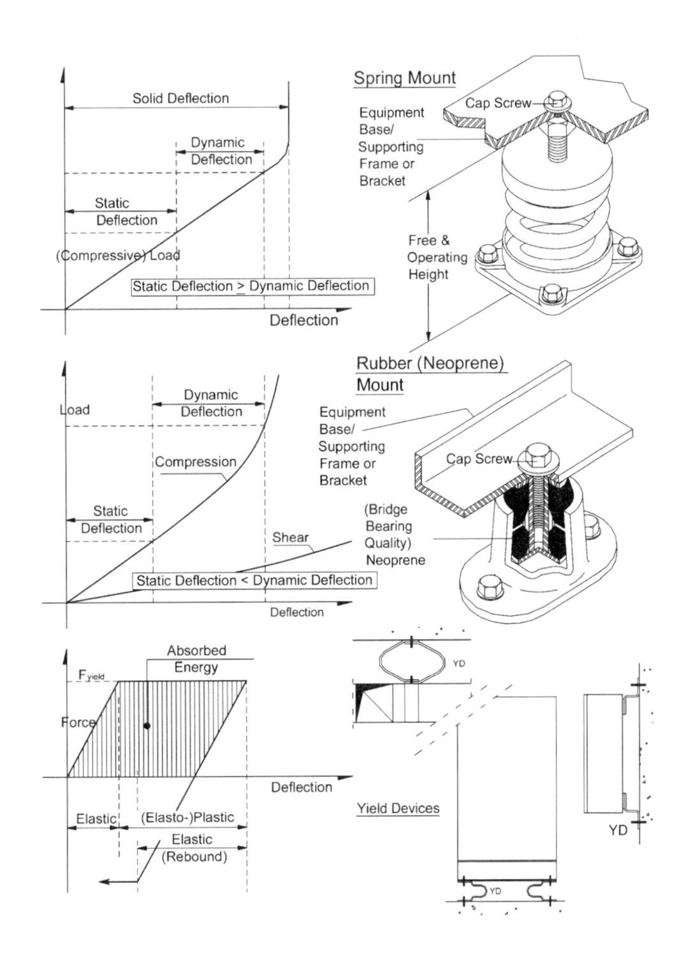

Figure 16-3 Shock isolation devices.

Steel Coil Spring Systems

Steel coil spring systems are one of the simplest and most widely used methods for shock, seismic, and vibration control. Their spring constants remain equal for all deflections to solid, and there is no dynamic stiffening effect.

The use of spring systems can be divided into two categories, springs subjected to compression forces only and springs subjected to tensile forces only. Most compression spring systems are noncaptive in the axial uplift direction; therefore, the maximum occurring dynamic deflection has to be less than the static deflection and only accelerations up to a maximum of 1 *g* are transmitted from outside sources into the supported system.

The common range of steel spring system natural frequencies is from 1.6 Hz to 3.6 Hz, which correspond to static deflections of 4 in. (100 mm) to 3/4 in. (19 mm), respectively.

Spring systems with a horizontal to vertical spring stiffness ratio ranging between 0.6 and 1.2 are suitable to absorb shock motions in all three axes, provided lateral stability is ensured. Otherwise, equipment could topple over. Lateral stability can either be achieved by means of a stable spring design, additional snubbing devices, or sway braces. These devices need to be designed so as to not engage during shock motions in order to ensure a low level of force transmission. Sudden contact or short-circuiting will transmit peak accelerations into the equipment.

Elastomeric Mounting Systems

These systems have the advantage of being molded into various shapes and stiffnesses and have more internal damping than steel spring systems.

By changing the hardness or durometer of the elastomeric compound, it is possible to vary the mounting stiffness without changing its geometry. Steel inserts and housings make it possible to provide a proper means of connection between structure and supported system which, therefore can endure a combination of compression, tensile, and shear forces without loosing stability or having the need for additional snubbing devices.

To avoid additional deflection through relaxation, elastomeric mounts should not be designed for large static deflections for their normal day-to-day operation. In general, the compressive static deflection should not be larger than 10 to 15% of the unloaded total height and should not exceed 25 to 40% for shear loading.

Designers must be aware that under dynamic loading, elastomeric materials act stiffer and show increased load-deflection ratios than when subjected to static loads. This effect, the dynamic to static stiffness ratio, increases with the elastomeric compound's hardness and has to be taken into account during the dynamic analysis.

YIELD DEVICES

These are characterized by their capability to absorb kinetic energy by elastoplastic behavior, or yielding at predesignated steel sections. The transmitted forces are limited to the internal forces necessary to exceed the yield stress of the material.

Yield devices are steel straps bent in various shapes dependent on the required shock deflections. By varying the width and thickness of the straps, the final level of elasticity, stiffness, and yielding force can be established.

The common range of the yield devise system's natural frequencies are from 10 to 20 Hz before yielding. Due to a lack of damping capacity in the elastic phase, amplification or resonance effects might occur. Yield devices are more suitable for highly fragile auxiliary systems, such as piping, air ducts, or cable trays, and for nonrotational equipment, such as electrical panel boards and telecommunication equipment. Composite yield devices, combining steel and elastomeric materials, have been designed and can reduce vibration transmission.

However, due to the yielding nature of these devices, misalignment of equipment and auxiliary systems might occur. Remedial action or even a full replacement of the supporting systems could become necessary.

GENERAL DESIGN PROCEDURE FOR SHOCK ISOLATION AND RESTRAINT SYSTEMS

The first step in the selection of any suitable isolation system is to check on internal sources of vibration and/or noise disturbances. For a required percentage of vibration isolation I against a certain disturbing frequency f_d, a suitable isolation system has to be implemented with a maximum allowable natural frequency $f_{n,max}$ equal to

$$f_{n,\,max} = f_d \times \sqrt{(1-\alpha)/(2-\alpha)}\ ,$$

where $\alpha = I/100\%$.

When the design requirements call for shock isolation, the next step is to verify the fragility level of the equipment or system to be supported against the occurring shock accelerations.

The required shock isolation system might be one and the same with the required vibration isolation system, provided there is enough deflection capacity left in the vibration isolation system to absorb the specified shock motions.

In securing equipment and systems to the structure, the following methods are available.

Hard-mounted systems. If the fragility level exceeds the proposed shock level, no flexible intermediary is required. The system can be connected to the structure directly by means of anchor bolts. Sliding systems can be used for certain types of systems with relatively high fragility levels. Here, the system can move freely with certain restrictions. This should only be considered in the case of small shock motions, not exceeding 1 in. (25 mm). Connections to piping and ducts have to be flexible enough to absorb these shock motions.

Hard-mounted systems are commonly used for auxiliary systems, tanks, and (expansion) vessels with high centers of gravity. Sliding systems are used for tanks and vessels with low centers of gravity, hence, without the danger of toppling over. Snubbing devices can be used to restrain excessive motions.

Flexible supporting systems. Where shock levels exceed the fragility levels, the use of a flexible intermediary becomes apparent. Following is an example of an isolation system selection procedure for a typical shock response spectrum, as shown in Figure 16-4.

EXAMPLE

An electrical panel board with a maximum fragility level of 4 g is to be installed with shock values as indicated in Figure 16-4. Using an inertia base lowers the overall center of gravity, which helps to reduce rocking modes. Note that this is a severe simplification of the

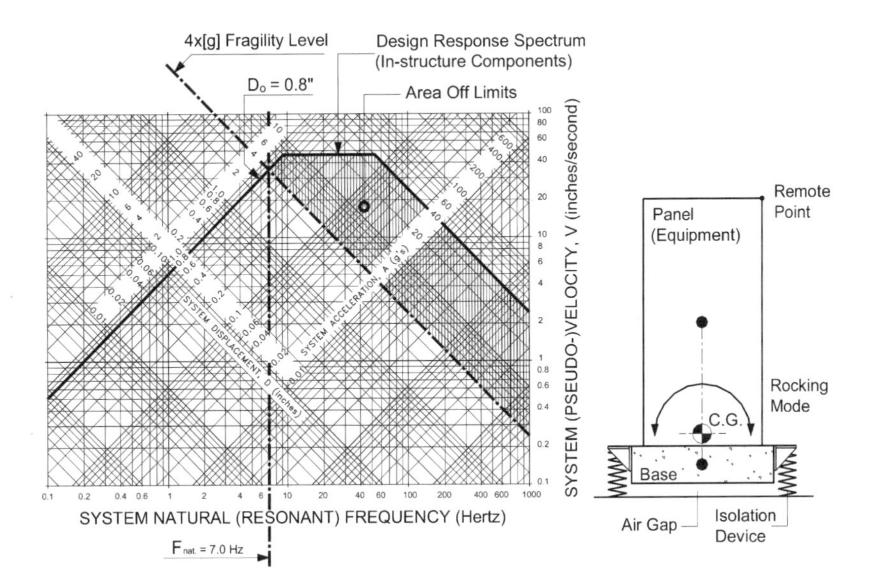

Figure 16-4 Typical shock response spectrum.

facts. Only through dynamic analyses is it possible to account for all translation, rotation, and rocking effects.

A shock isolation system with a natural frequency not exceeding 7.0 Hz would be suitable. Rubber mounts or steel springs may be used on this system and must be able to absorb the dynamic deflection of 0.80 in. (20 mm). In the upward and downward vertical directions, this must be added to or subtracted from the static deflection.

For rubber mounting systems, a dynamic to static stiffness ratio must be incorporated into the design. A 7 Hz dynamic system with a dynamic to static stiffness ratio of 1.4 requires the use of a 5.9 Hz static natural frequency in the vertical direction (7 Hz divided by $\sqrt{1.4}$). For a static deflection of 0.25 in. (6 mm), the minimum vertical deflection capacity has to be 1.05 in. (26 mm). For some rubber mounting systems commonly used for vibration isolation purposes, this could be excessive.

Rough design check:

$$\text{g-level} \approx \text{total deflection/static deflection} = 1.05\text{ in.}/0.25\text{ in.} \approx 4$$

A steel spring system might be more suitable for this case. The static vertical deflection has to be at least 0.80 in. (20 mm); hence, the system's maximum natural frequency is 3.5 Hz. The total deflection capability of the spring element has to exceed 1.60 in. (40 mm) to avoid the spring coils from running out of free space or "going solid."

Note that the components of shock isolation systems that are attached to the structure (i.e., the parts below the flexible intermediary) will be subjected to the full shock acceleration level applicable for hard-mounted systems. The weight of the isolation device can therefore play a considerable role in the final anchor bolt design.

As an example, a nonisolated device weighing 10 lb (4.5 kg) subjected to an acceleration level of 40 g supports a mass of 400 lb (181 kg), which is reduced to a level of 4 g. Thus, at least 20% $\{= (10 \times 40) / [(10 \times 40) + (400 \times 4)] \times 100\%\}$ of the total force acting on the anchor bolts is directly related to the nonisolated weight of the device.

The use of anchor bolt studs or sleeves in slotted or large holes should be avoided. The sudden movement of the device's base plate can snap the anchor bolt due to dynamic impact. When slotted or large holes cannot be avoided, an epoxy of sufficient compressive strength should be used to fill the gaps.

In any situation, vertical and horizontal motions could occur simultaneously. Results become more complicated when rotational or rocking modes are introduced. These occur due to eccentricities of the equipment's or auxiliary system's center of gravity in relation to the center of gravity of the shock isolation system. Any location on the isolated system is subjected to shock motions, which can be split into three translations and three rotations.

Sufficient air gaps and/or rattle space should be provided for the system to move freely through these shock motions without any contact with the structure and adjacent or overhead equipment and auxiliary systems. The displacement of connections to piping and air ducts through flexible joints and at remote points located farthest away from the overall center of gravity have to be determined and checked.

The use of dynamic analysis software has made it possible to design for all these effects quickly.

In general, shock isolation systems are very similar to systems applied for vibration isolation and seismic restraint. However, in shock environments, higher dynamic deflection capabilities in the system are required. Motion restraints used in the seismic field, where the accelerations are low and short-circuiting the system results in an acceptable increase of the transmitted accelerations, are to be avoided during shock, when acceleration levels are much higher, unless the dynamic shock system can handle it.

TYPICAL SUPPORTING SYSTEMS

Depending on their location, the proposed shock levels, and the concentration of equipment, isolation systems can be designed to support single or multiple pieces of equipment. New developments in protective structure design show shock isolation systems supplied as an intermediate component between the protective shell and the inner structure (see Figure 16-5).

Single Supporting Systems

Standard off-the-shelf equipment, originally designed to be simply floor, wall, or ceiling mounted, is often not capable of withstanding the forces set free during a shock impact. Local deformations might occur because of insufficient stiffness of the equipment's housing structure, even when supported by an isolation system. Using an additional supporting frame or inertia base can help to avoid local stresses.

In the case of lightweight equipment, the additional mass and stiffness supplied by a supporting frame can help achieve a more stable system. Tall, slender equipment, such as electrical panel boards, benefits from an additional concrete or steel base because of a lower overall center of gravity. With the additional base and with a possibly wider spread of the isolation devices, which results in a more stable layout, rocking effects can be reduced considerably.

However, awkward differential motions between personnel and panel boards may occur because of the flexibility of the isolation system. Sway braces or breaking pins, allowing the shock isolation devices to come into action only when a certain force is exceeded, are some of the possible solutions to restrain the flexibility during day-to-day operation.

Floating-Floor Systems

Wall-to-wall shock isolation systems can be used where equipment is concentrated in areas such as utility and plant rooms. These systems eliminate the differential motions between personnel and equipment. The equipment would be anchored to a steel or concrete subfloor, which would be isolated from the structure.

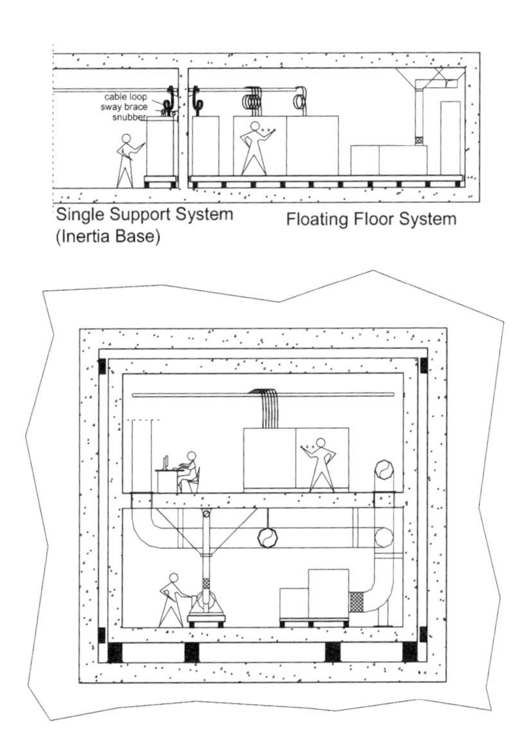

Figure 16-5 Typical supporting systems.

To avoid resonance during normal operation, flexible or vibration isolated systems should have natural frequencies over 3 Hz. This range includes low-deflection steel springs, rubber mounts, or even yielding devices, with or without breaking pins and/or snubbing devices.

To achieve a successful isolation system based on a floating floor design, it has to be implemented in the early stages of the structural design in order to account for the additional weight. Not having to deal with the installation of separate isolation systems for equipment in the mechanical and electrical installation phase of the building process can prove to be beneficial, both logistically and timewise.

Low-g Building Supporting Systems

One of the new developments in the shock isolation of protective structures is the low-*g* structure. This is a complete building section that contains the operational facilities and is isolated at its base level within a protective outer shell structure.

The design relies on the principle of achieving a long stand-off distance between the source of shock impact and the structure. Less economical because of the requirements of a double concrete structure, the advantage of creating a safe low-*g* environment for personnel, equipment, and auxiliary systems can be beneficial.

However, a low-*g* environment might still require equipment and systems to be flexibly mounted because of vibration and noise control requirements. Equipment must still be anchored to prevent overturning during shock impact.

A thorough dynamic analysis is required to help design for the combination of different flexible systems and their impacts on each other.

CONCLUSION

It is apparent that, although shock, seismic, and vibration load characteristics might be different, it is possible to use isolation devices, initially designed for vibration control purposes, to deal with shock impact as well, provided they incorporate sufficient dynamic deflection capabilities.

Due to the complexity of the dynamic sources and the various vibrational, rotational, and rocking modes of systems, the performance of a dynamic analysis is still required.

17 Residential

Residential equipment and associated systems are subject to the same seismic design methods as equipment in commercial buildings. In general, equipment should be attached to the structure. Although building codes include some exceptions based on weight, size, or location, some equipment may be considered important to the service of the residence after an earthquake and, therefore, should be anchored.

When designing seismic anchorage for residential equipment, it is important to meet the code requirements, which require the equipment to remain in place for safety purposes to prevent personal injury. The attachment of residential equipment can be performed at a relatively low cost. Equipment that is anchored properly will not only stay in place but will have a very good chance of remaining functional after an earthquake.

This chapter includes suggested requirements for several different types of equipment. Each section discusses the potential value of the equipment after an earthquake and recommended techniques for attachment.

WATER HEATERS AND STORAGE TANKS

Most equipment that stores water requires bracing for safety purposes because of its weight. In addition, the water heater or storage tank may be the only source of potable water available immediately after an earthquake if water mains are cut off from the residence or an entire neighborhood. For this reason, the water heater may be the most important piece of equipment in a residence.

Water heaters and storage tanks are most often located in a basement or garage. In either case, the floor is usually concrete with stud walls or a combination of stud walls and concrete walls on one, two, or three sides of the equipment.

Because the equipment tends to have a high center of gravity and a narrow base, the equipment is prone to swaying and falling over rather than sliding. Bracing of the equipment above and below the vertical center of gravity to the adjacent walls reduces the possibility of damage to piping, flue, or burner assemblies. Figures 17-1 and 17-2 show recommended bracing techniques.

BOILERS/FURNACES

The importance of boilers and furnaces in cold regions equals that of water storage. Heating system components may be located in the basement or garage and are usually set on a concrete floor. In some cases, they are located inside the residence within a closet or utility room and set directly on the wood floor.

The equipment should be anchored at the four corners of the base either directly through the equipment mounting holes or by attaching clips to the base of the equipment for anchorage, as shown in Figure 17-3.

AIR-CONDITIONING SYSTEMS

In earthquake regions where temperatures are excessive, air-conditioning systems can be important to the health of sick or elderly persons. Increased fatalities during a heat wave are a fact of life. An earthquake could damage cooling systems during a period of high temperatures and increase fatalities.

Components of a cooling system include condensing units, fan-coils, heat pumps, and packaged air-conditioning units. Condensing units are usually located outside the residence, set on a concrete pad or on the roof. Fan-coils and heat pumps are located inside the attic, closet space, utility room, basement, or garage of the residence. Packaged air-conditioning units are usually located on the roof, or smaller units are located in a wall or window.

Unlike water heaters, condensing units are not top heavy and only require attachments to reduce the possibility of sliding. In most cases they can be directly anchored to the concrete slab or to attachment clips anchored to the slab on all four sides of the unit, as shown in Figure 17-3.

Fan-coils and heat pumps on a wood-framed floor or roof should be directly anchored through the cabinet into a plywood platform or wood-frame support, as shown in

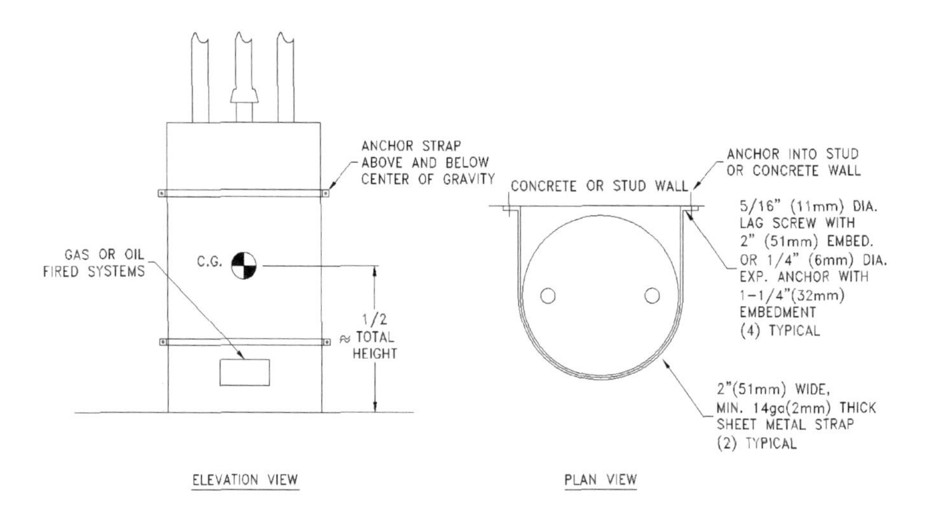

Figure 17-1 Water heater/storage tank anchored to wall.

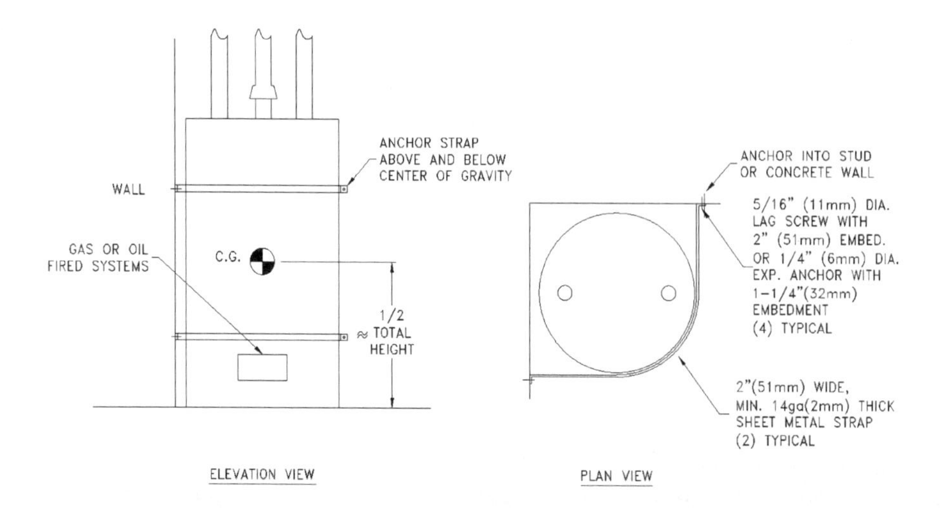

Figure 17-2 Water heater/storage tank anchored to corner.

Figure 17-3. Units in the basement or garage should be anchored to the concrete floor, as shown in Figure 17-3.

Wall or window air-conditioning units should be anchored to the wall frame and braced and/or supported outside the residence, as shown in Figure 17-4.

WATER PUMPS

Water pumps used for heating and/or cooling systems or well water should be anchored directly to a concrete slab or floor, as shown in Figure 17-5.

ELECTRIC GENERATORS

Loss of power can leave many residents stranded and isolated from the rest of society. Electric power has become essential for some residents for communication, where the advent of cordless phones and television have replaced traditional phones and transistor radios. In addition, power is required to operate heating and cooling systems.

Generators within residences are not common. However, if permanently installed, they should be properly anchored to a concrete slab or pad as shown in Figure 17-3.

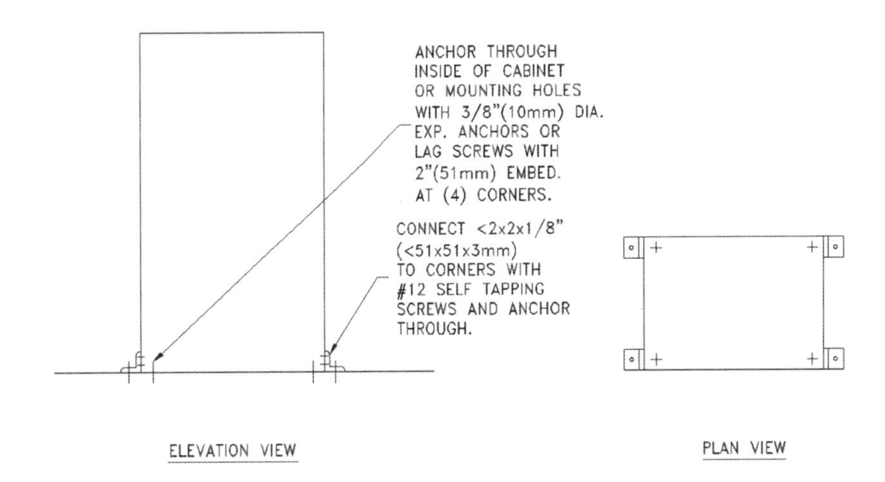

Figure 17-3 Water heater/storage tank anchored to corner.

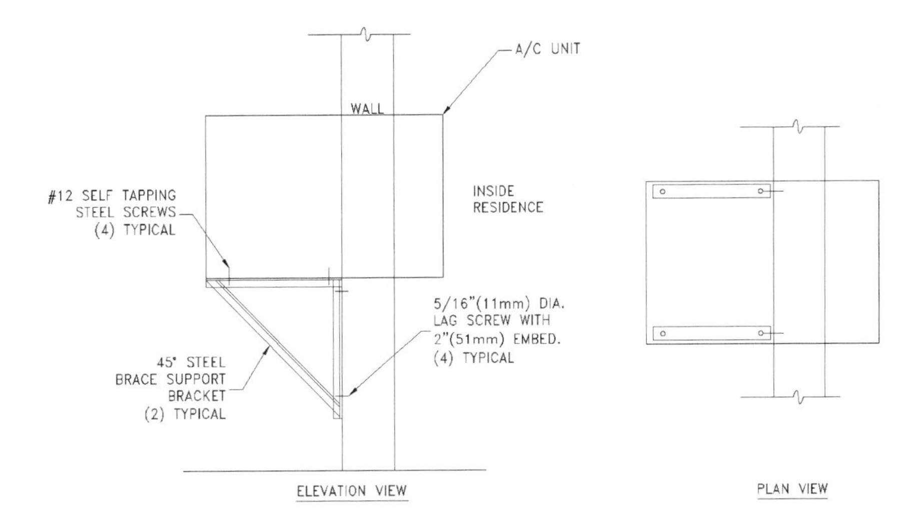

Figure 17-4 Window air conditioner to wall frame.

PROPANE TANKS

Propane tanks used for residential appliances, such as stoves, ovens, furnaces, and water heaters, are set outside on concrete pads.

The most common type of tank is a low-profile, horizontal cylinder. The cylinder is set within a pair of saddles. To prevent damage during an earthquake, the tank should be strapped down to the saddles and the saddles anchored to the concrete pad, as shown in Figure 17-6.

GAS, FUEL OIL, AND ELECTRIC LINES

Many water heaters, boilers, and furnaces produce heat with electricity or by burning gas or fuel oil. Anchoring of equipment will help to prevent failure at the connection to electric conduits. Anchoring the equipment will also prevent gas or fuel supply lines from breaking and leaking, eliminating a potential fire hazard. Installing flexible connectors between rigid piping or conduit and the appliance is also an effective way to prevent damage.

Since most residential gas lines or fuel oil lines are supported within the framework of the residence, additional bracing is not necessary. However, if the lines are exposed and susceptible to swaying, additional bracing is advisable, as shown in Figures 17-7 and 17-8, to avoid breaking the connections to the equipment.

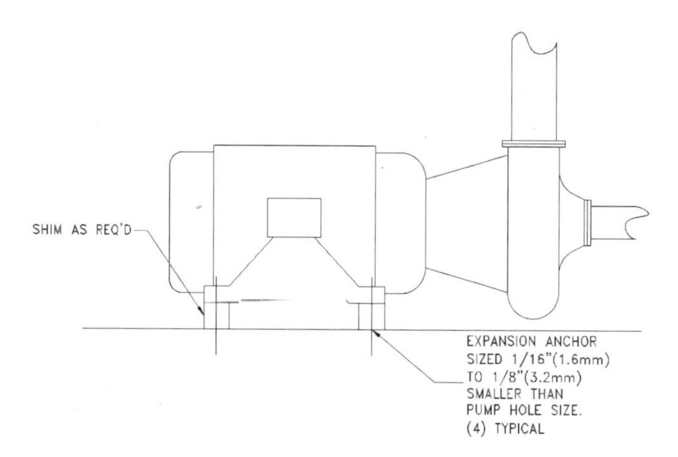

Figure 17-5 Pump anchored to floor or slab.

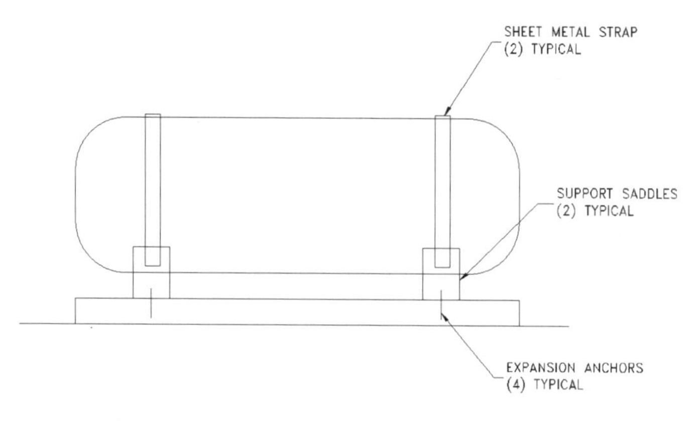

Figure 17-6 Propane tank anchored to floor or slab.

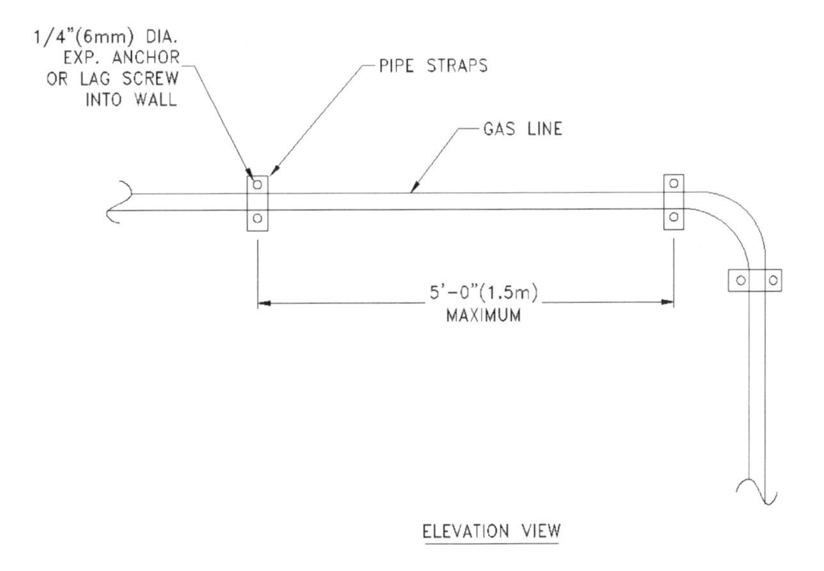

Figure 17-7 Gas pipelines anchored to wall.

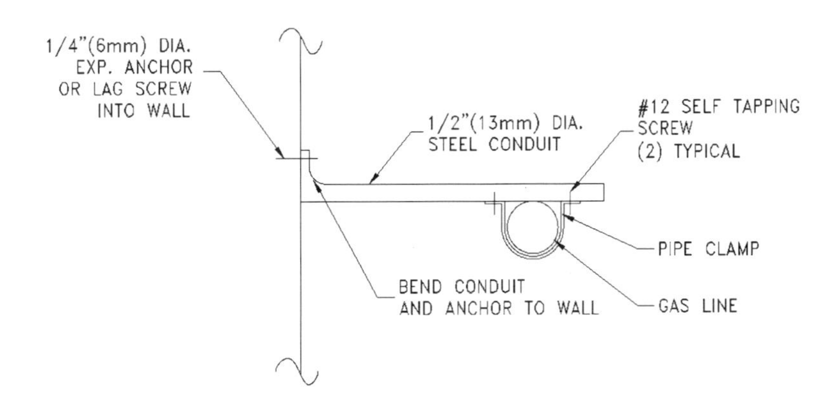

Figure 17-8 Gas pipelines anchored to wall.

18 Dos and Don'ts for Seismic Restraint Systems

The first seventeen chapters explain earthquakes and the restraining systems used on mechanical and electrical equipment. What happens when these systems are incorrectly specified or installed? This chapter contains photographs from the Northridge and Loma Prieta earthquakes in California.

HOUSEKEEPING PADS

Housekeeping pads and their performance during earthquakes were not considered until about 1995. The Northridge quake of 1994 opened many people's eyes to the role that housekeeping pads play in seismic restraint systems. Photographs 18-1 and 18-2 show what happens when the concrete in the pad is not reinforced and the pad is not attached to the structural slab. This pad shattered like a pane of glass. The chiller was on an upper story, and seismograph readings showed that this floor had over 1 *g* of vertical acceleration. The pad probably lifted off the ground with the chiller attached to it, came down, and shattered. Note in both photographs that the seismic snubbers are still attached to the shattered portions of the pad and to the equipment in other locations. While the seismic restraint system above the housekeeping pad was designed correctly, the housekeeping pad was not.

Another problem associated with housekeeping pads is the correct spacing and edge distances that are required for the use of post-installed anchors. As can be seen in Photographs 18-3 and 18-4, the post-installed anchors used to install the seismic restraint were installed too close to the edge of the slab, and the edge broke off. In Photograph 18-3, the

Photograph 18-1

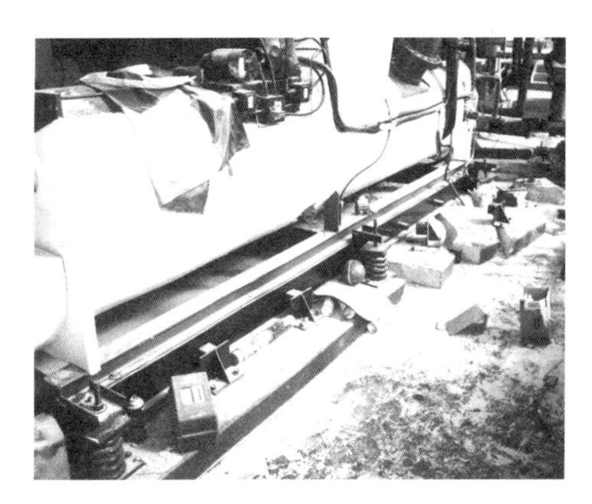

Photograph 18-2

combination isolator and restraint broke the housekeeping pad edge off on one side and pulled the anchor bolt cleanly out of the inner location. In Photograph 18-4, the seismic snubber did its job initially and held the equipment in place.

COOLING TOWERS AND CONDENSERS

Large, roof-mounted equipment is usually mounted on supplemental structural steel frames raised above the roof. As shown in Photographs 18-5 and 18-6, the supplemental structural steel frame to which the tower was attached failed. The failure was caused by an inadequately cross-braced structural steel frame. Without adequate cross-bracing, the main side members rotated. This added an additional degree of freedom to the system. Independent steel rails (whether inside or outside of the building) should not be used to support systems in seismic areas.

Another variation is the use of cooling tower antivibration rails. They usually consist of channels or angles with springs and bolts fitted between them and are seldom designed to resist horizontal seismic forces. As shown in Photographs 18-7 and 18-8, the rails failed and the tower moved laterally 12 in. (305 mm). Note in Photograph 18-7 that you cannot see any restraints. Photograph 18-8 shows a restraint at the end of the rail that was inadequate to resist the forces and the restraining bolt broke out the hole.

Photographs 18-9 and 18-10 show cooling towers that were mounted correctly on top of concrete piers. Note how the pier's length and width are sufficient to accept the post-installed anchors correctly and that they are both correctly cross-braced.

Photograph 18-11 shows a correctly mounted condenser on supplemental structural steel frame with combination spring isolator/seismic restraints.

Photograph 18-3

Photograph 18-4

Photograph 18-5

Photograph 18-6

Photograph 18-7

Photograph 18-8

Photograph 18-9

Photograph 18-10

Photograph 18-11

ROOFTOP AIR-HANDLING UNITS

Manufacturers of most rooftop units (RTUs) offer support curbs for their units. These standard sheet metal curbs with their configuration, cross-bracing, and duct support systems can be properly braced. Photographs 18-12, 18-13, and 18-14 show aluminum isolation curb caps on top of standard sheet metal curbs. These curb caps had inadequate restraints built into them and physically sheared themselves apart during the earthquake. Note in Photograph 18-12 that the aluminum curb's springs are nearly horizontal, a sign that the RTU has shifted to one side. Photographs 18-13 and 18-14 also show lateral shifting of the RTU to the sides. In addition to the damage to the curb, the weatherseals also failed, permitting rainwater to enter the building, causing more damage.

Photographs 18-15 and 18-16 show two styles of combination vibration isolation and structural curbs that fared much better than the aluminum curb caps.

Photograph 18-16 is a complete vibration isolation seismically restrained curb. Photograph 18-17 shows an RTU mounted on supplemental steel with combination spring isolators and seismic restraints.

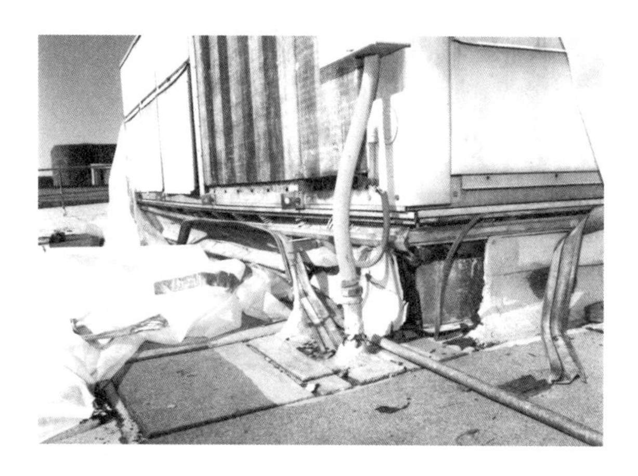

Photograph 18-12

Photograph 18-13

Photograph 18-14

Photograph 18-15

In Photograph 18-15, an RTU is mounted on a supplemental structural steel frame mounted on combination spring isolator and seismic restraints that fared well.

Photograph 18-18 shows a rooftop unit mounted on top of a supplemental structural steel frame with separate spring isolators and seismic restraints.

Note that the snubbers and supplemental steel are intact but the connection of the RTU to the steel failed, and the unit shifted laterally off the steel frame. In Photograph 18-19, a spring and snubber arrangement similar to Photograph 18-18 did not fare so well. The snubbers on this unit were not bolted to the lower structure as designed, and you can see that in Photograph 18-20, the bolt-hole in the snubber was never bolted down.

FLOOR-MOUNTED EQUIPMENT

Equipment adequately designed for direct support can use seismically rated vibration isolators, as shown in Photograph 18-21. The key to proper performance is a positive connection of the equipment frame to the isolator housing.

Unfortunately, this did not occur at the chiller installation shown in Photographs 18-22 and 18-23, where poor welded connections failed. Photograph 18-22 shows a chiller where restraints failed because of poor welding. Photograph 18-23 is a close-up view of the poor field weld at the chiller leg.

Photograph 18-16

Photograph 18-17

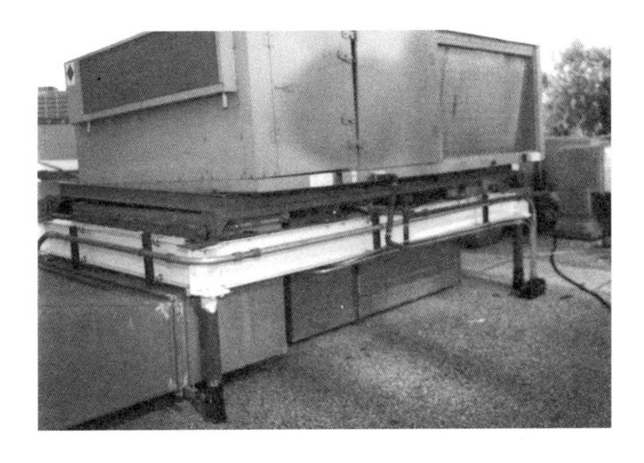

Photograph 18-18

Photograph 18-19

Photographs 18-24 to 18-27 illustrate what happens when equipment is mounted on vibration isolators with insufficient seismic ratings. Photograph 18-24 also shows that the equipment frame supplied with the equipment was not capable of supporting itself during an earthquake. It was not clear if this isolator was even bolted to the slab. Photograph 18-25 shows a isolator that has no seismic capabilities. Even though the lower half was bolted down, the upper half "jumped out" of the lower half.

Photographs 18-26 and 18-27 show two more pieces of equipment that were mounted on isolators that were not able to resist the seismic forces. In Photograph 18-26, this equipment could have been saved if there had been a flexible pipe connector. A connector either would have given the system enough play or it would have failed, leaving the equipment flanges intact. In Photograph 18-27, the isolator is a vertically restrained mount that was not able to resist the seismic forces. Note that the housing's lower baseplate is bent, and the upper portion of the housing has separated from the lower.

Photograph 18-20

Photograph 18-21

Photograph 18-22

Photograph 18-23

Some types of equipment require a supplemental base (i.e., a steel frame or concrete inertia base) before direct support on vibration isolators. They can use one of the following: an open spring and separate seismic snubber system, as shown in Photographs 18-28 and 18-29, or a seismically rated, combination vibration isolator and seismic restraint, as shown in Photographs 18-30 to 18-32. Both Photograph 18-28 and 18-29 show equipment with separate snubbers and isolators that are adequately restrained.

Photographs 18-30, 18-31, and 18-32 show combination isolator and seismic restraints. Photograph 18-30 is a correctly designed concrete inertia base. Note the edge distance between the housekeeping pad edge and the anchor bolts of the restraint. Photograph 18-31 shows a vertically restrained mount attached to a steel frame. Note that the frame is cross-braced.

Photograph 18-24

Photograph 18-25

Photograph 18-26

Photograph 18-27

Photograph 18-28

Photograph 18-29

Photograph 18-30

Photograph 18-31

Photograph 18-32 shows a piece of equipment with multiple combination isolator/restraints. Note the large steel frame to which they are attached.

Systems without any restraints are shown in Photographs 18-33, 18-34, and 18-35. In Photograph 18-33, the pump was installed on a rail with spring isolators under the rail without any restraints.

In Photographs 18-34 and 18-35, the pumps were installed on steel frames with spring isolators only. Note how the frames moved and bent the spring mounts. In Photograph 18-35, note that the frame has moved laterally about 6 in. (152 mm).

Systems with restraints but without an adequate supplemental support frame are shown in Photographs 18-36 to 18-40. These photographs clearly show the twisting that occurs when independent steel rails supported on spring isolators and without proper cross-bracing are used.

The units in Photographs 18-36 and 18-37 had seismic snubbers attached to the rails. Note that in both cases, the rails rolled away from the snubbers, which were not fastened to the rails—they were slipped over pipe sleeves.

Photograph 18-32

Photograph 18-33

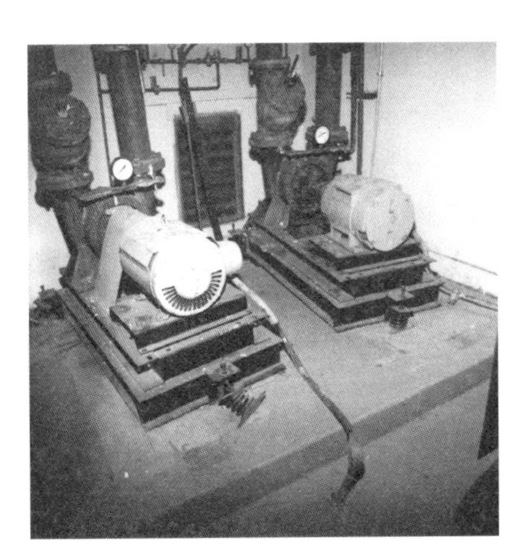

Photograph 18-34

Photograph 18-35

Photograph 18-36

Photograph 18-37

Two chillers are shown in Photograph 18-38. The one on the right was supported on independent steel rails, open springs, and seismic restraints (also shown in Photograph 18-39). The one on the left was supported on a steel frame with proper cross-bracing and seismically rated vibration isolators. The frame on the left was correctly designed and was able to resist the seismic forces, but the independent rails on the right were not. The rails rolled away from the seismic snubber and fell over, as shown in Photograph 18-39.

Once there is a properly designed vibration isolation and seismic restraint system, the next step is to consider the load transfer element (e.g., concrete housekeeping pads, piers, and pedestals, wood platforms, and steel frames) between the seismic restraints and the main building structure. Photographs 18-40 and 18-41 show equipment mounted on tall concrete piers that could not accept the horizontal seismic loads.

Photograph 18-38

Photograph 18-39

Photograph 18-40

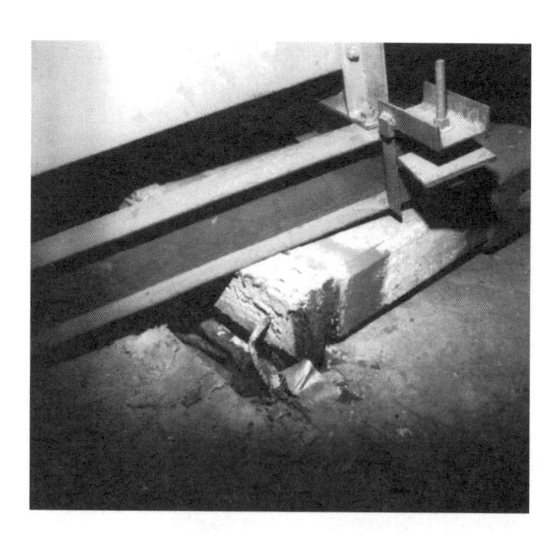

Photograph 18-41

Photograph 18-42 shows a fan on a tall supplemental structural steel frame. Note the amount of cross-bracing that was required to resist the seismic forces.

PIPING AND DUCTWORK

In most cases, damage in buildings caused by suspended systems is due to a complete lack of seismic sway bracing. Photographs 18-43 to 18-46 show the amount of movement that can occur on clevis and trapeze supported piping.

In Photograph 18-43, the clevis slipped along the pipe about 6 in. (152 mm).

In Photograph 18-44, a trapeze-supported single pipe moved back and forth about 4 in. (100 mm). This movement wore off the insulation on the pipe. Note that the sheet metal sleeve (installed to distribute the weight of the pipe on the insulation) fell out.

Photograph 18-45 shows a trapeze containing two pipes that suffered damage similar to that shown in Photograph 18-44.

Photograph 18-42

Photograph 18-43

Photograph 18-44

Photograph 18-45

In Photograph 18-46, an exhaust duct riser was crushed between two pipes. The unrestrained pipes survived but the exhaust duct did not.

Bracing or stiffening of the pipe clevis bolt is critical to prevent support failure. At the support shown in Photograph 18-47, the unbraced clevis hanger was severely distorted.

In Photograph 18-48, the angle brace that was welded to the clevis deformed and the welds broke. Pipe penetrations through walls, especially those constructed of gypsum board, should not be used as a seismic sway brace location, as shown in Photograph 18-49.

Photograph 18-50 shows the importance of strapping down a pipe to the trapeze even if the pipe is set in rollers.

No-hub pipe connections on risers, such as cast iron pipe connected with a shield and clamp assembly, should be protected against the differential motion between floors, as shown in Photograph 18-51.

Photograph 18-46

Photograph 18-47

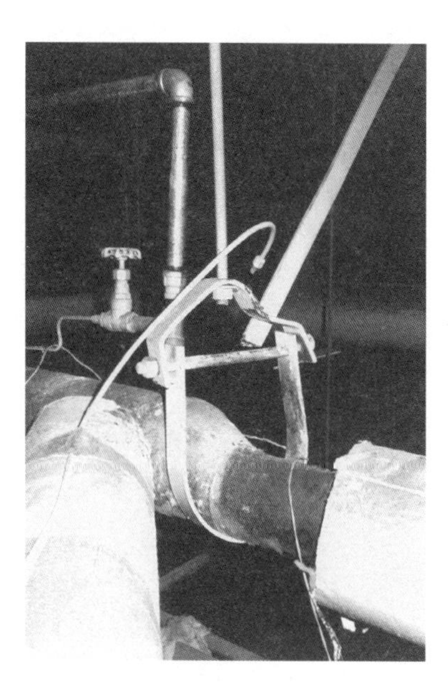

Photograph 18-48

Photograph 18-49

Pipe supports from wide flange steel beams should not be friction connections, such as the C-clamps shown in Photograph 18-52, unless they are provided with a restraining strap or hook to the opposite side of the beam flange.

Piping on the roof must also be designed for proper seismic bracing, as shown in Photograph 18-53.

Photographs 18-54 and 18-55 summarize all aspects of correct seismic sway brace installation, including proper installation angles, spring hangers located up against the structure, pipe sleeves over the clevis cross-bolts, and vertical rod stiffeners and longitudinal bracing connected directly to the pipe.

Damage from ductwork generally occurs when the connections of the sections fail and fall, as shown in Photographs 18-56 to 18-58.

Photograph 18-56 shows the straps that originally held a round duct. The duct shown in Photograph 18-57, ended up on the floor because it was not braced. The unbraced duct in Photograph 18-58 separated and fell to the floor.

Photograph 18-50

Photograph 18-51

Photograph 18-52

Photograph 18-53

Photograph 18-54

Photograph 18-55

Photograph 18-56

Photograph 18-57

Photograph 18-58

Photograph 18-59

Electrical conduit supported along walls and in ceiling space can contribute to damage within buildings, as shown in Photographs 18-59 and 18-60.

PIPE RISERS

Pipe risers' supports, anchors, and guides must be designed to allow for thermal growth and for seismic forces. Photographs 18-61 and 18-62 illustrate the use of separate spring supports and guides.

Photograph 18-61 shows a pipe riser with telescoping guides with specially designed attachment brackets. Photograph 18-62 shows spring-supported pipe risers with specially designed attachment brackets.

Photograph 18-63 shows risers installed on vertically restrained spring isolators. The brackets were improperly designed for the forces and bent during the earthquake.

Photograph 18-60

Photograph 18-61

Photograph 18-62

Photograph 18-63

INTERNAL ISOLATION

Equipment such as air-handling units can be purchased (or come standard) with internal vibration isolation. Photographs 18-64 and 18-65 show what happens to internal isolation during an earthquake.

Photograph 18-64 shows a housed isolator that is not vertically restrained. The equipment frame, which was made out of sheet metal, bent and twisted. Note the addition of a cable on the left. This was added after the earthquake to keep the fan section in place.

Photograph 18-65 shows the top of an unrestrained fan inside an air-handling unit. The fan section can move laterally and strike the copper pipes to the left.

FRAGILITY LEVELS

Photographs 18-66 and 18-67 show equipment whose fragility levels were not sufficient to prevent them from being damaged.

Photograph 18-66 shows a cooling tower whose legs were not capable of resisting the seismic forces. The legs buckled and the tower collapsed on itself.

Photograph 18-67 shows a condenser whose legs have bent and shifted off the isolator.

Photograph 18-64

Photograph 18-65

Photograph 18-66

Photograph 18-67

19 Seismic Testing of Restraints

ASHRAE has been the leader in the testing of seismic restraints for HVAC&R equipment. In 2008, ASHRAE, along with ANSI, published a testing standard, ANSI/ASHRAE Standard 171-2008, "Method of Testing Seismic Restraint Devices for HVAC&R Equipment" which describes the testing procedures for the following seismic restraint systems:

- Cable restraints
- Combination floor mounted isolator/restraints
- Seismic snubbers
- Seismic bumpers
- Solid/rigid bracing

The objective of these tests is to determine the capacity of seismic restraints/braces for HVAC&R equipment. The tests will determine the maximum force a restraint can withstand without breakage or permanent deformation.

Just as anchor bolts require testing to achieve allowable shear and tensile loads, so do seismic restraints and braces. Testing has historically been static force tests designed to find allowable forces. Depending on the type of restraint being tested, there could be as many as six different tests required.

For example, testing for figure 3-5C restraints as shown in Chapter 3 would include transverse and longitudinal horizontal tension tests, vertical tension, vertical compression, and 45° tension tests in both the transverse and longitudinal directions. Testing would include increasing cyclic loading based on a known failure load or pretesting to find expected failure load. During the cyclic loading, the restraint is checked for deformation at the beginning and end of each cycle.

See Figures 19-1 to 19-5 for the basic setups for these six tests.

BIBLIOGRAPHY

ASHRAE. 2008. *Method of testing seismic restraint devices for HVAC&R equipment.* ANSI/ASHRAE Standard 171-2008. Atlanta GA: ASHRAE.

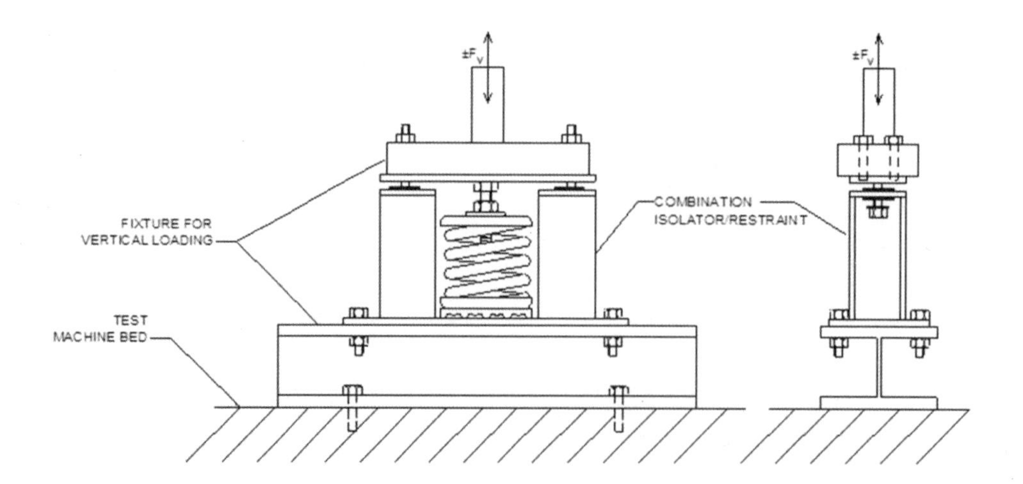

Figure 19-1 Vertical tension and compression F_V test setup.

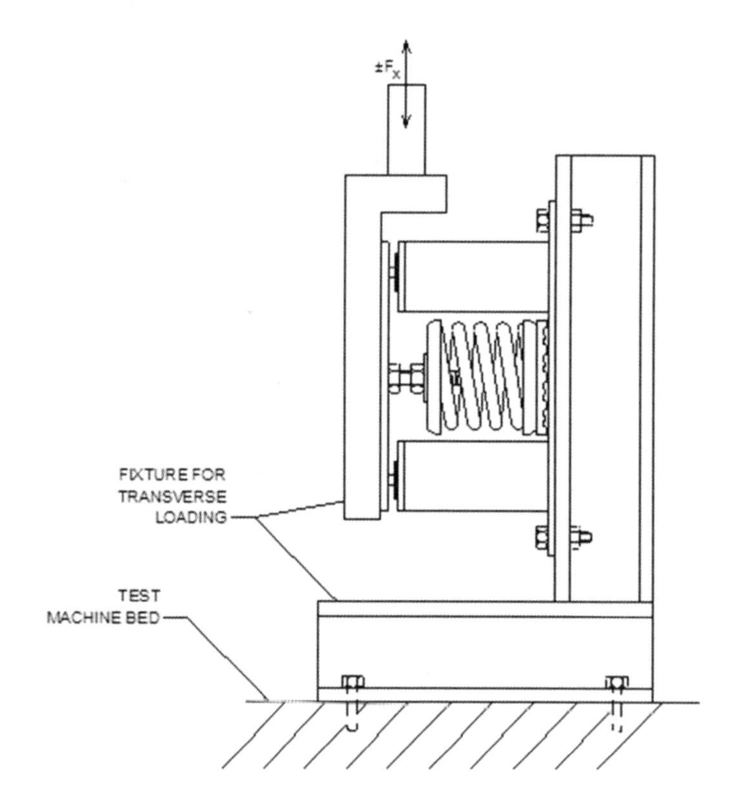

Figure 19-2 Horizontal longitudinal loading F_X test setup.

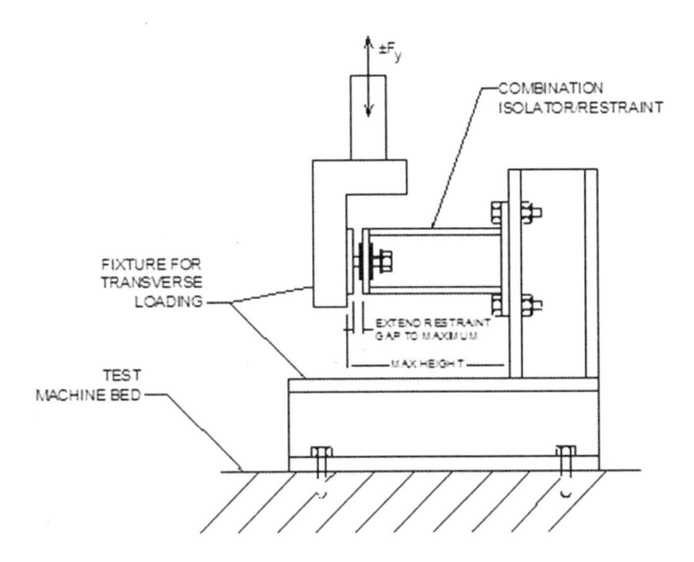

Figure 19-3 Horizontal transverse loading F_y test setup.

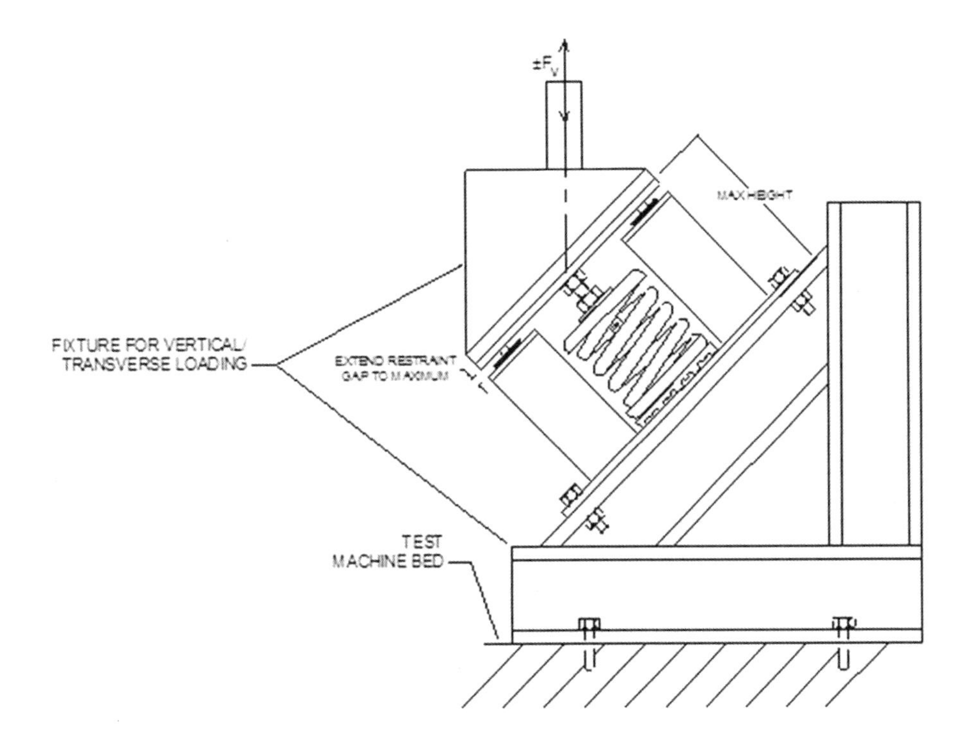

Figure 19-4 45° longitudinal-vertical load F_{xv} test setup.

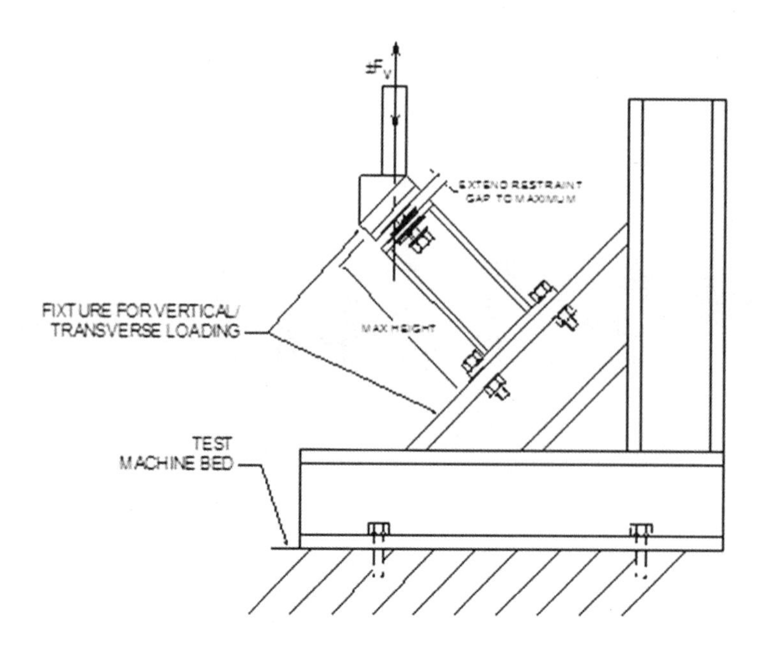

Figure 19-5 45° lateral-vertical load F_{yv} test setup.

Appendix A Abridged Modified Mercalli Intensity Scale

1. Not felt except by a few under especially favorable circumstances.
2. Felt only by a few persons at rest, especially on upper floors of buildings. Delicately suspended objects may swing.
3. Felt quite noticeably indoors, especially on upper floors of buildings, but many people do not recognize it as an earthquake. Standing automobiles may rock slightly. Vibration like passing truck. Duration estimated.
4. During the day, felt indoors by many, outdoors by a few. At night, some awakened. Dishes, windows, and doors disturbed; walls make creaking sound. Sensation like heavy truck striking building. Standing automobiles rock noticeably.
5. Felt by nearly everyone, many awakened. Some dishes, windows, etc. broken; cracked plaster in a few places; unstable objects overturned. Disturbances of trees, poles, and other tall objects sometimes noticed. Pendulum clocks may stop.
6. Felt by all, many frightened and run outdoors. Some heavy furniture moved; a few instances of fallen plaster and damaged chimneys. Damage slight.
7. Everybody runs outdoors. Damage negligible in buildings of good design and construction, slight to moderate in well-built ordinary structures, considerable in poorly built or badly designed structures; some chimneys broken. Noticed by persons driving cars.
8. Damage slight in specially designed structures, considerable in ordinary substantial buildings with partial collapse, great in poorly built structures. Panel walls thrown out of frame structures. Fall of chimneys, factory stacks, columns, monuments, and walls. Heavy furniture overturned. Sand and mud ejected in small amounts. Changes in well water. Persons driving cars disturbed.
9. Damage considerable in specially designed structures; well-designed frame structures thrown out of plumb; damage great in substantial buildings, with partial collapse. Buildings shifted off foundations. Ground cracked conspicuously. Underground pipes broken.
10. Some well-built wooden structures destroyed; most masonry and frame structures destroyed with foundations; ground badly cracked. Rails bent. Landslides considerable from river banks and steep slopes. Shifted sand and mud. Water splashed over banks.
11. Few, if any (masonry) structures remain standing. Bridges destroyed. Broad fissures in ground. Underground pipelines completely out of service. Earth slumps and land slips in soft ground. Rails bent greatly.
12. Damage total. Waves seen on ground surface. Lines of sight and level distorted. Objects thrown into air.

Appendix B Glossary

A

ACI

American Concrete Institute.

Active fault

Faults that are constantly moving, producing stresses that can give rise to earthquakes.

AISC

American Institute of Steel Construction, Inc. Publishers of the *Manual of Steel Construction*.

Allowable force or load

The working or required load of a material or component with built-in safety factors dependent upon the material or component application.

Allowable stress design (ASD)

Design practice within a defined allowable or working stress. For steel, this would be an appropriate factor of the yield stress.

B

Baseplate

A plate used for support and anchorage of a vibration isolator.

Bending moment

The result of a load applied on an axis parallel to and in the center of a support member.

BOCA

Building Officials Code Administrators.

Bounding spectrum

The maximum response that can be accepted by a piece of equipment as a result of an applied shock without loss of operation.

C

Cable brace

A steel cable designed for tension loads for use as a seismic sway brace for suspended equipment, piping, or ductwork.

Cantilevered

A support member connected at one end and unsupported at the other end.

Cast-in-place anchor

A headed steel bolt or equal deformation set within a concrete form before pouring is completed for use of anchoring equipment or anchor plates.

Center of gravity

The point within a piece of equipment or component about which it will balance.

Chemical anchor

An anchor designed to bond directly to concrete within a predrilled hole using a chemical compound.

Collapse earthquake

The sudden fall of the roof of an underground cavern or mine.

Compressive force or load

An axial force that produces a uniform compression of a material perpendicular to its cross section.

D

Deformability

A gage for a material's capability to deform or bend.

Design strength

A material's ultimate loading capability with appropriate reduction factors for specific applications (e.g., earthquake loads, gravity loads, live loads, etc.).

Ductile material

A material that will undergo large deflections under load before failure.

Ductile connection

A connection point between two ductile materials that behaves similarly to a ductile material.

DSA

Department of the State Architect in California. Reviews all public schools in California for compliance with state adopted codes.

E

Elastomeric

A material with flexibility in all directions that will recover to its original form if removed from its environment when deformed due to its environment.

Epicenter

Projected location on the surface above the focus of the earthquake.

Expansion anchor

A post-installed concrete anchor that uses some form of expansion of a wedge or its shell against the drilled hole in the concrete. Tensile forces are resisted by friction. This is not true in the case of an undercut anchor where tensile forces are resisted by mechanical means of a flared bottom.

Explosive earthquakes

Caused by underground human-made explosions.

F

Fault

An offset in a geological structure.

Flexible connector

A connector designed with an appropriate amount of flexibility between a piece of equipment, component, or system and another system to achieve the desired amount of separation.

Flexible equipment

A piece of equipment constructed so that it deforms more under load than that to which it is attached.

Flexibly mounted equipment

A piece of equipment supported on or from a vibration isolator.

Focus

Location deep inside the earth along the fault where the slip occurs and out of which the P and S waves emanate.

Fragility

The maximum shock in *g*s that a piece of equipment can take without suffering sufficient damage to render it inoperable.

G

Gravity load

The vertical load from a component's or system's weight.

Groove-fitted joint

A mechanical connection between two pipe sections using a tongue-and-groove concept.

Ground acceleration

The acceleration at ground or grade level due to an earthquake.

H

Housekeeping pads

Concrete pad that is used under equipment to raise it off the structural slab. Also called plynths, or plinths, in many countries.

I

IBC®

The *International Building Code*®, which combined the three model building codes in the United States—the Building Officials Code Administration (BOCA), the Southern Building Code Congress International (SBCCI), and the *Uniform Building Code* (UBC).

ICC

The International Code Council, publishers of the *International Building Code*® (*IBC*®).

Impact

The collision of one mass in motion with another mass.

Inactive fault

A fault that has not been active for thousands of years.

L

Limit stop

A design feature within a vibration isolator that limits upward motion due to earthquake loads or reduction of gravity loads.

Longitudinal brace

Brace that restrains pipes or ducts parallel to the longitudinal direction.

Love wave

Horizontal surface wave that mimics an S wave and moves residential buildings off their foundations.

M

Maximum unbraced rod length

The maximum length a threaded vertical support rod can be and accept a defined compressive force without buckling.

Modified Mercalli Intensity Scale

System to determine the strength of an earthquake. Verbal accounts and visual inspection using 12 values for damage. See Appendix A for an abridged version of the scale.

Moment magnitude

Based on the geometry of the fault plane after the seismic event; more accurate than the Richter magnitude, but takes weeks if not months to determine.

MSS-SP

Manufacturers Standardization Society.

N

National Design Specification for Wood Construction

Complete design manual for all types of wood and wood products used in buildings.

NBC

National Building Code of Canada.

No-hub pipe

Piping with connections that do not interlock or permanently join, such as shield and clamp assemblies for cast iron drainage pipe.

Normal fault

Vertical displacement where one side slides down away from the other side.

O

OSHPD

Office of Statewide Health Planning and Development in the state of California; reviews all hospital projects for compliance with state codes.

Oscillate

To allow a component to vibrate.

P

P waves

Movement waves with zones of compression and elongation that emanate from the focus of the earthquake. These waves can be refracted as they encounter different strata. Also known as primary waves.

Prestretched cable

Cable that has been stretched after manufacturing to reduce the constructional stretch that is inherent from winding and can provide certified modulus of elasticity.

R

Rayleigh wave

An elliptically rolling wave in the direction of travel with both horizontal and vertical movement.

Response spectrum

The maximum response (in acceleration, velocity, or displacement) experienced by a single degree-of-freedom system, as a function of its natural frequency, in response to a shock.

Restoring force

A force that can be used in seismic design to restore the loading conditions to their original condition (e.g., gravity loads).

Resultant force or load

The addition of all forces in the x, y, and z directions, resulting in one single force in a specific direction.

Retrofit

A new installation within an existing facility.

Reverse fault

Vertical fault where one side rises up past the other side.

Richter magnitude

Based on seismographs, the time delay between the P and S waves is plotted against the maximum amplitude of the S waves and a straight line drawn between them on special graph paper that has the Richter magnitude line also plotted on it. Where it crosses this line is the Richter magnitude. It is available immediately after the seismic event and is, therefore, quoted by the media. It is also used to pinpoint the focus.

Rigid equipment

A piece of equipment constructed so that it does not deform under load relative to that to which it is attached.

Rigidly mounted systems

Systems that are either solidly braced or bolted directly to structure without vibration isolators.

Rotating equipment

Mechanical equipment with rotating parts.

S

S waves

Vertical and horizontal movement waves that emanate from the focus and can be reflected as they encounter different strata. They do the most damage and are also known as secondary waves.

SBCCI

Southern Building Code Congress International.

Seismic capacity

The rated or allowable load of a prescribed restraint to accept seismic loads. For seismic sway brace systems, this could be in the form of a length of pipe or duct (e.g., the seismic capacity of the sway brace is 50 ft of 8 in. diameter pipe).

Seismic joint

A section of pipe routed from one building to another and designed to accept a predicted amount of differential motion between the two buildings.

Seismic force or load

The horizontal or vertical seismic force defined by code.

Shallow concrete anchor

From the 1997 *Uniform Building Code* (UBC) and 2009 *International Building Code* (IBC), any cast-in-place, chemical or expansion anchor with an embedment length-to-diameter ratio less than 8.

Shock loads

Short duration, high impact dynamic loads.

SMACNA

Sheet Metal and Air Conditioning Contractors' National Association.

Snubber

A device used to increase the stiffness of an elastic system whenever the displacement exceeds the design value; a seismic restraint used on isolated systems with an air gap and neoprene cushioning.

Solid brace

A steel angle or strut channel designed for tension and compression loads for use as a seismic sway brace for suspended equipment, piping, or ductwork.

Stiffness

The ratio of the change in force to the change in displacement in an elastic element.

Supplemental support frame

A separate steel frame to support equipment designed for both gravity and seismic loads.

Sway bracing

Generic term for systems that seismically restrain mechanical, fire protection, or electrical systems. There are two types of sway bracing: one is solid bracing using steel members such as angles or pipes, and the second is cable bracing using aircraft cable.

T

Tectonic plates

The outermost layer of rock that covers the entire surface of the earth.

Tectonic earthquake

The most common type of earthquake; caused by the shifting or sliding of tectonic plate edges past each other.

Tensile force or load

An axial force that produces a uniform stretching of a material perpendicular to its cross section.

Transcurrent fault

Predominately horizontal slippage defined as either left-lateral or right-lateral movement. For example, if you are standing on one side of a fault and the movement on the other side is from the right to the left, it is a left-lateral fault. Also known as a strike-slip fault.

Transverse brace

Brace that restrains pipes or ducts perpendicular to the longitudinal direction.

Twisting moment

The result of a load applied in an axis perpendicular to and off center from a support member.

V

Vibration isolated systems

Systems that incorporate resilient mountings between the equipment and the structure; must still be seismically restrained.

Volcanic earthquake

Earthquake that is caused by volcanic activity.